MOLECULAR BIOLOGY
OF
PLANT VIRUSES

MOLECULAR BIOLOGY
OF
PLANT VIRUSES

edited by

C. L. Mandahar

Botany Department
Panjab University, India

KLUWER ACADEMIC PUBLISHERS
Boston / Dordrecht / London

Distributors for North, Central and South America:
Kluwer Academic Publishers
101 Philip Drive
Assinippi Park
Norwell, Massachusetts 02061 USA
Telephone (781) 871-6600
Fax (781) 871-6528
E-Mail <kluwer@wkap.com>

Distributors for all other countries:
Kluwer Academic Publishers Group
Distribution Centre
Post Office Box 322
3300 AH Dordrecht, THE NETHERLANDS
Telephone 31 78 6392 392
Fax 31 78 6546 474
E-Mail <services@wkap.nl>

 Electronic Services <http://www.wkap.nl>

Library of Congress Cataloging-in-Publication Data

Molecular biology of plant viruses / edited by C.L. Mandahar.
    p.   cm.
  Includes bibliographical references and index.
  ISBN 0-7923-8547-0   (alk. paper)
  1. Plant molecular virology. I. Mandahar, C.L.
  QR351.M63   1999
  572.8'2928--DC21   99-28569
                                    CIP

Printed on acid-free paper.

Printed in the United States of America

CONTENTS

List of Figures · · · · · xiii
List of Tables · · · · · xv
List of Contributors · · · · · xvii
Preface · · · · · xxi
Acknowledgements · · · · · xxxi

1. Genome of RNA Viruses · · · · · 1-28
   François Héricourt, Isabelle Jupin, and Anne-Liss Haenni.

   Introduction · · · · · 1
   Specific Gene Classes · · · · · 1
   Polymerase-Associated Functions · · · · · 1
   Coat Protein · · · · · 9
   Cell-to-Cell Movement · · · · · 10
   Systemic Movement · · · · · 11
   Proteinases · · · · · 11
   Vector Transmission · · · · · 11
   Symptoms · · · · · 12
   Non-Coding Functions of Viral RNAs · · · · · 12
   Cis-Acting Elements Required for Translation · · · · · 12
   Cis-Acting Elements Required for Replication · · · · · 14
   Interactions with Coat Protein · · · · · 17
   Conclusions · · · · · 18
   References · · · · · 18

2. Genome of DNA Viruses · · · · · 29-46
   Thomas Frischmuth

   Introduction · · · · · 29
   Genome Organization · · · · · 30
   Family *Geminiviridae* · · · · · 30
   Genus *Nanovirus* · · · · · 31
   Family *Caulimoviridae* · · · · · 32
   Genus *Caulimovirus* · · · · · 32
   Genus *Badnavirus* · · · · · 33
   Gene Functions · · · · · 34
   Family *Geminiviridae* · · · · · 34
   Genus *Nanovirus* · · · · · 36
   Family *Caulimoviridae* · · · · · 36

Genus *Caulimovirus*   36
Genus *Badnavirus*   37
Gene Expression   38
Family *Geminiviridae* and Genus *Nanovirus*   38
Family *Caulimoviridae*   39
References   41

3.   Genome Organization in RNA Viruses   47-98
Sergey Morozov and Andrey Solovyev

Introduction   47
Genome Maps of Positive-Stranded RNA Viruses   48
Picorna-like Supergroup   48
Family *Potyviridae*   49
Family *Sequiviridae*   51
Family *Comoviridae*   52
Sobemo-like Supergroup   53
Genus *Sobemovirus*   53
Genus *Polerovirus*   55
Genus *Enamovirus*   55
Carno-like Supergroup   56
Genus *Umbravirus*   57
Beet Western Yellows Virus ST9-associated RNA   58
Genus *Luteovirus*   58
Family *Tombusviridae*   58
Alpha-like Supergroup   61
Family *Closteroviridae*   62
Family *Bromoviridae*   63
Family *"Tubiviridae"*   66
Order *"Tymovirales"*   73
Genome Maps of Plant Negative-Stranded RNA Viruses   77
Family *Rhabdoviridae*   78
Family *Bunyaviridae*   78
Genus *Tenuivirus*   79
Genome Maps of Plant Double-Stranded RNA Viruses   80
Family *Reoviridae*   80
Family *Partitiviridae*   81
Conclusions   81
Acknowledgements   82
References   82

4.   Gene Expression in Positive Strand RNA Viruses: 99-119
Conventional and Aberrant Strategies
Alexy Agranovsky and Sergey Morozov

Introduction   99
Viral Genomic RNA as a Single Translation Unit   100
  Initiation Codon Choice and Translation in Plants   100
    Conventional Scanning and Initiation   100
    Non-AUG Initiation Codons in Plant Virus RNAs   101
    Role of 5'- and 3'- Noncoding Regions in Initiation of Translation   101
Access to Internal Genes   103
  Divided versus Monopartite Genomes   104
  Transcriptional Control: Subgenomic RNAs   104
  Non-Orthodox Mechanisms of Initiation   107
    Leaky Scanning   107
    Internal Ribosome Entry Sites (IRES Elements)   107
  Non-Orthodox Mechanisms of Elongation and Termination   108
    Readthrough of Leaky Stop Codons   108
    Ribosomal Frameshifiting   109
Proteolytic Processing   110
  Chymotrypsin-like Serine or Cysteine Proteinases   112
  Papain-like Cysteine Proteinases   112
Conclusions   113
References   114

5. Molecular Basis of Genetic Variability in RNA Viruses   121-141
    Jozef J. Bujarski

Introduction   121
Genetic Mechanisms of Variability   121
  Mutation   121
    Point Mutations   122
    Insertions and Deletions   123
    Frameshift Mutations   123
    Mutant Stability   124
    Effects of Mutations on Host-Virus Interactions   124
  RNA Recombination   125
    Natural Sequence Rearrangement   126
    Recombination between Viral and Host RNAs   126
    Role of RNA Structure   127
    Role of Replicase Proteins   130
  Pseudorecombination   131
  Defective Interfering RNAs   132
Conclusions   134
Acknowledgements   136
References   136

6. Genetic Variability and Evolution   143-159
    F. Garcia-Arenal, A. Fraile, and J. M. Malpica

    Introduction   143
    Variability under Experimental Conditions   143
    Variability under Natural Conditions   145
    Factors Determining Genetic Structure of Virus Populations   149
        Founder Effects   149
        Selection   150
        Complementation   152
    Conclusions   153
    Acknowledgements   155
    References   155

7. Molecular Basis of Virus Transport in Plants   161-182
    Scott M. Leisner

    Introduction   161
    Pathways of Virus Movement   161
    Identification of Viral Movement Protein Genes   163
    Characteristics of Plant Viral Movement Proteins   164
    Tubules   169
    Models of Cell-to-Cell Movement   169
    Movement Protein-Host Interactions   174
    Viral Genes Influencing Vascular Movement   174
    Conclusions   176
    Acknowledgements   177
    References   177

8. Molecular Basis of Virus Transmission   183-200
    Johannes F. J. M. van den Heuvel, Alexander W. E. Franz,
    and Frank van der Wilk

    Introduction   183
    Transmission by Arthropod Vectors   183
        Noncirculative Virus Transmission by Insects   184
        Circulative Virus Transmission by Insects   185
            Putative Receptors   186
            Persistence in Hemolymph of Vectors   187
    Transmission by Soil-Inhabiting Nematodes   188
    Transmission by Zoosporic Fungi   190
        Persistent Transmission   190
        Non-persistent Transmission   191
    Transmission by Seed and Pollen   192

Viral Genetic Determinants 193
Host Genetic Determinants 195
References 195

9. Molecular Basis of Symptomatology 201-210
A. L. N. Rao.

Introduction 201
Genetic Basis of Symptom Expression 201
Molecular Biology of Symptom Expression 203
Role of Viral Coat Protein 203
Role of Movement Protein 206
Role of Viral Replicases 206
Other Factors 207
Conclusions 207
References 207

10. Gene-for-Gene Interactions 211-224
Christopher D. Dardick and James N. Culver

Introduction 211
Host Factors 213
Characterization of R Genes 214
Similarities to Pathogen Resistance in Animals 215
Organization of Pathogen Recognition Loci 216
Virus Factors 216
Capsid Protein as a Virulence Factor 216
Tobamovirus Capsid Proteins as Elicitors
of Hypersensitive Response 216
Potexvirus X Capsid Protein as Elicitor
of Hypersensitive Response 217
Virulence Generated by Different Virus Functions 217
Tobacco Mosaic Tobamovirus Movement Protein 217
Tobacco Mosaic Tobamovirus Replicase 218
Cauliflower Mosaic Caulimovirus (CaMV) Gene VI 218
Multiple Virus Determinants in Tomato Bushy Stunt Virus 218
Plant Viruses Evolve to Evade Host Recognition 219
Other Virus-Host Interactions Leading to Resistance 219
Applications 220
R Genes can be transferred across Species Barriers 220
Rational Design of R Genes 221
Engineering Broad-Spectrum Resistance 221
References 221

11. Molecular Biology of Viroids   225-239
Ricardo Flores, Marcos de la Peña, José Antonio Navarro,
Silvia Ambrós, and Beatriz Navarro

Introduction   225
Viroids : The Simplest Genetic Systems   225
    Nucleotide Sequence   225
    Domains in Proposed Secondary Structure and Elements
        of Tertiary Structure   226
    Role of Specific Domains   227
Genomic Diversity   228
    Variability and Viroid Quasispecies   228
    Selection Mechanisms   228
    Recombination   229
Origin and Evolution   229
    Viroids as Relics of RNA World   229
    Phylogenetic Reconstructions   230
    Classification   230
Replication   231
    Rolling Circle Model   231
    Transcription, Processing and Ligation, Hammerhead Ribozymes   232
Genetic Determinants of Pathogenesis and Symptomatology   234
    Pathogenic Domains   234
    Symptomatology   234
    Host Range   235
Genetic Determinants of Other Viroid Functions   236
    Transmission   236
    Viroid Transport in Host Plants   236
References   237

12. Molecular Biology of Transgenic Plants   241-254
Chuni L. Mandahar

Introduction   241
Genetic Variability of Challenge Virus   243
    Recombination   243
    Mutation   245
Heteroencapsidation   246
    Transmission of Non-transmissible Virus Isolates   247
    Complementation   247
    Extension of Host Range   248
Inheritance   248
Synergism   249
Other Effects   249
Conclusions   250

References   251

13. Gene Silencing   255-269
     Chuni L. Mandahar

     Introduction   255
     Types of Gene Silencing   257
     Mechanisms of Gene Silencing   258
       Transcriptional Gene Silencing   258
       Post-transcriptional Gene Silencing   258
         Threshold Model   259
         Ectopic Pairing Model   260
         Third Model   260
         Transmission Signal   261
     Effects of Gene Silencing   262
         Resistance in Transgenic and Non-transgenic Plants   263
         Recovery of Transgenic and Non-transgenic Plants from Disease 264
         Green Islands   265
     Conclusions   266
     References   266

Subject Index   271-281

LIST OF FIGURES

S. No.	Heading	Page No.
1.	Genome organization in geminiviruses	30
2.	Genome organization of nanoviruses	32
3.	Genome organization of caulimoviruses	33
4.	Genome organization of badnaviruses	34
5.	Linear map of begomovirus Rep protein	35
6.	Linear representation of rice tungro bacilliform, commelina yellow mottle, and cassava vein mosaic badnaviruses, and cauliflower mosaic caulimovirus genomes	38
7.	The RNAs of plant pararetroviruses	40
8.	Genome organization in genera of family *Potyviridae*	49
9.	Genome organization in genera of family *Sequiviridae*	51
10.	Genome organization in genera of family *Comoviridae*	52
11.	Genome organization in genera of sobemo-like supergroup	54
12.	Genome organization of pea enation mosaic enamovirus	56
13.	Genome organization in genera of carmo-like supergroup	57
14.	Genome organization in family *Tombusviridae*	60
15.	Genome organization of closteroviruses	62
16.	Genome organization in family *Bromoviridae*	65
17.	Genome organization in genera of proposed family *Tubiviridae*	69
18.	Genome organization in genera of proposed order *Tymovirales*	74
19.	Genome organization in viruses of proposed family *Potexviridae*	77
20.	Genome organization in genus *Nucleorhabdovirus*	78
21.	Genome organization in genus *Tospovirus* of family *Bunyaviridae*	79
22.	Genome organization in genus *Tenuivirus*	80
23.	Pairwise comparisons of expression strategies in alpha-like plant viruses	105
24.	Pairwise comparisons of expression strategies in sobemo-like, alpha-like, and picorna-like plant viruses	111
25.	A model of mechanism of heteroduplex-mediated recombination in brome mosaic virus	128
26.	Types of defective RNAs in plant viruses	132
27.	Two copy-choice models of mechanism leading to formation of deletions during replication of broad bean mottle bromovirus RNA	134
28.	Pathways of virus movement in plants	162
29.	Structure of movement proteins of tobacco mosaic tobamovirus, cauliflower mosaic caulimovirus, and cowpea mosaic comovirus	165
30.	A. Structure of brome mosaic bromovirus RNA3	205
	B. Symptom phenotype induced in *Chenopodium quinoa* by wild-type	

brome mosaic bromovirus and a variant   205
31. Induction of hypersensitive response and some biochemical
    responses involved   212
32. Schematic rod-like structure model for viroids of potato spindle
    tuber viroid family   227
33. Consensus phylogenetic tree for viroids   231
34. Schematic representation of hammerhead structures of avocado
    sunblotch, peach latent mosaic, and chrysanthemum chlorotic mottle
    viroids   233

LIST OF TABLES

S. No.	Heading	Page No.
1.	General organization of RNA deduced wholly or partly from nucleotide sequence	2
2.	Genome characteristics of viruses with single-stranded RNA	5
3.	General arrangement of genes in RNA genomes	6
4.	Functions of cauliflower mosaic virus genes	37
5.	Genome organization of rice dwarf virus, the genus *Phytoreovirus*	81
6.	Intrapopulation nucleotide diversities in some plant viruses	149
7.	Nucleotide diversities in some virus genes	151
8.	Symptom phenotypes specified by viral genes	202
9.	Identified *R* genes share similar features	213
10.	Viral proteins known to function as hypersensitive response elicitors	217
11.	Sequenced viroids with their abbreviations and sizes	226

LIST OF CONTRIBUTORS

Dr. Alexy A. Agranovsky
Department of Virology
A. N. Belozersky Institute of Physico-Chemical Biology
Moscow State University
Moscow 119899
Russia

Dr. Silvia Ambrós
Instituto de Biologia Molecular y Celular de Plantas (UPV-CSIC)
Universidad Politécnica de Valencia
46022 Valencia
Spain

Dr. Jozef. J. Bujarski
Presidential Research Professor
Plant Molecular Biology Center
Northern Illinois University
DeKalb, Illinois 60115
USA

Dr. James N. Culver
Center for Agricultural Biotechnology
Biotechnology Institute
University of Maryland
College Park, Maryland 20742
USA

Dr. Christopher D. Dardick
Molecular and Cell Biology Program
University of Maryland
College Park, Maryland 20742
USA

Dr. Ricardo Flores
Instituto de Biologia Molecular y Celular de Plantas (UPV-CSIC)
Universidad Politécnica de Valencia
46022 Valencia
Spain

Dr. A. Fraile
Departamento de Biotechnologia
Ciudad Universitaria
E. T. S. I. Agrónomos
28040 Madrid
Spain

Dr. Alexander W. E. Franz
Department of Virology
DLO Research Institute for Plant Protection (IPO-DLO)
6700 GW Wageningen
The Nethrelands

PD Dr. Thomas Frischmuth
Biologisches Institut
Abteilung für Molekularbiologie und Virologie der Pflanzen
Pfaffenwaldring 57
Universität Stuttgart
D-70550 Stuttgart
Germany

Dr. F. Garcia-Arenal
Departamento de Biotechnologia
Ciudad Universitaria
E. T. S. I. Agrónomos
28040 Madrid
Spain

Dr. Anne-Lise Haenni
Director of Research
Centre National de la Recherche Scientifique
Department Organisation et Expression du Genome
Institut Jacques Monod
CNRS/Universite Paris 7
2, place Jussieu – Tour 43
75251 Paris Cedex 05
France

Dr. François Héricourt
Centre National de la Recherche Scientifique
Department Organisation et Expression du Genome
Institut Jacques Monod
CNRS/Universite Paris 7
2, place Jussieu – Tour 43
75251 Paris Cedex 05
France

Dr. Isabelle Jupin
Chargee de Recherche
Centre National de la Recherche Scientifique
Department Organisation et Expression du Genome
Institut Jacques Monod
CNRS/Universite Paris 7
2, place Jussieu – Tour 43
75251 Paris Cedex 05
France

Dr. Scott M. Leisner
Department of Biology
University of Toledo
Toledo, OH 43606
USA

Dr. J. M. Malpica
Departamento de Protección Vegetal
CIT-INIA
28049 Madrid
Spain

Dr. Chuni L. Mandahar
Department of Botany
Panjab University
Chandigarh 160014
India

Dr. Sergy Yu. Morozov
Department of Virology
A. N. Belozersky Institute of Physico-Chemical Biology
Moscow State University
Moscow 119899
Russia

Dr. Beatriz Navarro
Instituto de Biologia Molecular y Celular de Plantas (UPV-CSIC)
Universidad Politécnica de Valencia
46022 Valencia
Spain

Dr. José-Antonio Navarro
Instituto de Biologia Molecular y Celular de Plantas (UPV-CSIC)
Universidad Politécnica de Valencia
46022 Valencia
Spain

Dr. Marcos de la Peña
Instituto de Biologia Molecular y Celular de Plantas (UPV-CSIC)
Universidad Politécnica de Valencia
46022 Valencia
Spain

Dr. A. L. N. Rao
Department of Plant Pathology
University of California
Riverside, California 92521
USA

Dr. Andrey Solovyev
Department of Virology
A. N. Belozersky Institute of Physico-Chemical Biology
Moscow State University
Moscow 119899
Russia

Dr. Frank van der Wilk
Department of Virology
DLO Research Institute for Plant Protection (IPO-DLO)
6700 GW Wageningen
The Nethrelands

Dr. Johannes F. J. M. van den Heuvel
Department of Virology
DLO Research Institute for Plant Protection (IPO-DLO)
6700 GW Wageningen
The Netherlands

PREFACE

Molecular biology is concerned with maintenance, transmission, and expression of genetic information at molecular level during replication, growth, and performance of different functions by an organism. It also seeks to explain the relationship between structure and function of biological molecules and how these relationships contribute to the operation and control of biochemical processes. Among others, structure and sequence analysis of nucleic acids, protein structure, molecular basis of gene expression involving the processes of transcription, translation and other steps, and determination of genetic maps are some of the important aspects of molecular biology.

Of the many technical advances that resulted in tremendous increase in our knowledge of molecular biology, the ability to remove a specific fragment of DNA from an organism, manipulate it in the test tube, and return it to the same or a different organism takes the preeminent place. This recombinant DNA technology or genetic engineering is central to molecular biology. Manipulation of RNA genome and its analysis for determining the functions of genes and gene products became much simpler, particularly due to two major advances. One, the conversion of single-stranded viral RNA to double–stranded cDNA clones by using reverse transcriptase and, second, the synthesis of infectious RNA transcripts *in vitro* from cloned cDNAs and their expression in bacteria. This could be possible due to the development of *in vitro* transcription systems. Development of techniques that specifically bring alterations in the desired gene sequences enable us to understand the role played by viral genes. Some other techniques like engineering chimeric proteins, genes, and viruses, use of reporter β-glucoronidase genes, use of RNase protection analysis and reverse transcriptase-polymerase chain reaction assay, use of transgenic plants and of cDNA and cRNA probes, and determination of nucleotide sequence of virus genomes have also substantially contributed to the development of molecular biology. The elucidation of complete genomic sequences of numerous viruses was particularly a great step forward. So are the isolation, sequencing, and expression of resistance genes.

Some other modern techniques are also being extensively used. Site-directed mutagenesis like substitution and deletion are routinely employed. This ability to make defined mutations in viral genomes at the DNA plasmid level and then to examine the effects of the mutation on the biology of viruses has allowed precise mapping of viral sequences involved in replication, regulation, and gene expression. In fact, it is becoming relatively simple to define precise viral sequences involved in these and other virus functions and in virus-host interactions. Use of molecular markers (random amplified polymorphic DNA, restriction fragment length polymorphism, and polymerase chain reaction) in tagging and mapping of different resistance genes, in construction of molecular maps, and in marker-assisted breeding has been very useful. Transfection and agroinfection as alternative means of infecting plant cells are important techniques. Availability of computer data base and computer analysis by different types of programs has proved of immense value.

The application of the DNA recombinant and other improved techniques ushered in a new era in the study of plant viruses so that the face of plant virology is now completely changed. Emphasis now is on the study of viral genomes, genetic maps, genes and gene expression, gene products, and genetic basis of virus function and biological properties. In short, the studies of the molecular and genetic aspects of plant viruses now get precedence. This is shown by the clearly observable shift towards genetic phenomena of plant viruses in reviews and research papers being published in the old traditional virology journals and annual review publications. In addition, progressively greater number of research papers are now appearing in new journals (*Cell, Gene, Plant Cell, Virus Genes and Development, Trends in Genetics,* and *Trends in Microbiology*) in which hardly any paper on molecular and genetic aspects of plant viruses was published earlier, apart from *Nucleic Acids Research* and *EMBO Journal* in which such papers had started appearing earlier.

But the take-off was slow. It is only about two decades hence that plant virologists started experimenting with the genetic basis of some viral phenomena and virus-induced reactions in plants. But now such studies are a raging torrent causing an information explosion in molecular and genetic aspects of plant viruses. However, there is hardly a book that emphasizes these aspects. A definite need, therefore, exists for a book that should ideally analyze, collate, and review such published information. This book seeks to do this and correct an aberration. It makes an effort to understand the complexity of plant viral genomes and their functioning, and deals exclusively with the molecular basis of nearly all plant virus-directed phenomena in host plants. It also shows the tremendous progress made in molecular biology of plant viruses.

The picture that emerges is fascinating, illuminating, as well as mind boggling since much has been made of the very limited information tapped on these infinitely small genetic materials. Nature seems to have plugged in some message and wired in some function in every fold, every nook, every cranny, every loop, and every turn of the genome. Evolution has ensured that not a nucleotide is out of place or out of context. In its proper environment each nucleotide, together with other nucleotides, does something essential for the virus to survive and replicate, for the various known or as yet unknown functions it performs, and for a virus to be what it is.

The progress in the study of plant viruses has always been correlated with the development of new techniques. This has been beautifully summed up in the following paragraphs that Morozov and Solovyev wrote in their chapter for this book but have instead been incorporated here.

"Twenty years have passed since a new era in studies of plant virus genomes started. Until late seventies, the main methods used for genome mapping were *in vitro* translation, electrophoretic analysis of virus-infected tissues, and studies of pseudorecombinants of multicomponent plant viruses and their deletion variants *in planta* and in isolated protoplasts (reviews: Atabekov and Morozov, 1979; Davies, 1979; Zaitlin, 1998). Direct RNA sequencing was efficient for rather short, usually terminal, RNA fragments only (reviews: Davies, 1979; Hirth and Richards, 1981). A breakthrough in virus genome studies in eighties came from elaboration of new methods for RNA reverse transcription and cDNA cloning, restriction enzyme mapping, and rapid sequencing as well as computer-assisted comparisons of the determined nucleotide and protein sequences. Pioneering advances of the new era were sequences of whole tobacco mosaic virus genome (Goelet *et al.*, 1982), RNA 3 of brome mosaic virus (Ahlquist *et al.*, 1981), and cucumber mosaic virus (Gould and Symons, 1982). Further progress in studies of viral RNA genomes revealed a broad diversity of the plant virus gene arrangements as well as the new principles of viral genome expression, evolution, and taxonomy (Reviews: Zaccomer *et al.*, 1995; Roossinck, 1997; Mayo and Pringle, 1998).

Molecular taxonomy of plant viruses is now not far from completion (Mayo and Pringle, 1998; Pringle, 1998a). During last 15 years, basic principles in delineation of RNA virus taxa shifted from formal characteristic (such as host specificity, virion morphology, size of RNA segments and level of genome fragmentation) to the evolutionary relationships based on the nucleotide and protein sequence similarities and delimitation of phylogenetically conserved gene blocks. Most representatives information pertinent to taxonomy comes from comparisons of replication related protein genes, especially RNA-dependent RNA polymerase gene (Koonin, 1991; Koonin and Dolja, 1993; Zanotto *et al.*; 1996). From phylogenetic point of view, these genes represent 'core' elements of RNA virus genomes whereas other gene modules are considered as 'accessory' ones (Koonin and Dolja, 1993). Various combinations of replicative and accessory genes (as well as different routes of their expression, see Chapter 4), arising mostly through gene shuffling and splitting of the preexisted modules, gave rise to the present diversity of virus RNA genomes (Gibbs, 1987; Morozov *et al.*, 1989; Simon and Bujarski, 1994; Roossinck, 1997)".

Since virus classification is still not complete, certain areas of confusion continue to persist. To cite only one example, in contrast to the placement of genus *Luteovirus* by Morozov and Solovyev, van den Heuvel *et al.* (Chapter 8), following Pringle (1998b), place it in the proposed family *Luteoviridae* to which two other genera of *Polerovirus* and *Enamovirus* are also assigned. A few other such unsettled cases of plant viruses, genera and families perforce occur in different chapters.

Genetics is the study of structure, function, and expression of genes and their inheritance and is often defined as a science that deals with heredity and variation. Plant viruses contain a very limited genetic information, which varies from four to ten products only (Davies and Hull, 1982; van Vloten-Doting and Neeleman, 1982). How could this meager information make the RNA plant viruses so successful as to be the dominant amongst viruses? Part of the answer could be that some of the plant viral proteins [like capsid protein, movement protein, and helper-component proteinase (HC-Pro)] are multifunctional and highly versatile. In all these cases, different domains of the multifunctional protein participate in different functions. This explains the presence of a few genes but the successful performance of lot many functions by plant viruses. Another part of the answer could be the presence in a genome of the non-coding regions that control certain viral activities.

RNA is labile; still it carries genetic information and does so successfully as more than half the genera and still more of the definitive virus species infecting eukaryote hosts contain RNA genomes (Murphy *et al.*, 1995). It is so despite the fact that, since copying is not proofread during replication as in DNA (Domingo and

Holland, 1994), errors rapidly build-up and accumulate. This is contrary to the selection of the fittest concept and has been called "Muller's ratchet" (Lai, 1995). This may be responsible for the small size (8-15 kb) of RNA genomes as well as for the several functions being performed by the same viral protein molecules. Chapter 1 deals with the RNA as the viral genetic material.

The chapter 2 deals with DNA as the viral genome. No effort is made here to fathom the reason as to why, surprisingly, the DNA viruses number far less than the RNA viruses. Only the characteristics of DNA as the viral genetic material are discussed.

Possibly the first report about the location of a gene on the genome of a plant virus pertains to tobacco mosaic virus (TMV). Gene responsible for local lesion on *Nicotiana sylvestris* was reported to be situated about 1,500 to 2,000 nucleotides from the 3` end of the RNA while the coat protein gene was placed in the 5` half of the genome (Kado and Knight, 1966,1968). Changes in cytological events and symptom expression in host were determined by still another gene (Kado *et al.,* 1968; Liu and Boyle, 1972). Since then, there is an unbelievable advancement, as shown by Morozov and Solovyev in chapter 3, in our knowledge about the gene organization in genomes of different RNA plant viruses.

An infecting virus, upon entering a cell, has to employ all the relevant genes for its survival and perpetuation. Most of the plant virus RNAs are polycistronic and have evolved an array of strategies to synthesize the translation products of the different genes in the polycistronic message. Translation of mRNA is important in the post-transcriptional and overall regulation of eukaryotic gene expression (Kozak, 1989; Gallie, 1993). Chapter 4 deals with these strategies that plant viruses adopt during gene expression.

The scanning model of translation regards the eukaryotic mRNA as a passive messenger, which is decoded linearly from 5` to 3` end. In the late 1980s, this concept began to change with the publication of reports concerning the existence of frameshifting and readthrough events during translation. Some viral messages seemed to be actively involved in forcing the ribosomes to perform certain complicated gymnastic steps (Atkins *et al.,* 1990). Moreover, structures and nucleotide sequences present in the 3` untranslated region of mRNAs (Gallie, 1993; Standart and Jackson, 1994; Jacobson, 1996) were suggested to regulate various cellular processes in response to environment and development. Additionally, translation initiation at the 5` end was controlled directly by specific regulatory proteins (Dubnau and Struhl, 1996) or indirectly by the length of poly (A) tail (Sheets *et al.,* 1994). Thus, mRNAs are not mere passive structures and neither are decoded in a simple linear fashion. All this and much else are treated in chapter 4.

Way back in 1926, McKinney reported the first strain of a virus. Jensen (1933, 1936) and Price (1934) followed him. After that there was no looking back till today the occurrence of strains is regarded as a self-evident truth. Genetic variability is now seen to have great importance in virus evolution, pathology, and adaptation to new hosts and changed environment, besides the several other advantages conferred on plant viruses. Bujarski discusses genetic basis of variability in plant viruses in chapter 5.

The extensive diversity of RNA arises due to a combination of several evolutionary factors, including high mutation rates, genome reassortment, genetic recombination, and gene duplication (Dolja and Carrington, 1992; Gibbs, 1995; Koonin and Dolja, 1993; Lai, 1995). Out of these, mutation and recombination are more important. The huge variability in biological and serological properties has been assigned to the frequent generation of mutants due to high error rate of viral RNA polymerase during replication because of the absence of proofreading activity of RNA-dependent RNA polymerase (Domingo and Holland, 1994). The mutation in itself is thus sufficient to explain all the known variation in plant viruses.

Such high mutational loads could have created several serious problems for the viruses. Viruses, therefore, evolved the means of correcting the high error rates by recombining mutant genomes with wild-type genomes. Recombination in this way eliminates detrimental mutations and maintains the functional integrity of a genome as well as ensures genome conservation. Bujarski and Kaesberg (1986) first reported genetic recombination in plant viruses. Since then it has been noted in several other cases but is not regarded to be common feature of plant virus replication. Yet it has been assigned some important functions like repair of genomes, contribution towards evolution of viruses and viral genomes, and in generation of new virus groups (Goldbach, 1986; Morozov et al., 1989; Keese and Gibbs, 1993; Koonin and Dolja, 1993).

A very efficient recombination process can be dangerous. It could lead to more common formation of parasitic RNA forms like satellites, defective interfering RNAs, and chimeric RNAs, which would compete with viral genome for replication but also be harmful pathologically. On the other hand, absence of or less efficient recombination process may significantly decrease the various advantages of recombination, including the capacity of viruses to adapt to evolving hosts by incorporating evolutionarily successful genes or modules from other viruses. Consequently, a balance must be existing between the two extremes in plant viruses. Mechanism of recombination has been extensively reviewed recently (MacFarlane, 1997; Simon and Bujarski, 1994; Nagy and Simon. 1997) and is also discussed in chapter 5.

In contrast, the opinion of Falk and Bruening (1994) does not fit into the above observations concerning the role of variability. They state: ' Even though ordinary infections in agricultural and natural settings continually provide viruses with multiple opportunities to interact, new viral diseases are usually due to minor variants of already known viruses and not to new viruses of recombinant origin. In fact, existing viruses are surprisingly stable, having evolved into fit competitors that evolve only slowly." Prof. Garcia-Arenal has further elaborated this theme in chapter 6, which is devoted to viral diversity. His conclusion: genetic stability is the rule rather than an exception and high potential to vary need not result in high variation rates of virus populations.

All the viral genetic variability as well as genetic diversity would have come to a naught if the viruses were to be confined to the initially infected cell and had not evolved the capacity to move out of and away from it. Studies with a *ts* TMV mutant indicated the involvement of a 30K protein, called movement protein and coded by viral genome, in its cell-to-cell movement (Nishiguchi et al., 1978). Since then the progress in the genetic basis of virus transport in plants has been remarkable. Leisner brings this out clearly in chapter 7. Efficiency of virus movement has far reaching consequences. It determines pathogenicity, virulence, and, in some case, host range of plant viruses (Atabekov and Taliansky, 1990; Maule, 1991).

There is practically an unlimited literature on virus transmission by vectors. However, its molecular basis has somehow remained a neglected field. The start was made in the last decade and soon sufficient literature had accumulated for Pirone to review in 1991 the genes and gene products involved in insect transmission of plant viruses. van den Heuvel does so again in this book in chapter 8. The genetic basis of virus transmission by seeds and pollen is also discussed.

Using a combination of protein stripping and nucleic acid mutagenesis, Kado and Knight (1966) mapped the location of the local lesion gene in tobacco mosaic virus. This was followed by the use of pseudorecombinants in multipartite plant viruses for garnering information on location of genes involved in symptomatology, particularly lesion size and lesion colour, in several hosts. Bruening (1977) and Van Vloten-Doting and Jaspars (1977) reviewed this. Still, van Vloten-Doting (1987) in the first review on the molecular basis of disease symptoms had to make some speculations about how virus gene products mediate plant disease symptoms because the existing knowledge about it at that time was very little. It still remains limited with the realization that disease symptoms in infected plants are caused by the activity of one or more viral genes and/or due to the activation of some host responses. Rao discusses the molecular basis of symptomatology in chapter 9, which in fact is one of the first real overview of this important topic.

The concept of gene-for-gene interaction envisages resistance/susceptibility genes in plants and the corresponding avirulence/virulence genes in the pathogen and that resistance is expressed only when a plant bears a resistance (R) gene that corresponds to an avirulence (avr) gene in a pathogen. Interaction between the products of these R and avr genes ultimately determines the resistance of a plant to virus infection (Bent, 1996; Baker et al., 1997). The number of characterized avr genes is more than 30. Several R genes have been isolated (Staskawicz et al., 1995), including the R gene from tobacco (which is the gene N operative against TMV) (Whitham et al., 1994). The interaction between R and avr genes in plant (tobacco)-virus (TMV) interaction in N gene-mediated resistance against TMV is the best studied case (Goodman and Novacky, 1994). However, the first example of gene-for-gene interaction is tomato carrying the Tm-1 resistance gene against TMV (Meshi et al., 1988). The other virus-host combinations in which such gene-for-gene interactions have been worked out and various other aspects of this interaction are dealt with by Culver in chapter 10.

Flores et al. briefly discuss in chapter 11 the molecular basis of various phenomena excited by the viroids.

Since the first report of the development of a transgenic plant (tobacco) expressing a virus (TMV) gene (capsid protein) (Powell et al., 1986), many other such transgenic plants have been developed. These plants are resistant to infection against the transgene-providing virus. In nature, a related strain of the transgene-providing virus may infect the transgenic plants so that this plant for all practical purposes contains two viruses. It could be a dangerous situation since recombination between these viruses may generate new strains/viruses with all the possible dangers. This and other genetic phenomena operative in transgenic plants are discussed in chapter 12.

Transgenic plants have been put to several uses. One of which is to make a transgenic plant produce a useful product in greater quantity by incorporating more than one copies of the relevant gene. But, instead, surprisingly the production of the concerned product got inhibited. It was interpreted as due to silencing of all the genes when they are present in more than one copy. Gene silencing has been invoked to explain some of the well-established phenomena in plant virus-host interaction. Chapter 13 for the first time deals with these.

REFERENCES

Ahlquist, P., Luckow, V., and Kaesberg, P. (1981). Complete nucleotide sequence of brome mosaic virus RNA 3. *J. Mol. Biol.* 153, 23-38.

Atabekov, J. G., and Morozov, S. Yu. (1979). Translation of plant virus messenger RNAs. *Adv. Virus Res.* 25, 1-91.

Atabekov, J. G., and Taliansky, M. E. (1990). Expansion of a plant virus-coded transport function by different viral genomes. *Adv. Virus Res.* 38, 201-248.

Atkins, J. F., Weiss, R. B., and Gesteland, R. F. (1990). Ribosome gymnastics–degree of difficulty 9.5, style 10.0. *Cell* 62, 413-423.

Baker, B., Zambryski, P., Staskawicz, B., and Dinesh-Kumar, S. P. (1997). Signalling in plant-microbe interactions. *Science* 276, 726-733.

Bent, A. F. (1996). Plant disease resistance genes: Function meets structure. *Plant Cell* 8, 1757-1771.

Bruening, G. (1977). Plant covirus systems: Two component systems. *Comp. Virol.* 11, 55-141.

Bujarski, J. J., and Kaesberg, P. (1986). Genetic recombination between RNA components of a multipartite plant virus. *Nature (London)* 321, 528-531.

Davies, J. W. (1979). The translation of plant viral nucleic acids in extracts from eukaryotic cells. *In* "Nucleic Acids in Plants" (T. C. Hall, a nd J. W. Davies, Eds.), Vol. II, pp. 111-149. CRC Press, Boca Raton. FL.

Davies, J. W., and Hull, R. (1982). Genome expression of plant positive-strand RNA viruses. *J. Gen. Virol.* 61, 1-14.

Dolja, V. V., and Carrington, J. C. (1992). Evolution of positive-strand RNA viruses. *Semin. Virol.* 3, 315-326.

Domingo, E., and Holland, J. J. (1994). Mutation rates and rapid evolution of RNA viruses. *In* "The Evolutionary Biology of Viruses" (S. S. Morse, Ed.), pp. 161-184. Raven Press, New York.

Dubnau, J., and Struhl, G. (1996). RNA recognition and translational regulation by a homeodomain protein. *Nature (London)* 379, 694-699.

Falk, B. W., and Bruening, G. (1994). Will transgenic crops generate new viruses and new diseases? *Science* 263, 1395-1396.

Gallie, D. R. (1993). Posttranslational regulation of gene expression in plants. *Annu. Rev. Plant Physiol Plant Mol. Biol.* 44, 77-105.

Gibbs, A. (1987). Molecular evolution of viruses: 'Trees', 'clocks', and 'modules'. *J. Cell. Sci. Suppl.* 7, 319-337.

Gibbs, M. (1995). *In* "Molecular Basis of Virus Evolution" (A. J. Gibbs, C. H. Calisher, and F. Garcia-Arenal, Eds.), pp. 351-368. Cambridge University Press, New York.

Goelet, P., Lomonossoff, G. P., Butler, P. J. G., Akam, M. E., Gait, M. J., and Karn. J. (1982). Nucleotide sequence of tobacco mosaic virus RNA. *Proc. Nat. Acad. Sci. USA* 79, 5818-1822.

Goldbach, R. (1986). Molecular evolution of plant RNA viruses. *Annu. Rev. Phytopathol.* 24, 289-310.

Goodman, R. N., and Novacky, A. J. (1994). The Hypersensitive Reaction in Plants to Pathogens, A Resistance Phenomenon. Amer. Phytopathol. Soc., St. Paul, MN.

Gould, A. R., and Symons, R. H. (1982). Cucumber mosaic virus RNA 3. Determination of the nucleotide sequence provides the amino acid sequences of protein 3A and viral coat protein. *Eur. J.*

Biochem. 126, 217-226.

Hirth, L., and Richards, K. E. (1981). Tobacco mosaic virus: Model for structure and function of a simple virus. *Adv. Virus Res.* 26, 145-199.

Jacobson, A. (1996). *In* "Translational Control." (J. W. B. Hershey, M. B. Mathews, and N. Sonenberg, Eds.), pp. 451-480. Cold Spring Harbor Lab. Press, Cold Spring Harbor, New York.

Jensen, J. H. (1933). Isolation of yellow mosaic viruses from plants infected with tobacco mosaic. *Phytopathology* 23, 964-974.

Jensen, J. H. (1936). Studies on the origin of yellow mosaic viruses. *Pytopathology* 26, 266-277.

Kado, C. I., and Knight, C. A. (1966). Location of a local lesion gene in tobacco mosaic virus RNA. *Proc. Natl. Acad. Sci. USA* 55, 1276-1283.

Kado, C. I., and Knight, C. A. (1968). The coat protein genes of TMV. I. Location of gene by mixed infection. *J. Mol. Biol.* 36,15-23.

Kado, C. I., van Regenmortel, M. H. V., and Knight, C. A. (1968). Studies on some strains of tobacco mosaic virus in orchids. I. Biological, chemical and serological studies. *Virology* 34. 17-24.

Keen, N. T. (1990). Gene-for-gene complementarity in plant-pathogen interactions. *Annu. Rev. Genet.* 34, 447-463.

Keen, N. T. (1992). The molecular biology of disease resistance. *Plant Mol. Biol.* 19, 109-122.

Keese, P., and Gibbs, A. (1993). Plant viruses: Master explores of evolutionary space. *Curr. Opin. Genet. Dev.* 3, 873-877.

Koonin, E. V. (1991). The phylogeny of RNA-dependent RNA polymerase of positive-strand RNA viruses. *J. Gen. Virol.* 72, 2197-2206.

Koonin, E. V., and Dolja, V. V. (1993). Evolution and taxonomy of positive-strand viruses: Implications of comparative analysis of amino acid sequences. *Crit. Rev. Biochem. Mol. Biol.* 28, 375-430

Kozak, M. (1989). The scanning model for translation :An update. *J. Cell Biol.* 108, 229-241.

Lai, M. M. C. (1995). Recombination and its evolutionary effect on viruses with RNA genomes. *In* "Molecular Basis of Virus Evolution" (A. J. Gibbs, C. H. Calisher, and F. Garcia-Arenal, Eds.), pp. 119-132. Cambridge University Press, New York.

Liu, K. C., and Boyle, J. C. (1972). Intracellular morphology of two tobacco mosaic virus strains in,and cytological responses of systemically susceptible potato plants. *Phytopathology* 62, 1303-1311.

MacFarlane, S. A. (1997). Natural recombination among plant virus genomes: Evidence from tobraviruses. *Semin. Virol.* 8, 25-31.

Maule, A. J. (1991). Virus movement on infected plants. *Crit. Rev. Plant Sci.* 9, 457-473.

Mayo, M. A., and Pringle, C. R. (1998). Virus taxonomy – 1997. *J. Gen. Virol.* 79, 649-657.

McKinney, H. H. (1926). Virus mixtures that may not be detected in young tobacco plants. *Phytopathology* 16, 893.

Meshi, Y., Watanabe, Y., Saito, T., Sugimoto, A., Maeda, T., and Okada, Y. (1987). Function of the 30kDa protein of tobacco mosaic virus: Involvement in cell-to-cell movement and dispensability for replication. *EMBO J.* 6, 2557-2563.

Morozov, S. Yu., Dolja, V. V., and Atabekov, J. G. (1989). Probable reassortment of genomic elements among elongated RNA-containing viruses. *J. Mol. Evol.* 29, 52-62.

Murphy, F. A., Fauquet, C. M., Bishop, D. H. L., Ghabrial, S. A., Jarvis, A. W., Martelli, G. P., Mayo, M. A., and Summers, M. D. (1995). Virus Taxonomy: Classification and Nomenclature. Sixth Rep. Inter. Committee on Taxonomy of Viruses. *Arch. Virol., Suppl.* 10. Springer-Verlag, New York. pp. 586.

Nagy, P. S., and Simon, A. E. (1997). New insights into the mechanisms of RNA recombination. *Virology* 235, 1-9.

Nishiguchi, M., Motoyoshi, F., and Oshima, N. (1978). Behavior of a temperature-sensitive strain of tobacco mosaic virus in tomato leaves and protoplasts. *J. Gen. Virol.* 39,53-61.

Pirone, T. P. (1991). Viral genes and gene products that determine insect transmissibility. *Semin. Virol.* 2, 81-87.

Powell, P. A., Nelson, R. S., De, B., Hoffmann, N., Rogers, S. G., Fraley, R. T., and Beachy, R. N. (1986). Delay of disease development in transgenic plants that express the tobacco mosaic virus coat protein gene. *Science* 232, 738-743.

Price, W. C. (1934). Isolation and study of some yellow strains of cucumber mosaic. *Phytopathology* 24, 743-761.

Pringle, C. R. (1998a). The universal system of virus taxonomy of the International Committee on

Virus Taxonomy (ICTV), including new proposals ratified since publication of the Sixth Report in 1995. *Arch. Virol.* 143, 203-210.

Pringle, C. R. (1998b). Virus taxonomy-San Diego 1998. *Arch. Virol.* 143, 1450-1459.

Roossinck, M. J. (1997). Mechanisms of plant virus evolution. *Annu. Rev. Phytopathol.* 35, 191-209.

Sheets, M. D., Fox, C. A., Hunt, T., Vandewoude, G., and Wickens, M. (1994). The 3`-untranslated regions of C-MOS and cyclin mRNAs stimulate translation by regulating cytoplasmic polyadenylation. *Gene Dev.* 8, 926-938.

Simon, A. E., and Bujarski, J. J. (1994). RNA-RNA recombination and evolution in virus infected plants. *Annu. Rev. Phytopathol.* 32, 337-362.

Standart, N., and Jackson, R. J. (1994). Regulation of translation by specific protein/mRNA interactions. *Biochemie* 76, 867-879.

Staskawicz, B. J., Ausubel, F. M., Baker, B. J., Ellis, J. G., and Jones, J. D. G. (1995). Molecular genetics of plant disease resistance. *Science* 268, 661-667.

Van Vloten-Doting, L. (1987). Symptoms of plant disease mediated by plant virus gene products. *In* "Molecular Basis of Virus Disease" (W. C. Russell, and J. W. Almond, Eds.). Cambridge Univ. Press, Cambridge.

Van Vloten-Doting, L., and Jaspars, E. M. J. (1977). Plant covirus systems: Three-component systems. *Comp. Virol.* 11, 1-53.

Van Vloten-Doting, L., and Neeleman, L. (1982). Translation of plant virus RNAs. *In* "Nucleic Acids and Proteins in Plants" (B. Parthier and D. Boulter, Eds), Vol. II. Structure, Biochemistry and Physiology of Nucleic Acids, pp.337-367. Springer-Verlag, Berlin.

Whitham, S., Dinesh-Kumar, S. P., Choi, D., Hehl, R., Corr, C., and Baker, B. (1994). The product of tobacco mosaic virus resistance gene *N*: Similarity to Toll and the interleukin-1 receptor. *Cell* 78, 1101-1115.

Zaccomer, B., Haenni, A.-L., and Macaya, G. (1995). The remarkable variety of plant RNA virus genomes. *J. Gen. Virol.* 76, 231-247.

Zaitlin, M. (1998). Elucidation of the genome organization of tobacco mosaic virus. *Phil. Trans. Biol. Sci.* (*London*), in press

Zanotto, P. M., Gibbs, M. J., Gould, E. A., and Holmes E. C. (1996). A reevaluation of the higher taxonomy of viruses based on RNApolyymerases. *J. Virol.* 70, 6083-6096.

ACKNOWLEDGEMENTS

This book could not have been possible without the essential inputs from four quarters: the library of the Institute of Microbial Technology (IMTECH), Chandigarh, the various contributors, my family, and the publishers. I am thankful to the Director Dr. Amit Ghosh, Dr. Naresh, and the staff of the extremely well maintained, managed, and stocked library of the IMTECH. My fullest thanks are to them. This library is the base upon which all the superstructure of this book has been built. Drs. S. M. Paul Khurana and I. D. Garg of the Central Potato Research Institute, Shimla were also of great help for collection of literature.

Several of the contributors are very distinguished plant virologists who have made basic contributions to the molecular aspects of plant virology. It is on the foundations laid by such people that molecular plant virology rests. Their readily agreeing to contribute to this book is in itself an indication of the importance of this project. It also gives a solid base to it. Research contributions of the other authors are also very important to some of the just developing aspects of molecular plant virology so that its future is tied down and secure due to the efforts of such workers. Each chapter sums up the present state of our knowledge of the respective field. Together, all the authors have made this book a valuable source of information on nearly all aspects of molecular plant virology. My gratitude is to all of them.

Nothing could have been accomplished without a working knowledge of the computer. Several persons provided me all the desired inputs. They include Profs. K. K. Dhir, R. K. Kohli, and S C. Kaushal, Dr. Daisy Rani, Dr. Kanwaljit Singh, and Messers Rajiv, Amarjit Singh, Sushil Sharma, H. S. Kang, and the people at the Hightech Computers. I freely asked for and always got all the needed help and directions from them. The support extended by my cousin, Mr. P. K. Sharma, and his colleagues is gratefully acknowledged. Mr. Manish Gupta was always of immense help, which I often sought. Without him, I would have floundered for a much longer period. The cooperation and assistance extended by Mr. S. K. Sharma and Mr. Jitendra Maan during final preparation of the manuscript was an asset. I profusely thank all of them. Absence of any one of them at the particular time may have spelled much trouble for me.

Vinod, Shelly, Atul, Abha, Akash, Abhishek, and Arshya – they are all so dear – allowed me all the time and affection that I needed. But above all, I am beholden to Aruna, my wife, who has always been a support in all ways imaginable. All of them together created an ambience conducive to my work. My debt to Dr. Karl Maramorosch will ever stand. I got his guidance whenever I sought it. My friend and colleague Prof. S. S. Kumar was always a source of strength.

My sincere thanks are due to the Publishers of this book. Ultimately, it is they who have made this book possible and believed in its merit to agree to publish it.

1 GENOME OF RNA VIRUSES

François Héricourt, Isabelle Jupin
and Anne-Lise Haenni

INTRODUCTION

The vast majority of plant virus groups contain an RNA genome, most frequently of positive polarity. This chapter provides an updated overview of genome of such plant viruses stressing the functions of coding and non coding regions of genome in virus amplification. Three concise tables accompany the text. In each table, the viruses are presented in alphabetical order by genera as officially recognized in the Plant Virus Classification of the International Committee on Taxonomy of Viruses (1996). Table 1 presents general features of RNA genomes and their associated RNAs like satellite RNA (sat-RNA) and defective interfering RNA (DI). Table 2 shows further major characteristics of viruses with a single-strand RNA (ssRNA) genome. Table 3 classifies the viruses in supergroups where these have been established, outlines gene arrangements (also consult chapter 3), and gives the main translation strategies (chapter 4) used by plant RNA viruses. Recent key references as well as certain other references, not mentioned in the review by Zaccomer et al. (1995), are provided. A few virus groups are not presented, due to lack of sufficient sequence data concerning them. These are the fabaviruses, betacryptoviruses and cytorhabdoviruses. Virus-related RNA molecules (DI and sat-RNA) and viroids are not discussed here.

SPECIFIC GENE CLASSES

Polymerase-Associated Functions

In all plant viruses, replication of viral genome is believed to involve at least one or two viral proteins containing specific motifs, whose structure and function appear very similar between plant and animal viruses (reviewed in Buck, 1996).

An *in vitro* polymerase activity associated with an enzyme complex isolated from plants operates in various viruses (reviewed in de Graaf and Jaspars, 1994). Although there were some early debates as to the origin of this enzymatic activity, it is now clear that the RNA-dependent RNA (RdRp) activity is encoded by a specific ORF present in viral genome. In a few cases, the RdRp activity of this ORF product has been demonstrated biochemically (Hong and Hunt, 1996). In most cases, putative RdRps have been identified based on the presence of conserved amino acid sequence motifs, including a GDD triplet (Argos, 1988; Koonin and Dolja, 1993),

Table 1. General organization of RNA deduced wholly or partly from nucleotide sequence

Genus [a]	Virus	ss/ds	RNA segments	RNA (nucleotides) [b]	Sat/DI	Accession number	Notes
Alfamovirus F: *Bromoviridae*	AMV[c]	ss	3	1: 3644 2: 2593 3: 2142		L00163 K02702 K03542	
Alphacryptovirus F: *Partiviridae*	BCV3	ds	2	2: 1607		S63913	
Bromovirus F: *Bromoviridae*	BMV	ss	3	1: 3234 2: 2865 3: 2117		X02380 X01679 V00099	RNA3: oligo(A) in intergenic region; DI in BBMV
Bymovirus F: *Potyviridae*	BaYMV	ss	2	1: 7632 2: 3585		D01091 D01092	
Capillovirus	ASGV	ss	1	6496		D14995	
Carlavirus	PVM	ss	1	8535		X53062	
Carmovirus F: *Tombusviridae*	CarMV	ss	1	4003		X02986	Sat-RNA and DI in TCV
Closterovirus F: *Closteroviridae*[d]	BYV	ss	1	15480		X73476	DI in CTV[1]
Comovirus F: *Comoviridae*	CPMV	ss	2	B: 5889 M: 3481		X00206 X00729	
Cucumovirus F: *Bromoviridae*	CMV	ss	3	1: 3357 2: 3050 3: 2216	Sat-RNA, DI	D00356 D00355 D10538	
Dianthovirus F: *Tombusviridae*[d]	CRSV	ss	2	1: 3756 2: 1394		L18870 M88589	
Enamovirus	PEMV	ss	2	1: 5706 2: 4253	Sat-RNA	L04573 S53233	
Fijivirus F: *Reoviridae*	MRDV	ds	10	S6: 2193		X55701	
Furovirus	SBWMV	ss	2	1: 7099 2: 3593	DI	L07937 L07938	
	BNYVV	ss	4	1: 6746 2: 4612 3: 1774 4: 1108	Sat-RNA, DI	D00115 X04197 M36894 M36896	
Hordeivirus	BSMV	ss	3	1: 3768 (α) 2: 3289 (β) 3: 3164 (γ)		J04342 X03854 X52774	Internal poly(A) of 21 nt (RNA1) and 20 nt (RNA2 and 3); RNA3 in-frame direct tandem repeat of 366 nt in ORF1
Idaeovirus	RBDV	ss	2	1: 5449 2: 2231		S51557 S55890	
Ilarvirus F: *Bromoviridae*	TSV	ss	3	1: 3491[2] 2: 2926[2] 3: 2205		U80934 U75538 X00435	
Luteovirus	BYDV	ss	1	5677	Sat-RNA	X07653	Three subgroups: A: BYDV (PAV-isolate); B: PLRV; C: SbDV

Machlomovirus *F:* *Machlomoviridae*[d]	MCMV	ss	1	4437		X14736	
Marafivirus	OBDV[3]	ss	1	6509		U87832	
Necrovirus *F: Tombusviridae*[d]	TNV	ss	1	3759	Sat-virus	D00942	
Nepovirus *F: Comoviridae*	TBRV	ss	2	1: 7356 2: 4662	Sat-RNA	D00322 X04062	
Nucleorhabdovirus *F: Rhabdoviridae*	SYNV	ss	1	13720	DI	L32603	
Oryzavirus *F: Reoviridae*	RRSV	ds	10	S4: 3823[4] S7: 1938[5] S9: 1132[6] S10: 1162[5]		U66714 U66713 L38900 U66712	S1-S3, S5, S6: sequences established, accession numbers available; unpublished
Phytoreovirus *F: Reoviridae*	WTV	ds	12	S4: 2565 S5: 2613 S6: 1700 S7: 1726 S8: 1472 S9: 1182 S10: 1172 S11: 1128 S12: 851	DI	M24117 J03020 M24116 M77019 J04344 M24115 M24114 M77020 M77021	S1-S12 sequence of RDV established[7]
Potexvirus	PVX	ss	1	6435		D00344	DI in ClYMV; Sat- RNA in BaMV
Potyvirus *F: Potyviridae*	PVY	ss	1	9704		D00441	
Rymovirus *F: Potyviridae*	BrSMV[8]	ss	1	9672		Z48506	
Sequivirus *F: Sequiviridae*	PYFV	ss	1	9871		D14066	
Sobemovirus	SBMV	ss	1	4194	Sat-RNA (viroid- like)	M23021	Sat-virus in PMV, a possible member of the group
Tenuivirus	RStV	ss	4	1: 8970[9] 2: 3514 3: 2475 4: 2137		D31879 D13176 D01094 D01039	RNA5 (1317 nt) in MStV; RNA5 (2704 nt) and RNA6 (2584 nt) in RGSV[10]
Tobamovirus	TMV	ss	1	6395		J02415	Sat-virus in TMGMV
Tobravirus	TRV	ss	2	1: 6791 2: 1799 1905 3389		D00155 X03241 X03686 X03955	Widely different lengths of RNA2 depending on the strain
Tombusvirus *F: Tombusviridae*	TBSV	ss	1	4776	Sat-RNA, DI RNA	M21958	
Tospovirus *F: Bunyaviridae*	TSWV	ss	3	L: 8897 M: 4821 S: 2916	DI	D10066 S48091 D00645	
Trichovirus	ACLSV	ss	1	7555		M58152	
Tymovirus	TYMV	ss	1	6318		X07441	

| *Umbravirus* | GRV[11] | ss | 1 | 4019 | Sat-RNA[12] | Z69910 | |
| *Waikavirus* F: *Sequiviridae* | RTSV | ss | 1 | 12484 | | M95497 | |

a: For genera that have been associated into families, the name of the family (F) is added
b: Excluding poly(A) tails   c: Underlined virus abbreviations indicate type-member
d: Plant Virus Classification of the International Committee on Taxonomy of Viruses (1996)
References: *1*, Mawassi *et al.* (1995); *2*, Scott *et al.* (1998); *3*, Edwards *et al.* (1997); *4*, Upadhyaya *et al.* (1998); *5*, Upadhyaya *et al.* (1997); *6*, Upadhyaya *et al.* (1995); *7*, Uyeda *et al.* (1995); *8*, Götz & Maiss, (1995); *9*, Toriyama *et al.* (1994); *10*, Toriyama *et al.* (1998); *11*, Taliansky *et al.* (1996); *12*, Murant *et al.*(1988)

ABBREVIATIONS

Viruses and corresponding genus: ACLSV = apple chlorotic leafspot trichovirus; AMV = alfalfa mosaic alfamovirus; ASGV = apple stem grooving capillovirus; BaMV = bamboo mosaic potexvirus; BaYMV = barley yellow mosaic bymovirus; BBMV = broad bean mottle bromovirus; BCV1 = beet cryptic alphacryptovirus 1; BCV3 = beet cryptic alphacryptovirus 3; BlSV = bluberry scorch carlavirus; BMV = brome mosaic bromovirus; BNYVV = beet necrotic yellow vein furovirus; BrSMV = brome streak mosaic rymovirus; BSMV = barley stripe mosaic hordeivirus; BVQ = beet furo-like virus Q; BWYV = beet western yellows luteovirus; BYDV = barley yellow dwarf luteovirus; BYV = beet yellows closterovirus; CarMV = carnation mottle carmovirus; CCMV = cowpea chlorotic mottle bromovirus; CCSV = cucumber chlorotic spot closterovirus; CfMV = cocksfoot mottle sobemovirus; CLRV = cherry leafroll nepovirus; ClYMV = clover yellow mosaic potexvirus; CMV = cucumber mosaic cucumovirus; CNV = cucumber necrosis tombusvirus; CPMV = cowpea mosaic comovirus; CRSV = carnation ringspot dianthovirus; CTLV = citrus tatter leaf capillovirus; CTV = citrus tristeza closterovirus; CymRSV = cymbidium ringspot tombusvirus; GFLV = grapevine fanleaf nepovirus; GRV = groundnut rosette umbravirus; GVB = grapevine trichovirus B; LIYV = lettuce infectious yellows closterovirus; MCMV = maize chlorotic mottle machlomovirus; MRDV = maize rough dwarf fijivirus; MStV = maize stripe tenuivirus; OBDV = oat blue dwarf marafivirus; PCV = peanut clump furovirus; PDV = prune dwarf ilarvirus; PEBV = pea early browning tobravirus; PEMV = pea enation mosaic enamovirus; PLRV = potato leafroll luteovirus; PMV = panicum mosaic sobemovirus; PPV = plum pox potyvirus; PSbMV = pea seedborne mosaic potyvirus; PVM = potato carlavirus M; PVX = potato potexvirus X; PVY = potato potyvirus Y; PYFV = parsnip yellow fleck sequivirus; RBDV = rasberry bushy dwarf idaeovirus; RCNMV = red clover necrotic mosaic dianthovirus; RDV = rice dwarf phytoreovirus; RGSV = rice grassy stunt tenuivirus; RRSV = rice ragged stunt oryzavirus; RStV = rice stripe tenuivirus; RTSV = rice tungro spherical waikavirus; RYMV = rice yellow mottle sobemovirus; SbDV = soybean dwarf luteovirus; SBMV = southern bean mosaic sobemovirus; SBWMV = soil-borne wheat mosaic furovirus; SYNV = sonchus yellow net nucleorhabdovirus; TBRV = tomato black ring nepovirus; TBSV = tomato bushy stunt tombusvirus; TCV = turnip crinkle carmovirus; TEV = tobacco etch potyvirus; TMGMV = tobacco mild green mosaic tobamovirus; TMV = tobacco mosaic tobamovirus; TNV = tobacco necrosis necrovirus; TRV = tobacco rattle tobravirus; TSV = tobacco streak ilarvirus; TSWV = tomato spotted wilt tospovirus; TuMV = turnip mosaic potyvirus; TVMV = tobacco vein mottling potyvirus; TYMV = turnip yellow mosaic tymovirus; WClMV = white clover mosaic potexvirus; WTV = wound tumor phytoreovirus.
Others. CI = cytoplasmic inclusion; CP = coat protein; DI = defective interfering RNA; ds = double-strand; ER = endoplasmic reticulum; f/s = frameshift; HC-Pro = helper component proteinase; HSP = heat shock protein; ICR = internal control region; IRES = internal ribosome entry site; K = kilodalton; MP = movement protein; NI = nuclear inclusion; NMR = nuclear magnetic resonance; ns = non structural; nt = nucleotide; NTP = nucleotide triphosphate; ORF = open reading frame; RdRp = RNA-dependent RNA polymerase; r/t = readthrough; sat = satellite; sg = subgenomic; SGP = subgenomic promoter; ss = single-strand; UTR = untranslated region; vc = viral complementary; VPg = genome-linked virus protein; 3D = three dimensional.

Table 2 . Genome characteristics of viruses with single-stranded RNA

Genus	Virus	Polarity	Structure		Notes
			5' end	3' end	
Alfamovirus	AMV	+	Cap	No poly(A), no tRNA-like	
Bromovirus	BMV	+	Cap	tRNA-like	
Bymovirus	BaYMV	+	VPg proposed	Poly(A)	
Capillovirus	ASGV	+	Cap?	Poly(A)	Cap proposed in CTLV[1]
Carlavirus	PVM	+		Poly(A)	Cap in BISV[2]
Carmovirus	CarMV	+	Cap	No poly(A), no tRNA-like	
Closterovirus	BYV	+	Cap proposed	No poly(A), no tRNA-like	
Comovirus	CPMV	+	VPg	Poly(A)	
Cucumovirus	CMV	+	Cap	tRNA-like	
Dianthovirus	CRSV	+		No poly(A)	Cap in RCNMV
Enamovirus	PEMV	+	VPg	No poly(A), no tRNA-like	In RNA2, no termination codon, no 3' UTR
Furovirus	SBWMV	+	Cap	tRNA-like	
	BNYVV	+	Cap	Poly(A)	
Hordeivirus	BSMV	+	Cap	tRNA-like	
Idaeovirus	RBDV	+	Cap proposed in RNA 1	No poly(A), no tRNA-like	
Ilarvirus	TSV	+		No poly(A), no tRNA-like	
Luteovirus	BYDV	+	VPg	No poly(A), no tRNA-like	No VPg in BYDV (PAV isolate[3])
Machlomovirus	MCMV	+	Cap	No poly(A)	
Marafivirus	OBDV[4]	+	Cap proposed	Poly(A)	
Necrovirus	TNV	+	ppA	No poly(A), no tRNA-like	
Nepovirus	TBRV	+	VPg	Poly(A)	
Nucleorhabdovirus	SYNV	-			
Potexvirus	PVX	+	Cap	Poly(A)	
Potyvirus	PVY	+	VPg	Poly(A)	
Rymovirus	BrSMV[5]	+		Poly(A)	
Sequivirus	PYFV	+	VPg proposed	No poly(A), no tRNA-like	
Sobemovirus	SBMV	+	VPg	No poly(A), no tRNA-like	
Tenuivirus	RStV	- and ambisense	No cap	No poly(A)	RNA1: (-) polarity[6]
Tobamovirus	TMV	+	Cap	tRNA-like	
Tobravirus	TRV	+	Cap	tRNA-like	tRNA-like folding but can- not be aminoacylated
Tombusvirus	TBSV	+		No poly(A), no tRNA-like	
Tospovirus	TSWV	- and ambisense	(p)ppA on S RNA	No poly(A), no tRNA-like	
Trichovirus	ACLSV	+	Cap proposed	Poly(A)	Cap in GVB[7]
Tymovirus	TYMV	+	Cap	tRNA-like	
Umbravirus	GRV[8]	+		No poly(A), no tRNA-like	
Waikavirus	RTSV	+		Poly(A)	

References: *1*, Ohira *et al.* (1995); *2*, Cavileer *et al.* (1994); *3*, Shams-bakhsh *et al.* (1997); *4*, Edwards *et al.* (1997); *5*, Götz & Maiss (1995); *6*, Toriyama *et al.* (1994); *7*, Saldarelli *et al.* (1996); *8*, Taliansky *et al.* (1996).

Table 3. General arrangement of genes in RNA genomes

Genus	Virus	Super-group [a]	Protein motifs			MP	CP gene Position	sg RNA	Readthrough (r/t)	Frameshift (f/s)	Proteolytic cleavage
			Methyl-transferase	Helicase	RdRp						
Alfamovirus	AMV	A	126K (1a) of RNA1	126K (1a) of RNA1	90K (2a) of RNA2	32K (3a) of RNA3	3' on RNA3	Yes			
Alpha-cryptovirus	BCV3				54.9K (only prot. of RNA2)		52K (only prot. of RNA2 in BCV1)				
Bromovirus	BMV	A	106K (1a) of RNA1	106K (1a) of RNA1	90K (2a) of RNA2	32K (3a) of RNA3	3' on RNA3	Yes			
Bymovirus	BaYMV	P		270K (polyprot.) of RNA1	270K (polyprot.) of RNA1	270K (polyprot.) of RNA1	3' on RNA1				Proposed
Capillovirus	ASGV	A	241K (polyprot.)	241K (polyprot.)	241K (polyprot.)		3'	Proposed			Proposed
Carlavirus	PVM	A	223K (polyprot. in BlSV[1]	223K (polyprot. in BlSV[1]	223K (polyprot.)	Triple gene block[2]	Internal	Proposed			Yes in BlSV[3]
Carmovirus	CarMV	C			86K (r/t)	7K (prot. 5)	3'	Yes	Yes		
Closterovirus	BYV	A	295K (polyprot.; 1a)	295K (polyprot.; 1a)	348K (f/s polyprot.; 1a/1b)	65K[4]	Internal (duplicated)	Yes		Proposed	Yes
Comovirus	CPMV	P		58K (of 202K polyprot.) of B-RNA	87K (of 202K polyprot.) of B-RNA	48K (of 105/95K polyprot.) of M-RNA	3' on M-RNA				Yes
Cucumovirus	CMV	A	109K (1a) of RNA1	109K (1a) of RNA1	94K (2a) of RNA2	30K (3a) of RNA3; 15K (2b) of RNA2[5]	3' on RNA3	Yes			
Dianthovirus	CRSV	C			88K (f/s) of RNA1	33K prot. of RNA2	3' on RNA1	Yes in RCNMV		Yes in RCNMV[6]	
Enamovirus	PEMV			84K (prot. 2) and 130K (f/s prot. 2/3) of RNA1	130K (f/s prot. 2/3) of RNA1; 65K of RNA2		Internal on RNA1	Yes	Proposed	Proposed	Proposed

Table 3. continued

Furovirus	SBWMV	A	150K of RNA1	150K of RNA1	209K (r/t) of RNA1	Triple gene block	5' on RNA2	Proposed	Yes	Yes	Yes[7]
	BNYVV	A	237K of RNA1	237K of RNA1	237K of RNA1	Triple gene block	5' on RNA2	Yes	Yes	Yes	
Hordeivirus	BSMV	A	130K of RNA1	130K of RNA1	87K of RNA3	Triple gene block	5' on RNA2	Yes with poly(A) tail			
Idaeovirus	RBDV	A	190K of RNA1	190K of RNA1	190K of RNA1		3' on RNA2	Yes			
Ilarvirus	TSV	A		123K (1a) of RNA1[8]	91.6K (2a) of RNA2[8]	32K of RNA3 in PDV[9]	3' on RNA3	Yes			
Luteovirus	BYDV	S	111K (r/t)		99K (f/s)	17K in PLRV[10]	Internal	Yes			
Machlomovirus	MCMV	C					3'	Yes	Proposed	Yes	Proposed
Marafivirus	OBDV[11]	A	227K	227K			C-terminus of 227K; duplicated	Yes	Proposed		Yes
Necrovirus	TNV	C			82K (r/t)	7.5K (7a) and 7.3K (7b) in TNV-D^H (ref. 12)	3'	Yes	Proposed		
Nepovirus	TBRV	P		72K (of 254K polyprot.) of RNA1	92K (of 254K polyprot.) of RNA1	Postulated (on N-terminus of 150K polyprot.) of RNA2	3' on RNA2	Yes with poly(A) tail in CLRV[13]			Yes
Nucleo-rhabdovirus	SYNV				241.7K (L Prot.)		5' on (+) RNA	Yes with poly(A) tail			Proposed[14]
Oryzavirus	RRSV				141K (4a) proposed on S4 RNA[15]		S9 RNA[16]				
Phytoreovirus	WTV		89K (P5) on S5 RNA of RDV[17]		164K (P1) on S1 RNA of RDV[18]	42K (P8) on S8 and 41.5K (P9) on S9 RNAs; 36K (P11) on S11 RNA (capsomer)					

Table 3. continued

Potexvirus	PVX	A	166K	166K	166K	Triple gene block	3'	Yes			
Potyvirus	PVY	P	(CI) of 336K polyprot.	(CI) of 336K polyprot.	(NIb) of 336K polyprot.		3'				Yes
Rymovirus	BrSMV[1][9]	P	(CI) of 340K polyprot.	(CI) of 340K polyprot.	(NIb) of 340K polyprot.		3'				Proposed
Sequivirus	PYFV	P		336K (polyprot.)	336K (polyprot.)		Internal				Proposed
Sobemovirus	SBMV	S		105K	105K	17.8K (P1) in RYMV[20]	3'	Yes with VPg		Yes in CfMV[21]	Proposed
Tenuivirus	RStV				RNA1[22]		32K (NC) on vcRNA3	Yes			
Tobamovirus	TMV	A	126K	126K	183K (r/t)	30K	3'	Yes	Yes		
Tobravirus	TRV	A	134K	134K	194K (r/t)	29K (prot. 3)	Only gene on RNA2	Yes	Yes		
Tombusvirus	TBSV	C			92K (r/t)	22K (and 19K[23])	Internal	Yes	Proposed		
Tospovirus	TSWV				331.5K (L) of L vcRNA	33.6K (NSm) of M RNA	29K (N) of S vcRNA	Yes	Proposed		
Trichovirus	ACLSV	A	216K	216K	216K	36.5K in GVB[24]	3'; internal in GVB[24]				Proposed in ACLSV and GVB[24]
Tymovirus	TYMV	A	140K (of 206K polyprot.)	140K (of 206K polyprot.)	66K (of 206K polyprot.)	69K	3'	Yes			Yes
Umbravirus	GRV[25]	C			94K (f/s prot. 1/2)	28K (prot. 3)		Yes		Proposed	
Waikavirus	RTSV	P	390K (polyprot.)	390K (polyprot.)	390K (polyprot.)		Internal	Yes			Yes[25]

a: Key to supergroups: A, alpha-like; P, picorna-like; C, carmo-like; S, sobemo-like.
References: *1*, Cavileer *et al.* (1994); *2*, Zavriev *et al.* (1991); *3*, Lawrence *et al.* (1995); *4*, Agranovsky *et al.* (1998); *5*, Ding *et al.* (1995); *6*, Kim & Lommel (1994); *7*, Hehn *et al.* (1997); *8*, Scott *et al.* (1998); *9*, Bachman *et al.* (1994); *10*, Schmitz *et al.* (1997); *11*, Edwards *et al.* (1997); *12*, Molnár *et al.* (1997); *13*, Brooks & Bruening (1995); *14*, Scholthof *et al.* (1994); *15*, Upadhyaya *et al.* (1998); *16*, Upadhyaya *et al.* (1995); *17*, Suzuki *et al.* (1996); *18*, Suzuki (1995); *19*, Götz & Maiss; *20*, Bonneau *et al.* (1998); *21*, Mäkinen *et al.* (1995); *22*, Toriyama *et al.* (1994); *23*, Scholthof *et al.* (1995); *24*, Taliansky *et al.* (1996); *25*, Thole & Hull (1998).

and from mutations of the ORFs containing these motifs which abolished viral replication (PVX: Longstaff *et al.*, 1993; TYMV: Weiland and Dreher, 1993; BMV: Traynor *et al.*, 1991; Table 3).

With the exception of certain virus groups (Table 3), most positive-strand RNA viruses contain ORFs encoding proteins with amino acid sequence motifs characteristic of well-defined RNA helicases (reviewed in Kadaré and Haenni, 1997). Some of the better characterized members possess nucleotide binding and hydrolysis activities and the capacity to bind nucleic acids (TMV: Evans *et al.*, 1985; PPV: Laín *et al.*, 1991; TYMV: Kadaré *et al.*, 1996). For only a few members of the potyvirus family, a NTP-dependent RNA helicase activity has been demonstrated (Laín *et al.*, 1990; Eagles *et al.*, 1994). In many cases, such ORFs are essential for RNA replication (BMV: Kroner *et al.*, 1990; TYMV: Weiland and Dreher, 1993; CPMV: Peters *et al.*, 1994; PPV: Fernandez *et al.*, 1997), presumably since helicases could function to unwind duplexes formed during RNA replication, or to remove secondary structures from RNA templates. In BMV, this protein domain also appears to influence RNA recombination (Nagy *et al.*, 1995). The BMV helicase and polymerase domains form a complex (Kao *et al.*, 1992; O'Reilly *et al.*, 1997, 1998) and colocalize on ER (Restrepo-Hartwig and Ahlquist, 1996).

A second NTP-binding and putative helicase motif is also present in hordei-, furo-, carla- and potexviruses, in the so called "triple gene block" involved in viral cell-to-cell movement (Mushegian and Koonin, 1993). In BNYVV, it binds nucleic acids (Bleykasten *et al.*, 1996) but its precise function in cell-to-cell movement remains to be clarified.

The positive-strand RNA viruses possessing capped genomic RNAs (Table 2) include members of the alpha-like and carmo-like virus supergroups. The addition of a cap structure at 5` end of viral RNA occurs in cytoplasm and utilizes virus-encoded enzymes (reviewed in Buck, 1996). A guanylyltransferase activity occurs in the TMV 126K protein (Dunigan and Zaitlin, 1990), while a methyltransferase domain was identified near N-terminus of replication proteins of all members of the alpha-like virus supergroup (Rozanov *et al.*, 1992).

VPgs are virus-encoded proteins covalently linked to 5`-terminal nucleotide of virus genomic RNA and are found in viruses of the picorna-like and sobemo-like plant virus supergroups (Table 2). VPg is processed from a polyprotein in nepo-, poty-, como- and luteoviruses (GFLV: Pinck *et al.*, 1991; TEV: Dougherty and Parks, 1991; CPMV: Peters *et al.*, 1992; PLRV: van der Wilk *et al.*, 1997). The 3D structure of PVY VPg is known (Plochocka *et al.*, 1996). The TVMV VPg interacts with viral polymerase, stimulates its activity *in vitro* (Hong *et al.*, 1995; Fellers *et al.*, 1998), and is likely to be involved in initiation of both the negative- and positive-strand RNA synthesis, as in poliovirus (Paul *et al.*, 1998).

Targeting of polymerase-associated viral proteins to ER appears to be linked to assembly of replication complexes on the cellular membranes (Restrepo-Hartwig and Carrington, 1994; Schaad *et al.*, 1997).

Coat Protein

The CP is multifunctional as it operates at different levels in the viral life cycle Besides its structural role, its other roles in different virus families include participation in viral replication (BMV: Marsh *et al.*, 1991; Flasinski *et al.*, 1995), cell-to-cell and systemic movement in infected host (BMV: Sacher and Ahlquist, 1989; CPMV: Wellink and van Kammen, 1989; WClMV: Forster *et al.*, 1992; TEV: Dolja *et al.*, 1994; CMV: Canto *et al.*, 1997), host range determination and symptom expression (AMV: Neeleman *et al.*, 1991; TCV: Heaton *et al.*, 1991; CMV: Suzuki *et al.*, 1995), and vector transmission (CNV: McLean *et al.*, 1994; Robbins *et al.*, 1997; CMV: Chen and Francki, 1990; Perry *et al.*, 1998). The case of alfamo- and ilarviruses is peculiar because CP of these viruses is also required for the early events of replication cycle, a phenomenon called "genome activation" (van der Vossen *et al.*, 1994). Different domains of the CP seem to contribute to these different functions (CCMV: Schneider *et al.*, 1997; AMV: van der Vossen *et al.*, 1994).

Most plant virus particles contain only one type of protein, but como-, faba- and marafiviruses encode two CP species. Sequence comparison between the CP of different RNA viruses has led to their classification in two possible families, related to their capsid structure (Dolja *et al.*, 1991). In the closterovirus family, a quasi duplication of CP gene encoding a protein of unknown function is expressed *in vivo* but not detected in viral particles (Boyko *et al.*, 1992; Febres *et al.*, 1994). The members of the *Sequiviridae* possess three distinct CP species (Shen *et al.*, 1993; Turnbull-Ross *et al.*, 1993) expressed from a polyprotein. More complex particles, such as enveloped reoviruses and rhabdoviruses, contain several proteins.

In addition to the CP, luteoviruses and some furoviruses also synthesize an elongated version of this protein by readthrough of termination codon of CP gene. This readthrough protein is present in low amounts in viral capsid and is involved in transmission by insect or fungal vector (see below).

Cell-to-Cell Movement

Unlike animal viruses, plant viruses encounter a cell wall and, therefore, their local spread occurs via intercellular cytoplasmic connections, the plasmodesmata. Cell-to-cell virus movement is mostly controlled by a single movement gene (MP) product, in TCV it requires two gene products, necroviruses apparently encode 2 or 3 MPs, while movement function in carla-, furo-, hordei- and potexviruses is controlled by three gene products encoded by the "triple gene block" (chapter 7). In contrast, potyviruses do not encode a specific MP but rely on the use of several multifunctional proteins such as the CP, HC-Pro and CI to achieve cell-to-cell movement (Rojas *et al.*, 1997, Rodriguez-Cerezo *et al.*, 1997).

Sequence comparisons revealed that no clear conserved motifs exist among MPs of different virus families, although sequence homologies exist between MPs of closely related virus families. In spite of sequence divergences, different MPs possess a number of common properties: they can often complement each other or allow the movement of viruses otherwise limited to particular tissues; bind nucleic acids in a cooperative manner; localize inside or in the vicinity of plasmodesmata and increase their size exclusion limit. The MPs also perform certain other functions: CMV MP promotes movement of nucleic acids from one cell to the neighbouring one (Ding *et al.*, 1995); MP also colocalizes with microtubules in TMV-infected cells (McLean *et al.*, 1995; Heinlein *et al.*, 1995), a feature that suggests the involvement of the cytoskeleton for intracellular viral movement; a number of MPs are phosphorylated in plants possibly by a cell wall-associated kinase; and BMV MP also participates in the control of host range (de Jong *et al.*, 1995; Fujita *et al.*, 1996) and in symptom formation (Rao and Grantham, 1995).

All the above functions are carried by different domains in the MPs, as analyzed in several viruses (CMV, CPMV, RCNMV, TMV: Kahn *et al.*, 1998; Vaquero *et al.*, 1997).

Systemic Movement

Less is known about the molecular mechanism of systemic movement of plant viruses (reviewed in Séron and Haenni, 1996; Gilbertson and Lucas, 1996). Most viruses are believed to circulate through the phloem, although movement through xylem is reported (Opalka *et al.*, 1998). A function in vascular movement is most frequently ascribed to CP. However, replication-related proteins (BSMV: Weiland and Edwards, 1994; BMV: Traynor *et al.*, 1991; CMV: Gal-On *et al.*, 1994) or specific non-structural viral proteins (TBSV: Scholthof *et al.*, 1995; CMV: Ding *et al.*, 1995) are also required for vascular movement of viruses. In other cases, vascular movement seems to rely on *cis*-acting sequences in the RNA rather than on a specific gene product (BSMV: Petty *et al.*, 1990; BNYVV: Lauber *et al.*, 1998).

Proteinases

Proteolytic processing of a precursor polyprotein is a translation strategy employed by many RNA viruses (chapter 4). It involves virus-encoded proteinases which are related in sequence and activity to known cellular proteinases (reviewed in Dougherty and Semler, 1993; and chapter 4). Poty-, como- and nepoviruses encode chymotrypsin-like proteinases (TEV: Verchot *et al.*, 1992; CPMV: Dessens *et al.*, 1991; GFLV: Margis and Pinck, 1992), while papain-like proteinases are encoded by poty-, tymo-, clostero- and carlaviruses (TEV: Carrington *et al.*, 1989; TYMV: Rozanov *et al.*, 1995; BYV: Agranovsky *et al.*, 1994, BlSV: Lawrence *et al.*, 1995). Sequence alignments of sobemo- and luteovirus genomes suggest that serine proteinases are encoded by viral genome (Gorbalenya *et al.*, 1988), but to date, no biochemical evidence for their activity has appeared, even though in luteoviruses, the VPg is cleaved from a protein precursor (van der Wilk *et al.*, 1997).

Vector Transmission

The natural spread of viruses is most frequently accomplished by insect, nematode or fungal vectors (chapter 8). A high degree of virus-vector specificity exists and in several cases CP appears to determine virus transmissibility by its respective vector. Alteration of a single amino acid on the CP surface can render a virus non-transmissible by its vector (Robbins *et al.*, 1997), and vector specificity can be altered when viral nucleic acid is heterologously encapsidated by a CP specific for a particular vector (Perry *et al.*, 1998).

In luteoviruses and some furoviruses, an elongated version of CP synthesized by readthrough of termination codon of CP gene is present in low amounts in viral capsid and is involved in transmission by the insect or fungal vector (BNYVV, BWYV, BYDV; Chay *et al.*, 1996; Wang *et al.*, 1995;Tamada *et al.*, 1996). In luteoviruses, interaction between the readthrough domain and the chaperonin encoded by an endosymbiont bacterium is essential for virus persistence in the insect (van den Heuvel *et al.*, 1997).

Virus-encoded non-structural proteins (helper components) are necessary for transmission of some viruses. In aphid-transmitted potyviruses, this function is ensured by the multifunctional HC-Pro protein (reviewed in Maia *et al.*, 1996), which is proposed to bind both to the aphid stylet and to the viral CP (Blanc *et al.*, 1997). In PEBV and TRV tobraviruses, at least one non-structural gene product appears to be involved in nematode-transmission (MacFarlane and Brown, 1995; Hernandez *et al.*, 1997).

Symptoms

Extensive work has been undertaken to investigate the molecular basis of symptom development (reviewed in van Loon, 1987; Daubert, 1989; chapter 9). Such analyses have identified viral genes or nucleotide sequences influencing symptoms, but multiple and independent loci also appear to jointly determine symptomatology. Modifications in TMV genome show that CP and RdRp genes are both implicated in symptom development (Banerjee *et al.*, 1995; Shintaku *et al.*, 1996). In BNYVV, two distinct proteins strongly and independently influence leaf symptomatology (Jupin *et al.*, 1992).

However, in certain systems, symptom expression is not associated with the expression of a viral gene product. The sequence and/or structure of their genome would instead be involved in determining their pathogeny (Rodriguez-Cerezo *et al.*, 1991). Viroids (Owens *et al.*, 1996) and certain satellite RNAs (Sleat and Palukaitis, 1990; Taliansky and Robinson, 1997) represent these systems.

NON-CODING FUNCTIONS OF VIRAL RNAs

In addition to their coding functions, viral RNAs contain signals that are necessary for their interaction with other components implicated in life cycle of a virus. These signals can be involved in translation, replication, and encapsidation processes.

Cis-Acting Elements Required for Translation

The RNA genomes of certain viruses have the same characteristics as cellular mRNAs, i.e. a 5` terminal cap structure and a poly(A) tail which are usually considered to have the same functions during translation both in cellular and viral RNAs. With few exceptions, full-length *in vitro*-synthesized viral transcripts require the cap structure for infectivity (reviewed in Boyer and Haenni, 1994).

The 5` UTR has been assigned two functions in different plant viruses: an enhancer effect in translation and an internal initiation of translation. The 5` leader sequence of plant virus RNA genome can contain *cis* elements involved in translation. It is well studied in the omega fragment, the 68 nucleotide-long 5` UTR of TMV RNA, which has a high enhancer effect in translation (Gallie *et al.*, 1987a). A (CAA)n repeat and, to a lesser extent, a short direct-repeat motif and the primary rather than the secondary structure (Gallie and Walbot, 1992) appear to be responsible for this enhancer effect of the omega fragment. A similar effect has been demonstrated for BMV RNA4 (Gallie *et al.*, 1987b), AMV RNA4 (Jobling and Gehrke, 1987), PSbMV (Nicolaisen *et al.*, 1992), TEV (Carrington and Freed, 1990; Nicolaisen *et al.*, 1992) and PVX (Zelinina *et al.*, 1992; Tomashevskaya *et al.*, 1993). In constrast, an inhibitory effect on translation was shown for the 5` UTR of AMV RNA3 (van der Vossen *et al.*, 1993).

Internal initiation (chapter 4) has been proposed for the 5` UTRs of certain members of the picorna-like supergroup such as comoviruses (CPMV: Thomas *et al.*, 1991; Verver *et al.*, 1991) or potyviruses (TEV: Carrington and Freed, 1990). These 5` UTRs are long and such a function has been etablished for picornaviruses. However, in CPMV, contradictory results were obtained using bicistronic constructs (Belsham and Lomonossoff, 1991; Riechmann *et al.*, 1991). Recent studies on a potyvirus and a tobamovirus have identified the viral 5` UTR sequences functioning as IRES. In TuMV, ribosomes are suggested to bind to an internal site within the 5` UTR of RNA and then presumably scan the sequence to reach the initiator AUG (Basso *et al.*, 1994). Similarly, translation of 3'-proximal CP gene of a crucifer-infecting TMV (crTMV) appears to occur *in vitro* by an internal ribosome entry mechanism. Use of a dicistronic RNA transcript system showed that integrity of the 148-nucleotide long region upstream of the CP gene of crTMV RNA is essential for internal initiation of translation (Ivanov *et al.*, 1997).

Except PEMV RNA2 (Demler *et al.*, 1993), all RNA viruses contain a 3' UTR. In TMV, the 3'-coterminal pseudoknots upstream of the tRNA-like structure, greatly increase translation rate of a chimeric mRNA coding for a reporter gene. It appears that the structure of the 3'-located pseudoknot, and the primary sequence, are important for this effect. Together with the omega fragment and the cap structure, it has a synergistic effect suggesting that omega, the cap and the 3' UTR interact to reach an efficient level of translation (Leathers *et al.*, 1993). The 3' UTR of BMV RNA3 has a similar effect to that of the 3' UTR of TMV RNA on translation, as opposed to the 3' UTR of the TYMV RNA and AMV RNA4 which has little impact on translation and RNA stability (Gallie and Kobayashi, 1994). Recently, a translational determinant important for the competitive translational activity of AMV RNA4 was identified in the 3' UTR. Site-directed mutagenesis demonstrated that mutations in one of the numerous AUG triplets downstream of the CP gene specifically reduce CP RNA translation under competitive conditions (Hann *et al.*, 1997). Interestingly, the 3' UTR of BYDV (PAV isolate) contains a translation enhancer sequence able to promote a cap-independent translation initiation (Wang *et al.*, 1997). A role in translation has also been proposed for the 3' UTR of WTV where alterations or extensions of this region can modify translation efficiency *in vitro* of a genomic transcript. This effect is postulated to result from intramolecular interactions between 5` and 3' UTRs (Xu *et al.*, 1989).

Rohde *et al.* (1994), Brierley (1995), and chapter 4 of this book review the frameshift expression mechanism in viruses. Frameshift results from movement of ribosomes either in 5` (-1 frameshift) or 3' (+1 frameshift) direction on mRNA; the -1 frameshift event being far more common than the +1 frameshift event. In almost all cases, a -1 frameshift requires different *cis*-acting elements: a heptanucleotide sequence known as a "slippery" sequence. This sequence is downstream of the point of frameshift which corresponds to a hairpin structure associated in many instances with a pseudoknot structure and a spacer region of 4 to 9 nucleotides between the slippery sequence and the hairpin structure.

Luteovirus are the best-studied plant viruses with respect to frameshift. A -1 frameshift phenomenon was reported for expression of polymerase gene of BYDV in a protoplast system (Brault and Miller, 1992). A pseudoknot is not required for frameshift in a German isolate of PLRV (Prüfer *et al.*, 1992) but is required in a Polish isolate of PLRV (Kujawa *et al.*, 1993). Both genomic RNAs of PEMV possess an ORF containing RdRp-related motifs that appears to be produced by -1 frameshift. A pseudoknot structure following the slippery site has been suggested for RNA1 (ten Dam *et al*, 1990). In contrast, translation studies performed *in vitro* on RNA of CRSV (Ryabov *et al.*, 1994) and RCNMV (Xiong *et al.*, 1993; Kim and Lommel, 1994) show that a hairpin structure downstream of the sliperry sequence can be formed, but no potential pseudoknot has been detected.

A +1 frameshift is rarely met with in plant viruses except the closteroviruses which seem to employ this strategy. The requirement of *cis* elements for this mechanism seems to be less strict than that required for -1 frameshift. Usually, a slippery run of bases and a rare or a termination codon appear to be necessary. For example, the BYV genome contains a shifty sequence followed by a hairpin structure that could potentially form a pseudoknot (Agranovsky *et al.*, 1994). On the other hand, a +1 frameshift has also been proposed for CTV (Karasev *et al.*, 1995) and LIYV (Klaasen *et al.*, 1995) but no downstream structure that might suggest frameshift, nor even a shifty sequence, could be detected.

Readthrough mechanism, suppression of a UAG or a UGA termination codon leading to the synthesis of two proteins from a cistron, occurs among some plant RNA viruses (reviewed in Maia *et al.*, 1996). Synthesis of the RdRp in tobamo-, tobra-, tombus-, and carmoviruses takes place by this mechanism. In BNYVV, PEMV, PCV and luteoviruses, it leads to fusion of CP to the protein produced by the readthrough domain. In SBWMV (Shirako and Wilson, 1993) and BVQ (Koenig *et al.*, 1998), both genomic RNAs possess a readthrough domain: RNA1 contains a readthrough domain encoding RdRp motifs and RNA2 carries the CP readthrough domain. In BYDV, *in vitro* analysis revealed that two regions 3' of the termination codon, one of which is a long-range signal, are required for efficient readthrough (Brown *et al.*, 1996). CarMV is unique since its 98K protein appears to result from a double readthrough event (Harbison *et al.*, 1985).

Besides the termination codon, features in the RNA are also required in *cis* for efficient readthrough. The two downstream codons appear to be crucial for readthrough in TMV RNA *in vivo* (Skuzeski *et al.*, 1991) and *in vitro* (Valle *et al.*, 1992).

Cis-Acting Elements Required for Replication

It has long been known that 3' end of positive-strand RNA genomes of certain plant viruses have tRNA-related properties (reviewed in Mans *et al.*, 1991; Florentz and Giegé, 1995). The 3D folding of these regions in the viral RNA presents certain features reminiscent of the 3D structure of canonical tRNAs. These tRNA-like structures are involved in negative-strand synthesis in TMV (Dawson *et al.*, 1986),

BMV (Miller *et al.*, 1986), and TYMV (Morch *et al.*, 1987; Tsai and Dreher, 1991). A possible function of this tRNA-like structure in TYMV may be to direct the replication complex to the correct 3' end of viral RNA and therefore to prevent internal initiation of negative-strand RNA synthesis (Deiman *et al.*, 1998).

Deiman and Pleij (1997) hint to the importance of pseudoknots in viral RNAs. The structure of TYMV RNA pseudoknot in solution was determined by NMR spectroscopy revealing its exact folding (Kolk *et al.*, 1998). In addition to the pseudoknots present in the tRNA-like structures, a few viruses have pseudoknots upstream of these structures which appear to participate in RNA replication. One of the three pseudoknots immediately upstream of the tRNA-like structure in TMV is necessary for RNA replication (Takamatsu *et al.*, 1990). In addition, analysis of disruption mutants in the upstream pseudoknot structure of TYMV RNA has demonstrated a role of this structure in replication (Deiman *et al.*, 1997).

Several lines of evidence suggest that the 3' poly(A) tract protects the RNA from degradation and also provides an essential *cis*-acting sequence for RNA replication. In CPMV, part of the poly(A) tail of the B-RNA can adopt a hairpin structure with a heptanucleotide sequence just upstream of the poly(A) sequence. Deletions from the 3' end of this RNA that prevent hairpin formation dramatically interfere with RNA replication (Eggen *et al.*, 1989a). Deletion of the poly(A) tail from BNYVV RNA3 causes a great reduction in the ability of RNA to replicate and poly(A) tail is restored in progeny (Jupin *et al.*, 1990). Addition of poly(A) also occurs during infection with 3' poly(A)-deficient transcripts of WClMV (Guilford *et al.*, 1991), CPMV (Eggen *et al.*, 1989b) and PPV (Riechmann *et al.*, 1990).

Similarities exist between viral RNA sequences (AMV, bromo-, cucumo-, tobamo-, tobra- and tymoviruses) and the ICR2 of eukaryotic promoters recognized by RNA polymerase III (Marsh *et al.*, 1989). These sequences have a role in replication of BMV RNA (Pogue *et al.*, 1990, 1992) and such a role is also proposed for the RNA of CMV (Boccard and Baulcombe, 1993) and AMV (van der Vossen *et al.*, 1993). In AMV, mutations in the ICR2 motif abolish RNA3 accumulation, which supports the idea that this structure is important for positive-strand promoter activity (van der Vossen and Bol, 1996). In BMV, deletion analysis revealed that the ICR2 motif and the tRNA-like structure are the minimal requirements for *in vivo* formation of a functional RdRp (Quadt *et al.*, 1995).

In many viruses, still other *cis*-acting elements are involved in replication even though these sequences do not appear to possess easily defined sequence motifs such as the ones discussed above.

Mutations that disrupt a putative hairpin at the 5` UTR end of a mutant derived from BMV RNA2 greatly reduce replication. The role of this structure (whose formation could be favoured by host factors) might be to liberate the 3' end of the negative-strand in the replicative form, thereby allowing initiation of positive-strand synthesis by RdRp (Pogue and Hall, 1992). *Cis*-acting functions in the 5` region of BNYVV RNA3 are located within the first 292 nucleotides from the 5` end. A secondary structure is proposed for this region and indeed efficient accumulation of RNA3 during infection requires base-pairing between sequence elements in this structure (Gilmer *et al.*, 1993). For PVX, mutation analysis has revealed that *cis*-acting elements from different regions of 5` UTR are required for positive-strand RNA synthesis but are not essential for negative-strand production (Kim and Hemenway, 1996). Studies on 5` UTR of AMV RNAs indicated that although elements between nucleotides 80 and 345 of 5` UTR of RNA3 are sufficient for replication, a specific sequence of 3 to 5 nucleotides is required to target the replicase to an initiation site corresponding to the 5` end of the RNA (van Rossum *et al.*, 1997a).

The 5` and 3' ends of the genome parts of viruses of cytorhabdovirus family (Wetzel *et al.*, 1994) and of tospovirus (reviewed in Elliot *et al.*, 1991) and tenuivirus (reviewed in Ramirez and Haenni, 1994) groups can potentially form stable base-paired panhandle structures that are believed to play a role in replication and encapsidation.

A sequence of 105 nucleotides in 3' end of CP gene of TEV is a *cis*-acting element required for genome replication (Mahajan *et al.*, 1996). This *cis*-acting element and the 3' UTR contain putative base-paired secondary structures necessary for viral amplification (Haldeman-Cahill *et al.*, 1998). A sequence of 8 U-rich nucleotides in PVX 3' UTR is necessary for viral replication (Sriskanda *et al.*, 1996). This sequence is the binding site for host proteins. In CymRSV (Havelda and Burgyan, 1995), the putative stem-loop structures composed of three hairpins at 3' end of genomic RNA participate in negative-strand synthesis. A similar situation operates in one of the five stem-loop structures of 3' UTR of AMV RNAs. Mutations that disturb this structure reduce or abolish RNA replication (van Rossum *et al.*, 1997b). Moreover, the region required for RdRp recognition of AMV RNAs was longer than the region required for CP binding *in vitro*. The 3' UTR of AMV RNA3 also contains a core promoter for negative-strand synthesis, as well as an enhancer element (van Rossum *et al.*, 1997c).

Synthesis of sgRNA occurs by internal initiation on negative-strand templates. This strategy is used by viruses of the alpha-like supergroup (reviewed in Goldbach *et al.*, 1991), and also of the carmo- and sobemo-like supergroups (reviewed in Buck, 1996). The SGP of BMV RNA3 is located in the 250 nucleotide-long intergenic region between MP and CP genes. Both *in vitro* and *in vivo* studies have identified a core SGP extending about 20 bases upstream of the sgRNA initiation site (Marsh *et al.*, 1988; French and Ahlquist, 1988). This core promoter is able to direct low levels of sgRNA synthesis. In addition, upstream sequences are required for full SGP activity, including a poly(A) tract immediately upstream of the core promoter and a further AU-rich sequence. Moreover, the activity of this SGP is position-dependent, since the promoter closest to the 3' end of positive-strand is the most active. The intergenic region of BMV RNA3 also contains a ICR2 motif but this structural element appears to be dispensable for sgRNA synthesis (Smirnayagina *et al.*, 1994). In contrast, the 286 nucleotide-long intergenic region of CMV RNA3 contains an ICR2 motif which is a part of the SGP (Boccard and Baulcombe, 1993). The same positional effect, as in BMV SGP, was observed for CMV SGP. In AMV RNA3, the SGP contains a core promoter, comparable to the BMV core region, and two enhancer regions located on either side of the transcriptional start site (van der Vossen *et al.*, 1995). Unlike BMV and CMV, higher levels of sgRNA synthesis were obtained when the SGP was closer to 5` end of genomic RNA. The SGP of BNYVV RNA3 differs significantly from those described above in that most of the promoter is located downstream of the transcription initiation site (Balmori *et al.*, 1993). A sequence of 8 nucleotides within the putative SGP of PVX as well as the spacing between this element and the start site for sgRNA synthesis are critical for accumulation of the two major sgRNA species (Kim and Hemenway, 1997).

The SGP is mostly identical to the genomic sequence from which it originated. However, for other viruses, high sequence similarities between the 5' termini of genomic and sgRNA can be encountered, suggesting that these sequences may be part of the genomic and SGPs. For RCNMV, the 14 nucleotide-long putative SGP is almost identical to 5' terminus of genomic RNA1 (Zavriev *et al.*, 1996). It is predicted that a stable hairpin could be folded in this putative sgRNA promoter element. TCV possesses two sgRNAs; computer analysis of their SGPs has revealed an extensive hairpin just upstream of the transcription start site (Wang and Simon, 1997). Similarity in the 5'-terminal sequences of genomic and sgRNAs has also been demonstrated for several luteovirus (reviewed in Miller *et al.*, 1995).

Interactions with Coat Protein

The mechanism of specific encapsidation of the viral RNA into its cognate capsid has been examined for several viruses. This is well exemplified by TMV in which a CP disk aggregate specifically recognizes an internal RNA sequence known as the assembly origin (reviewed in Butler, 1984). Besides encapsidation, interaction between viral RNA and CP can also be considered as a regulating element of viral life cycle as in TCV and to play an important role in viral replication as in AMV.

In AMV, the 3'-terminal 145 nucleotides of the three RNAs are homologous and can be folded into a structure consisting of a series of hairpins, separated by AUGC motifs, which possess high-affinity CP-binding sites (Houser-Scott *et al.*, 1994). These sites are involved in the "genome activation" process (reviewed in Jaspars, 1985). Recent data strongly suggest that the AUGC repeats provide sequence-specific determinants for CP/RNA interaction (Houser-Scott *et al.*, 1997). The requirement of CP for replication was supported by the detection of CP in purified RdRp preparations from AMV-infected plants (Quadt *et al.*, 1991). Futhermore, mutations in the CP gene reduce the accumulation of positive-strands by 100-fold, with little effect on negative-strand synthesis, indicating that the CP plays an additional role in asymmetric positive-strand accumulation (van der Vossen *et al.*, 1994). Other major CP-binding sites also exist in internal positions on the RNAs that can likewise form stable stem-loop structures (Zuimeda and Jaspars, 1984); these sites might be involved in virus assembly.

In TCV, two regions of genome interact specifically with CP. One region consists of two RNA motifs within the 3'-proximal CP gene. The other region consists of three RNA motifs that cover about 300 nucleotides in the gene encoding the putative RdRp and spans a suppressible termination codon. Binding of the CP subunits to this region might regulate the level of readthrough and hence of RdRp formation. Four of the five RNA motifs can adopt hairpin structures. It is proposed that both regions could be involved in virus assembly and that binding to CP could switch viral RNA from translation/replication functions to encapsidation (Wei *et al.*, 1990; Wei and Morris, 1991). A 186 nucleotide-long region at 3' end of CP gene acts as a specific packaging signal (Qu and Morris, 1997).

The 5' UTR of RNA of several tymoviruses contains conserved hairpins with protonatable internal loops, consisting of C-C and C-A mismatches (Hellendoorn *et al.*, 1996). The 5' UTR of TYMV RNA contains two such hairpins. Mutations within this region point to a functional role of the C-C and C-A mismatches in replication of the genomic RNA. Futhermore, a deletion of 75% of the 5' UTR, including the two hairpins, results in a drastically altered ratio between filled and empty capsids, suggesting a role of this 5` UTR in viral packaging (Hellendoorn *et al.*, 1997).

The BMV RNA1 contains at least two regions within the protein 1a coding sequence that are capable of high affinity, sequence-specific interaction with CP, suggesting that these binding sites probably initiate the encapsidation process (Duggal and Hall, 1993). In g, the minimal sequence required for *in vitro* initiation of assembly is mapped in the first 38 to 47 nucleotides at 5`-terminus of RNA. This region is A-C rich and lacks any discernible secondary structure (Sit *et al.*, 1994).

CONCLUSIONS

This chapter centers on certain major characteristics of viral RNA genomes. Yet, it is clear that at all stages, interaction of the virus with the host cell is mandatory for virus development. However, information on these interactions and on the mechanisms whereby they modulate virus multiplication and symptomatology (chapter 9) is still fragmentary. An important target of future research will be to investigate these interactions more thoroughly. There is no doubt that the information reaped from studies on virus genome organization and expression will constitute a vital stepping stone for such investigations.

REFERENCES

Agranovsky, A. A., Koonin, E.V., Boyko, V. P., Maiss, E., Frötschl, R., Lunina, N. A., and Atabekov, J. G. (1994). Beet yellows closterovirus: Complete structure and indentification of a leader papain-like thiol protease. *Virology* 198, 311-324.

Agranovsky, A. A., Folimonov, A.S., Folimonova, S.Y., Morozov, S.Yu., Schiemann, J. , Lesemann, D., and Atabekov, J. G. (1998). Beet yellows closterovirus HSP70-like protein mediates the cell-to-cell movement of a potyvirus transport deficient mutant and a hordeivirus based chimeric virus. *J. Gen. Virol.* 79, 889-895.

Argos, P. (1988). A sequence motif in many poly merases. *Nucleic Acids Res.* 16, 9909-9916.

Atkins, D., Hull, R., Wells, B., Roberts, K., Moore, P., and Beachy, R.N. (1991). The tobacco mosaic virus 30K movement protein in transgenic tobacco plants is localized to plasmedesmata. *J. Gen. Virol.* 72, 209-211.

Bachman, E. J., Scott, S.W., Xin, G., and Vance, V.B. (1994) . The complete nucleotide sequence of prune dwarf ilarvirus RNA 3: Implications of coat protein activation of genome replication in ilarviruses. *Virology* 201, 127-131.

Balmori, E., Gilmer, D., Richards, K., Guilley, H., and Jonard, G. (1993). Mapping the promoters for subgenome RNA syntheris on beet necrotic yellow vein virus RNA 3. *Biochimie* 75, 517-521.

Banerjee, N., Wang, J. Y., and Zaitlin, M. (1995). A single nucleotide change in the coat protein gene of tobacco mosaic virus is involved in the induction of severe chlorosis. *Virology* 207, 234 -239.

Basso, J., Dallaire, P., Charest, P. J., Devantier, Y., and Laliberté, J.-F. (1994). Evidence for an internal ribosome entry site within the 5` non-translated region of turnip mosaic potyvirus RNA. *J. Gen. Virol.* 75, 3157-3165.

Belsham, G. J., and Lomonossoff, G. P. (1991). The mechanism of translation of cowpea mosaic virus middle component RNA: No evidence for internal initiation from experiments in an animal cell transient expression system. *J. Gen. Virol.* 72, 3109-3113.

Blanc, S., Lopez -Moya, J. J., Wang, R., Garcia-Lampasona, S., Thornbury, D. W. , and Pirone, T. P. (1997). A specific interaction between coat protein and helper component correlates with aphid transmission of a potyvirus. *Virology* 231, 141-147.

Boccard, F., and Baulcombe, D. (1993). Mutational analysis of *cis*-acting sequences and gene function in RNA 3 of cucumber mosaic virus. *Virology* 193, 563 -578.

Bonneau, C., Brugidou, C., Chen, L., Beachy, R. N. and Fauquet, C. (1998). Expression of the rice yellow mottle virus P1 protein *in vitro* and *in vivo* and its involvement in virus spread. *Virology* 244, 79-86.

Boyer, J.-C., and Haenni, A.-L. (1994). Infectious transcripts and c DNA clones of RNA viruses. *Virology* 198, 415 - 426.

Boyko, V. P., Karasev, A. V., Agranovsky, A. A., Koonin, E. V., and Dolja, V. V. (1992). Coat protein gene duplication in a filamentous RNA virus of plants. *Proc. Natl. Acad. Sci. U.S.A.* 89, 9156-9160.

Brault, V. and Miller, W. A. (1992). Translational frameshifting mediated by a viral sequence in plant cells. *Proc. Natl. Acad. Sci. U.S.A.* 89, 2262-2266.

Brault, V., van den Heuvel, J. F., Verbeek, M., Ziegler-Graff, V., Reutenauer, A., Herrbach, E., Garaud, J.-C., Guilley, H., Richards, K., and Jonard, G. (1995) Aphid transmission of beet western yellows luteovirus requires the minor capsid readthrough protein P74. *EMBO J.* 14, 650 - 659.

Brierley, I. (1995). Ribosomal frameshifting on viral RNAs. *J. Gen. Virol.* 76, 1885 -1892.

Brooks, M., and Bruening, G. (1995). A subgenomic RNA associated with cherry leafroll virus infections. *Virology* 211, 33-41.

Brown, C. M., Dinesh-Kumar, S. P. and Miller W. A. (1996). Local and distant sequences are required for efficient readthrough of the barley yellow dwarf virus-PAV coat protein gene stop codon. *J. Virol.* 70, 5884-5892.

Bruyère, A., Brault, V., Ziegler-Graff, V., Simonis, M. T., van den Heuvel, J. F.J. M., Richards, K., Guilley, H., Jonard, G., and Herrbach, E. (1997). Effect of mutations in the beet western yellows virus readthrough protein on its expression and packaging and on virus accumulation, symptoms and aphid transmission. *Virology* 230, 323-334.

Buck, K. W. (1996). Comparison of the replication of positive-stranded RNA viruses of plants and animals. *Adv. Virus Res.* 47, 159-251.

Butler, P. J. G. (1984). The current picture of the structure and assembly of tobacco mosaic virus. *J. Gen..Virol.* 65, 253-279.

Canto, T., Prior, D. A. M., Hellwald, K.-H., Oparka, K. J., and Palukaitis, P. (1997). Characterization of cucumber mosaic virus IV. Movement protein and coat protein are both essential for cell-to-cell movement of cucumber mosaic virus. *Virology* 237, 237-248.

Carrington, J. C., and Freed, D. D. (1990). Cap-independent enhancement of translation by a plant potyvirus 5`- nontranslated region. *J. Virol.* 64, 1590-1597.

Carrington, J. C., Cary, S. M., Parks, T. D., and Dougherty, W. G. (1989). A second proteinase encoded by a plant potyvirus genome. *EMBO J.* 8, 365-370.

Cavileer, T. D., Halpern, B. T., Lawrence, D. M., Podleckis, E. V., Martin, R. R., and Hillman, B. I. (1994). Nucleotide sequence of the carlavirus associated with blueberry scorch and similar diseases. *J. Gen.Virol.* 75, 711-720.

Chay, C. A., Gunasinge, U. B., Dinesh-Kumar, S. P., Miller, W. A., and Gray, S. M. (1996). Aphid transmission and systemic plant infection determinants of barley yellow dwarf luteovirus-PAV are contained in the coat protein readthrough domain and 17 kDa protein, respectively. *Virology* 219, 57- 65.

Daubert, S. D. (1989). Sequence determinants of symptoms in the genomes of plant viruses, viroids and satellites. *Mol. Plant-Microbe Interact.* 1, 317-325.

Dawson, W. O., Beck, D. L., Knorr, D., A. and Grantham, G. L. (1986). c DNA cloning of the complete genome of tobacco mosaic virus and production of infectious transcripts. *Proc. Natl. Acad. Sci. U.S.A.* 83, 1832-1836.

de Graaf, M., and Jaspars, E. M. J. (1994). Plant viral RNA synthesis in cell-free systems. *Annu. Rev. Phytopathol.* 32, 311-335.

De Jong, W., Chu, A. ,and Ahlquist, P. (1995). Coding chamers in the 3a cell-to-celll movement gene can extend the host range of brome mosaic virus systemic infection. *Virology* 214, 464 - 474.

Deiman, B. A. L. M., and Pleij, C. W. A. (1997). Pseudoknots: A vital feature in viral RNA. *Semin. Virol.* 8, 166-175.

Deiman, B. A. L. M., Kortlever, R. M. and Pleij, C. W. A. (1997). The role of the pseudoknot at the 3` end of turnip yellow mosaic virus RNA in minus strand synthesis by the viral RNA-dependent RNA polymerase. *J. Virol.* 71, 5990-5996.

Deiman, B. A. L. M., Koenen, A. K., Verlaan, P. W. G. , and Pleij, C. W. A. (1998). Minimal template requirements for initiation of minus-strand synthesis *in vitro* by the RNA-dependent RNA polymerase of turnip yellow mosaic virus. *J. Virol.* 72, 3965-3972.

Demler, S. A., Rucker, D. G., and de Zoeten, G. A. (1993). The chimeric nature of the genome of pea enation mosaic virus: The independent replication of RNA 2. *J. Gen. Virol.* 74, 1-14.

Dessens, J. T., and Lomonossoff, G. P. (1991). Mutational analysis of the putative catalytic triad of the cowpea mosaic virus 24K protease. *Virology* 184, 738-746.

Ding, B., Li, Q., Nguyen, L., Palukaitis, P., and Lucas, W. J. (1995). Cucumber mosaic virus 3a protein potentiates cell-to-cell trafficking of CMV RNA in tobacco plants. *Virology* 207, 345-353.

Ding, S. W., Li, W. X., and Symons, R. H. (1995). A novel naturally occurring hybrid gene encoded by a plant RNA virus facilitate long distance virus movement. *EMBO J.* 14, 5762 -5772.

Dolja, V. V., Boyko, V. P., Agranovsky, A. A., and Koonin, E. V. (1991). Phylogenies of capsid proteins of rod-shaped and filamentous RNA plant viruses: Two families with distinct patterns of sequenceand probably structutre conservation. *Virology* 184, 79-86.

Dolja, V. V., Haldeman, R., Robertson, N. L., Dougherty, W. G., and Carrington, J. C. (1994). Distinct functions of capsid protein in assembly and movement of tobacco etch potyvirus. *EMBO J.* 13, 1482-1491.

Dougherty, W. G., and Parks, T. D. (1991). Post-translational processing of the tobacco etch virus 49 kDa small nuclear inclusion polyprotein: Identification of an internal cleavage site and delimitation of VPg and proteinase domains. *Virology* 183, 449-456.

Dougherty, W. G., and Semler, B. L. (1993). Expression of virus-encoded proteinases: Functional and structural similarities with cellular enzymes. *Microbiol. Rev.* 57, 781- 822.

Duggal, R. , and Hall, T. C. (1993). Identification of domains in brome mosaic virus RNA-1 and coat protein necessary for specific interaction and encapsidation. *J. Virol.* 67, 6406-6412.

Dunigan, D. D., and Zaitlin, M. (1990). Capping of tobacco mosaic virus RNA: Analysis of viral-coded guanylyltransferase-like activity. *J. Biol. Chem.* 265, 7779-7786.

Eagles, R. M., Balmori-Melian, E., Beck, D. L., Gardner, R. C., and Forster, R. L. (1994). Characterization of NTPase, RNA-binding, and RNA-helicase activities of the cytoplasmic inclusion protein of tamarillo mosaic potyvirus. *Eur. J. Biochem.* 224, 677 - 684.

Edwards, M. C., Zhang, Z., and Weiland, J. J. (1997). Oat blue dwarf marafivirus resembles the tymoviruses in sequence, genome organisation and expression strategy. *Virology* 232, 217-229.

Eggen, R., Verver, J., Wellink, J., Pleij, K., van Kammen, A., and Goldbach, R. (1989a). Analysis of sequences invloved in cowpea mosaic virus RNA replication using site specific mutants. *Virology* 173, 456-464.

Eggen, R., Verver, J., Wellink, J., de Jong, A., Goldbach, R. and van Kammen, A. (1989b). Improvements of the infectivity of *in vitro* transcripts from cloned cowpea mosaic virus c DNA: Impact of terminal nucleotide sequences. *Virology* 173, 447-455.

Elliott, R. M., Schmaljohn, C. S., and Collett, M. S. (1991). *Bunyaviridae* genome structure and gene expression. *Curr. Topics Microbiol. Immunol.* 169, 91-141.

Evans, R. K., Haley, B. E., and Roth, D. A. (1985). Photoaffinity labeling of a viral induced protien from tobacco: Characterization of nucleotide-binding properties. *J. Biol. Chem.* 260, 7800-7804.

Febres, V. J., Pappu, H. R., Anderson, E. J., Pappu, S. S., Lee, R. F. ,and Niblett, C. L. (1994). The diverged copy of the citrus tristeza virus coat protein is expressed *in vivo*. *Virology* 201, 178-181.

Fellers, J., Wan, J., Hong, Y., Collins, G., and Hunt, A. (1998). *In vitro* interactions between a Potyvirus encoded genome-linked protein and RNA-dependent RNA polymerase. *J. Gen. Virol.* 79, 2043-2049.

Fernández, A., Guo, H. S., Saenz, P., Simon-Buela, L., Gómez de Cedrón, M., and García, J. A. (1997). The motif V of plum pox potyvirus CI RNA helicase is involved in NTP hydrolysis and is essential for virus RNA replication. *Nucleic Acids Res.* 25, 4474-4480.

Flasinski, S., Dzianott, A., Pratt, S. and Bujarski, J. J. (1995). Mutational analysis of the coat protein gene of brome mosaic virus : Effects on replication and movement in barley and *Chenopodium hybridum*. *Mol. Plant-Microbe Interact.* 8, 23-31.

Florentz, C., and Giegé, R. (1995). *In* "tRNA: Structure, Biosynthesis and Function" (D. Söll, and U. RajBhandary, Eds.), pp.141-163. Am. Soc. Microbiol., Washington.

Forster, R. L., Beck, D. L., Guilford, P. J., Voot, D. M., van Dolleweerd, C. J., and Andersen, M. T.

(1992). The coat protein of white clover mosaic potexvirus has a role in facilitating cell-to-cell transport in plants. *Virology* 191, 480-484.

French, R., and Ahlquist, P. (1988). Characterization and engineering of sequences controlling *in vitro* synthesis of brome mosaic virus subgenomic RNA. *J. Virol.* 62, 2411-2420.

Fujita, Y., Mise, K., Okuno, T., Ahlquist, P., and Furusawa, I. (1996). A single codon change in a conserved motif of a bromo virus movement protein gene confers compatibility with a new host. *Virology* 223, 283-291.

Gallie, D. R., and Kobayashi, M. (1994). The role of the 3` untranslated region in non-polyadenylated plant viral m RNAs in regulating translational efficiency. *Gene* 142, 159 -165.

Gallie, D. R., and Walbot, V. (1992). Identification of the motifs within the tobacco mosaic virus 5` leader responsible for enhancing translation. *Nucleic Acids Res.* 17, 463-4638.

Gallie, D. R., Sleat, D. E., Watts, J. W., Turner, P. C., and Wilson, T. M. A. (1987a). The 5` leader sequence of tobacco mosaic virus RNA enhances the expression of foreign gene transcripts *in vitro* and *in vivo*. *Nucleic Acids Res.* 15, 3257-3273.

Gallie, D. R., Sleat, D. E., Watts, J. W., Turner, P. C., and Wilson, T. M. A. (1987b). A comparison of eukaryotic viral 5` leader sequences as enchances of m RNA expression *in vivo*. *Nucleic Acids Res.* 15, 8693-8711.

Gal-On, A., Kaplan, I., Roossinck, M. J., and Palukaitis, P. (1994). The kinetics of infection of zucchini squash by cucumber mosaic virus indicate a function for RNA 1 in virus movement. *Virology* 205, 280-289.

Gilbertson, R. L., and Lucas, W. J. (1996). How do viruses traffic on the 'Vascular Highway' ? *Trends Plant Sci.* 1, 260 -267.

Gilmer, D., Allmang, C., Ehresmann, C., Guilley, H., Richards, K., Jonard, G., and Ehresmann, B. (1993). The secondary structure of the 5` noncoding region of beet necrotic yellow vein virus RNA 3: Evidence for a role in viral RNA replication. *Nucleic Acids Res.* 21, 1389 -1395.

Goldbach, R., Le Gall, O., and Wellink, J. (1991). Alpha-like viruses in plants. *Semin. Virol.* 2, 19-25.

Gorbalenya, A. E., Koonin, E. V., Blinov, V. M., and Donchenko, A. P. (1988). Sobemovirus genome appears to encode a serine protease related to cysteine proteases of picornaviruses *FEBS Lett.* 236, 287-290.

Götz, R., and Maiss, E. (1995). The complete nucleotide sequence and genome organization of the mite-transmitted brome streak mosaic rymovirus in comparison with those of potyviruses. *J. Gen. Virol.* 76, 2035-2042.

Guilford, P. J., Beck, D. L., and Forster, R. L. S. (1991). Influence of the poly (A) tai l and putative polyadenylation signal on the infectivity of white clover mosaic potexvirus. *Virology* 182, 61-67.

Haldeman-Cahill, R., Daros, J. A., and Carrington, J. C. (1998). Secondary structures in the capsid protein coding sequence of 3` nontranslated region involved in amplification of the tobacco etch virus genome J. *Virol.* 72, 4072-4079.

Halveda, Z., and Burgyan, J. (1995) . 3` terminal putative stem loop structure required for the accumulation of cymbidium ringspot viral RNA. *Virology* 214, 269-272.

Hann, L. E., Webb, A. C., Cai, J. M., and Gehrke, L. (1997). Identification of a competitive translation deteminant in the 3` untranslated region of alfalfa mosaic virus coat protein mRNA. *Mol. Cell. Biol.* 17, 2005-2013.

Harbison, S. A., Davies, J. W., and Wilson, T. M. A. (1985). Expression of high molecular weight polypeptides by carnation mottle virus RNA. *J. Gen. Virol.* 66, 2597-2604.

Heaton, L. A., Lee, T. C., Wei, N., and Morris, T. J. (1991). Point mutations in the turnip crinkle virus capsid protein affect the symptoms expressed by *Nicotiana benthamiana*. *Virology* 183, 143-150.

Hehn, H., Fritsch, C., Richards, K., Guilley, H., and Jonard, G. (1997). Evidence for *in vitro* and *in vivo* autocatalytic processing of the primary translation product of beet necrotic yellow vein virus RNA 1 by a papain-like proteinase. *Arch. Virol.* 142, 1051-1058.

Heinlein, M., Epel, B. L., Padgett, H. S., and Beachy, R. N. (1995). Interaction of tobamovirus movement proteins with the plant cytoskeleton. *Science* 270, 1983-1985.

Hellendoorn, K., Michiels, P. J., Buitenhuis, R., and Pleij, C. W. A. (1996). Protonatable hairpins are conserved in the 5` untranslated region of tymovirus RNAs. *Nucleic Acids Res.* 24, 4910 -4917.

Hellendoorn, K., Verlaan, P. W., and Pleij, C. W. A. (1997). A functional role for the conserved protonable hairpins in the 5` untranslated region of turnip yellow mosaic virus RNA. *J. Virol.* 71, 8774 - 8779.

Hernandez, C., Visser, P. B., Brown, D. J., and Bol, J.F. (1997). Transmission of tobacco rattle virus isolate PpK20 by its nematode vector requires one of the two non-structural genes in the viral RNA2. *J. Gen. Virol.* 78, 465-467.

Hong, Y., and Hunt, A. G. (1996). RNA polymerase acitivity by a potyvirus-encoded RNA-dependent RNA polymerase. *Virology* 226, 146-151.

Hong, Y., Levay, K., Murphy, J. F., Klein, P. G., Shaw, J. G., and Hunt, A . G. (1995). A potyvirus polymerase interacts with the viral coat protein and VPg in yeast cells. *Virology* 214, 159 – 166.

Houser-Scott, F., Baer, M. L., Liem, K. F., Cai, J. M., and Gehrke, L. (1994). Nucleatide sequence and structural determinants of specific binding of coat protein or coat protein peptides to the 3` untranslated region of alfalfa mosaic virus RNA 4. *J. Virol.* 68, 2194 -2205.

Houser-Scott, F., Ansel-McKinney, P., Cai, J. M., and Gehrke, L. (1997). *In vitro* genetic selection analysis of alfalfa mosaic virus coat protein binding to 3` terminal AUGC repeats in the viral RNAs. *J.Virol.* 71, 2310-2319.

Ivanov, P. A., Karpova, O. V., Skulachev, M. V., Tomashevskaya, O. L., Rodionova, N. P., Dorokhov, Yu. L. , and Atabekov, J. G. (1997). A tobamovirus that contains an internal ribosome entry site functional *in vitro. Virology* 232, 32-43.

Jaspars, E. M. J. (1985). Interaction of alfalfa mosaic virus nucleic acid and protein. *In:* " Molecular Plant Virology " (J. W. Davies, Ed.,), Vol. I, pp. 155 - 221, CRC Press, Boca Raton, FL.

Jobling, S. A., and Gehrke, L. (1987). Enhanced translation of chemeric messenger RNAs containing a plant viral untranslated leader sequence. *Nature* 325, 622-625.

Jupin, I., Bouzoubaa, S., Richards, K., Jonard, G., and Guilley, H. (1990). Multiplication of beet necrotic yellow vein virus RNA 3 lacking a 3` poly (A) tail is accompanied by reappearance of the poly (A) tail and a novel short U-rich tract preceding it. *Virology* 178, 281-2 84.

Jupin, I., Guilley, H., Richards, K. E., and Jonard, G. (1992). Two proteins encoded by beet necrotic yellow vein virus RNA 3 influence symptom phenotype on leaves. *EMBO J.* 11, 479-488.

Kadaré, G., and Haenni, A.-L. (1997). Virus encoded helicases. *J. Virol.* 71, 2583-2590.

Kadaré, G., David, C. , and Haenni, A.-L. (1996). ATPase, GT Pase, and RNA binding activities associated with the 206 kilodalton protein of turnip yellow mosaic virus. *J. Virol.* 70, 8169 – 8174.

Kahn, T. W., Lapidot, M., Heinlein, M., Reichel, C., Cooper, B., Gafny, R., and Beachy, R. N. (1998). Domains of the TMV movement portein involved in subcellular localization. *Plant J.* 15, 15-25.

Kao, C. C., Quadt, R., Hershberger, R. P., and Ahlquist, P. (1992). Brome mosaic virus RNA replication proteins 1a and 2a form a complex *in vitro. J. Virol.* 66, 6322-6329.

Karasev, A. V., Boyko, V. P., Gowda, S., Nikolaeva, O. V., Hilf, M. E., Koonin, E. V., Niblett, C. L., Cline, K., Gumpf, D. J., Lee, R. F., Garnsey, S. M., Lewandowski, D. J., and Dawson, W. O. (1995). Complete sequence of the citrus tristeza virus RNA genome. *Virology* 208, 511-520.

Kim, K. H., and Lommel, S. A. (1994). Identification and analysis of the site of -1 ribosomal frameshifting in red clover necrotic mosaic virus.*Virology* 200, 574-582.

Kim, K. H., and Hemenway, C. (1996). The 5` nontranslation region of potato virus X RNA affects both genomic and sub-genomic RNA synthesis. *J. Virol.* 70, 5533-5540.

Ki m, K. H., and Hemenway, C. (1997). Mutations that alter a conserved element upstream of the potato virus X triple gene black and coat protein genes affect subgenomic RNA accumulation. *Virology* 232, 187-197.

Klaassen, V. A., Boeshore, M. L., Koonin, Ḻ. V., Tian, T., and Falk, B. W. (1995). Genome structure and phylogenetic analysis of lettuce infectious yellows virus, a whitefly-transmitted, bipartite closterovirus. *Virology* 208, 99-110.

Koenig, R., Pleij, C. W. A., Beier, C., and Commandeur, U. (1998). Genome properties of beet virus Q, a new furo-like virus from sugarbeet, determined from unpurified virus. *J. Gen. Virol.* 79, 2027 – 2036.

Kolk, M. H., van der Graaf, M., Wijmenga, S. S., Pleij, C. W. A., Heus, H. A., and Hilbers, W. C. (1998). NMR structure of a classical pseudoknot: Interplay of single- and double-stranded RNA,

Science 280, 434-438.

Koonin, E. V., and Dolja, V. V. (1993). Evolution and taxonomy of positive-strand RNA viruses: Implications of comparative analysis of amino acid sequences. *Crit. Rev. Biochem. Mol. Biol.* 28, 375-430.

Kroner, P. A., Young, B. M., and Ahlquist, P. (1990). Analysis of the role of brome mosaic virus 1a protein domains in RNA replication, using insertion mutagenesis. *J. Virol.* 64, 6110-6120.

Kujawa, A. B., Drugeon, G., Hulanicka, D., and Haenni, A.-L. (1993). Structural requirements for efficient translation frameshifting in the synthesis of the putative viral RNA-dependent RNA polymerase of potato leaf roll virus. *Nucleic Acids Res.* 21, 2165-2171.

Laín, S., Riechmann, J. L., and García, J. A. (1990). RNA helicase: A novel activity associated with a protein encoded by a positive-strand RNA virus. *Nucleic Acids Res.* 18, 7003-7006.

Laín, S., Martin, M. T., Riechmann, J. L., and García, J. A. (1991). Novel catalytic activity associated with positive-strand RNA virus infection: Nucleic acid stimulated AT Pase activity of the plum pox potyvirus helicase-like protein. *J. Virol.* 65, 1- 6.

Lauber, E., Guilley, H., Tamada, T., Richards, K. E., and Jonard, G. (1998). Vascular movement of beet necrotic yellow vein virus in *Beta macrocarpa* is dependent on an RNA 3 sequence domain rather than a gene product. *J. Gen. Virol.* 79, 385-393.

Lawrence, D. M., Rozanov, M. N., and Hillman, B. I. (1995). Autocatalytic processing of the 223-kDa protein of blue berry scorch carlavirus by a papain-like proteinase. *Virology* 207, 127-135.

Leathers, V., Tanguay, R., Kobayashi, M., and Gallie, D. R. (1993). A phylogenetically conserved sequence within viral 3` untranslated RNA pseudoknots regulates translation. *Mol .Cell. Biol.* 13, 5331-5347.

Longstaff, M., Brigneti, G., Boccard, F., Chapman, S., and Baulcombe, D. (1993). Extreme resistance to potato virus X infection in plants expressing a modified component of the putative viral replicase. *EMBO J.* 12, 379-386.

MacFarlane, S. A., and Brown, D. J. (1995). Sequence comparison of RNA2 of nematode-transmissible and nematode-nontransmissible isolates of pea early browing virus suggests that the gene encoding the 29kDa protein may be involved in nematode transmission. *J. Gen. Virol.* 76, 1299-1304.

Mahajan, S., Dolja, V. V., and Carrington, J. C. (1996). Roles of the sequence encoding tobacco etch virus capsid protein in genome amplification: Requirements for the translation process and a *cis*-active element. *J. Virol.* 70, 4370-4379.

Maia, I. G., Séron, K., Haenni, A.-L., and Bernardi, F. (1996). Gene expression from viral RNA genomes. *Plant Mol. Biol.* 32, 367-391.

Maia, I. G., Haenni, A.-L., and Bernardi, F. (1996). Potyviral HC-Pro: A multifunctional protein. *J. Gen. Virol.* 77, 1335-1341.

Mäkinen, K., Næss, V. , Tamm, T., Truve, E., Aaspõllu, A. and Saarma, M. (1995). The putative replicase of cocksfoot mottle sobemovirus is translated as a part of the polyprotein by -1 ribosomal frameshift. *Virology* 207, 566-571.

Mans, R. M. W., Pleij, C. W. A., and Bosch, L. (1991). Transfer RNA-like structures: Structure, function and evolutionary signficance. *Eur. J. Biochem.* 201, 303 -324.

Margis, R., and Pinck, L. (1992). Effects of site directed mutagenesis on the presumed catalytic triad and substrate binding pocket of grapevine fanleaf nepovirus 24 kDa proteinase. *Virology* 190, 884-888.

Marsh, L. E., Dreher, T. W., and Hall, T. C. (1988). Mutational analysis of the core and modulator sequences of BMV RNA 3 subgenomic promoter. *Nucleic Acids Res.* 16, 981-995.

Marsh, L. E., Pogue, G. P., and Hall, T. C. (1989). Similarities among plant virus (+) and (-) RNA termini imply a common ancestry with promoters of eukaryotic tRNAs. *Virology* 172, 415-427.

Marsh, L. E., Huntley, C. C., Pogue, G. P., Connell, J. P., and Hall, T.C. (1991). Regulation of (+): (-) strand asymmetry in replication of brome mosaic virus RNA. *Virology* 182, 76- 83.

Mawassi, M., Karasev, A. V., Mietkiewska, E., Gafny, R., Lee, R. F., Dawson, W. O., and Bar-Joseph, M. (1995). Defective RNA molecules associated with citrus tristeza virus. *Virology* 208, 383-387.

McLean, B. G., Zupan, J., and Zambryski, P. C. (1995). Tobacco mosaic virus movement protein associates with the cytoskeleton in tobacco cells. *Plant Cell* 7, 2101-2114.

Miller, W. A., Bujarski, J. J., Dreher, T. W., and Hall, T. C. (1986). Minus-strand initiation by brome mosaic virus replicase within the 3' tRNA-like structure of native and modified RNA templates *J. Mol. Biol.* 187, 537-546.

Miller, W. A., Dinesh-Kumar, S. P., and Paul, C. P. (1995). Luteovirus gene expression. *Crit. Rev. Plant Sci.* 14, 179-211.

Molnár, A., Havelda, Z., Dalmay, T., Szutorisz, H., and Burgyan, J. (1997). Complete nucleotide sequence of tobacco necrosis virus strain D^H and genes required for RNA replication and virus movement. *J. Gen.Virol.* 78, 1235-1139.

Morch, M. -D., Joshi, R. L., Denial, T. M., and Haenni, A.-L. (1987). A new 'sense' RNA approach to block viral RNA replication *in vitro*. *Nucleic Acids Res.* 15, 4123-4130.

Murant, A. F., Rajeshwari, R., Robinson, D. J., and Raschké, J. H. (1988). A satellite RNA of groundnut rosette that is largerly responsible for symptoms of groundnut rosette disease. *J. Gen. Virol.* 69, 1479-1486.

Nagy, P. D., Dzianott, A., Ahlquist, P., and Bujarski, J. J. (1995). Mutation in the helicase-like domain of protein 1a alter the sites of RNA-RNA recombination in brome mosaic virus. *J. Virol.* 69, 2547-2556.

Neeleman, L., van der Kuyl, A. C., and Bol, J. F. (1991). Role of alfalfa mosaic virus coat protein gene in symptom expression. *Virology* 181, 687 - 693.

Nicolaisen, M., Johansen, E., Poulsen, G. B., and Borkhardt, B. (1992). The 5' untranslated region from pea seedborne mosaic potyviurs RNA as a translational enhancer in pea and tobacco protoplasts. *FEBS Lett.* 303, 169-172.

O'Reilly, E. K., Paul, J. D., and Kao, C. C. (1997). Analysis of the interaction of viral RNA replication protein by using the yeast two-hybrid assay. *J. Virol.* 71, 7526-7532.

O'Reilly, E. K., Wang. Z., French, R., and Kao, C. C. (1998). Interactions between the structural domains of the RNA replication proteins of plant-infecting RNA viruses. *J. Virol.* 72, 7160-7169.

Ohira, K., Namba, S., Rozanov, M., Kusumi, T., and Tsuchizaki, T. (1995). Complete sequence of an infectious full-length cDNA clone of citrus tatter leaf capillovirus: Comparative sequence analysis of capillovirus genome. *J. Gen. Virol.* 76, 2305-2309.

Opalka, N., Burgidou, C., Bonneau, C., Nicole, M., Beachy, R. N., Yeager, M., and Fauquet, C. (1998). Movement of rice yellow mottle virus between xylem cells through pit membranes. *Proc. Natl. Acad., Sci. U.S.A.* 95, 3323 -3328.

Owens, R. A., Steger, G., Hu, Y., Fels, A., Hammond, R. W., and Riesner, D. (1996). RNA structural features responsible for potato spindle tuber viroid pathogenicity. *Virology* 222, 144 -158.

Paul, A. V., van Boom, J. H., Filippov, D., and Wimmer, E. (1998). Protein-primed RNA synthesis by purified poliovirus RNA polymerase. *Nature* 393, 280-284.

Perry, K. L., Zhang, L., and Palukaitis, P. (1998). Amino acid changes in the coat protein of cucumber mosaic virus differentially affect transmission by the aphids *Myzus persicae* and *Aphis gossypii. Virology* 242, 204-210.

Peters, S. A., Voorhorst, W. G, Wellink, J., and van Kammen, A. (1992). Processing of VPg-containing polyproteins encoded by the B-RNA from cowpea mosaic virus. *Virology* 191, 90 -97.

Peters, S. A., Verver, J., Nollen, E. A., van Lent, J. W., Wellink, J., and van Kammen, A. (1994). The NTP-binding motif in cowpea mosaic virus B polyprotein is essential for viral replication. *J. Gen. Virol.* 75, 3167-3176.

Petty, I. T., Edwards, M. C., and Jackson, A. O. (1990). Systemic movement of an RNA plant virus determined by a point substitution in a 5' leader sequence. *Proc. Natl. Acad. Sci.U.S.A.* 87, 8894-8897.

Pinck, M., Reinbolt, J., Loudes, A. M., Ret, M. L., and Pinck, L. (1991). Primary structure and location of the genome-linked protein (VPg) of grapevine fanleaf nepovirus. *FEBS Lett.* 284, 117-119.

Plochocka, D., Welnicki, M., Zielenkiewicz, P., and Ostoja-Zagorski, W. (1996). Three dimensional model of the potyviral genome-linked protein. *Proc. Natl. Acad. Sci. U.S.A.* 93, 12150-12154.

Pogue, G. P., and Hall, T. C. (1992). The requirement for a 5' stem-loop structure in brome mosaic virus replication supports a new model for viral positive-strand RNA initiation. *J. Virol.* 66, 674-684.

Pogue, G. P., Marsh, L. E., and Hall, T. C. (1990). Point mutations in the ICR2 motif of brome mosaic virus RNAs debilitate (+) strand replication. *Virology* 178, 152-160.

Pogue, G. P., Marsh, L. E., Connell, J. P., and Hall, T. C. (1992). Requirement for ICR-like sequences in the replication of brome mosaic virus genomic RNA. *Virology* 188, 742-753.

Prüfer, D., Tacke, E., Schmitz, J., Kull, B., Kaufmann, A., and Rohde, W. (1992). Ribosomal frameshifting in plants: A novel signal directs the -1 frameshift in the synthesis of the putative viral replicase of potato leafroll luteovirus. *EMBO J.* 11, 1111-1117.

Qu, F., and Morris, T. J. (1997). Encapsidation of turnip crinkle virus is defined by a specific packaging signal and RNA size. *J. Virol.* 71, 1428-1435.

Quadt, R., Rosdorff, H. J., Hunt, T. W., and Jaspars, E. M. J. (1991). Analysis of the protein composition of alfalfa mosaic virus RNA-dependent RNA polymerase. *Virology* 182, 309-315.

Quadt, R., Ishikawa, M., Janda, M., and Ahlquist, P. (1995). Formation of brome mosaic virus RNA-dependent RNA polymerase in yeast requires coexpression of viral proteins and viral RNA. *Proc. Natl. Acad. Sci. U.S.A.* 92, 4892-4896.

Ramirez, B.-C., and Haenni, A.-L. (1994). Molecular biology of tenuiviruses, a remarkable group of plant viruses. *J. Gen. Virol.* 75, 467-475.

Rao, A. L. N., and Grantham, G. L. (1995). A spontaneous mutation in the movement protein gene of brome mosaic virus modulates symptom phenotype in *Nicotiana benthamiana..J. Virol.* 69, 2689 – 2691.

Restrepo-Hartwig, M. A., and Carrington, J. C. (1994). The tobacco etch potyvirus 6 kilodalton protein is membrane associated and involved in viral replication. *J. Virol.* 68, 2388-2397.

Restrepo-Hartwig, M. A., and Ahlquist, P. (1996). Brome mosaic virus helicase- and polymerase-like proteins colocalize on the endoplasmic reticulum at sites of viral RNA synthesis. *J. Virol.* 70, 8908-8916.

Riechmann, J. L., Laín, S., and García, J. A. (1990). Infectious *in vitro* transcripts from a plum pox potyvirus cDNA clone. *Virology* 177, 710-716.

Riechmann, J. L., Laín, S., and García, J. A. (1991). Identification of the initiation codon of plum pox potyvirus genomic RNA .*Virology* 185, 544-552.

Robbins, M. A., Reade, R. D., and Rochon, D. M. (1997). A cucumber necrosis virus variant deficient in fungal transmissibility contains an altered coat protein shell domain. *Virology* 234, 138-146.

Rodríguez-Cerezo, E., Klein, P. G., and Shaw, J. G. (1991). A determinant of disease symptom severity is located in the 3`- terminal noncoding region of the RNA of a plant virus. *Proc. Natl. Acad. Sci. U.S.A.* 88, 9863-9867.

Rodríguez-Cerezo, E., Findlay, K., Shaw, J. G., Lomonossoff, G. P., Qiu, S. G., Linstead, P., Shanks, M., and Risco, C. (1997). The coat and cylindrical inclusion proteins of a potyvirus are associated with connections between plant cells. *Virology* 236, 296-306.

Rohde, W., Gramstat, A., Schmitz, J., Tacke, E., and Prüfer, D. (1994) . Plant viruses as model systems for the study of non-canonical translation mechanisms in higher plants.*J. Gen. Virol.* 75, 2141 – 2149.

Rojas, M. R., Zerbini, F. M., Allison, R. F., Gilbertson, R. L., and Lucas, W. J. (1997). Capsid protein and helper component-proteinase function as potyvirus cell-to-cell movement proteins. *Virology* 237, 283-296.

Rozanov, M. N., Koonin, E. V., and Gorbalenya, A. E. (1992). Conservation of the putative methyl-transferase domain: A hall mark of the 'sindbis-like' supergroup of positive-strand RNA viruses. *J. Gen. Virol.* 73, 2129-2134.

Rozanov, E. M., Drugeon, G., and Haenni, A.-L. (1995). Papain-like protease of turnip yellow mosaic virus: A prototype of a new viral proteinase group. *Arch. Virol.* 140, 273-288.

Ryabov, E. V., Generozov, E. V., Kendall, T. L., Lommel, S. A., and Zavriev, S. K. (1994). Nucleotide sequence of carnation ringspot dianthovirus RNA-1. *J. Gen. Virol.* 75, 243-247.

Sacher, R., and Ahlquist, P. (1989). Effects of deletions in the N-terminal basic arm of bnrome mosaic virus coat protein on RNA packing and systemic infection. *J. Virol.* 63, 4545-4552.

Saldarelli, P., Minafra, A., and Martelli, G. P. (1996). The nucleotide sequence and genomic organization of grapevine virus B. *J. Gen. Virol.* 77, 2645-2652.

Schaad, M. C., Jensen, P. E., and Carrington, J. C. (1997). Formation of plant RNA virus replication complexes on membranes: Role of an endoplasmic reticulum-targeted viral protein. *EMBO. J.* 16, 4049-4059.

Schmitz, J., Stussi-Garaud, C., Tacke, E., Prüfer, D., Rhode, W., and Rohfritsch, O. (1997). *In situ* localization of the putative movement protein (pr 17) from potato leafroll luteovirus (PLRV) in infected and transgenic potato plants. *Virology* 235, 311-322.

Schneider, W. L., Greene, A. E., and Allison, R. F. (1997). The carboxy-terminal two-thirds of the cowpea chlorotic mottle bromovirus capsid protein is incapable of virion formation yet supports systemic movement. *J. Virol.* 71, 4862-4865.

Scholthof, H. B., Scholthof, K.-B. G., Kikkert, M., and Jackson, A. O. (1995). Tomato bushy stunt virus spread is regulated by two nested genes that function in cell-to-cel l movement and host-dependent systemic invasion. *Virology* 213, 425-438.

Scholthof, K.-B. G., Hillman, B. I., Modrell, B., Heaton, L. A., and Jackson, A. O. (1994). Characterization and detection of sc4 : A sixth gene encoded by sonchus yellow net virus. *Virology* 204, 279-288.

Scott, S. W., Zimmerman, M. T., and Ge, X. (1998). The sequence of RNA 1 and RNA 2 of tobacco streak virus: Addtional evidence for the inclusion of alfalfa mosaic virus in the gennus *Ilarvirus. Arch. Virol.* 143, 1187-1198.

Séron, K., and Haenni, A.-L. (1996). Vascular movement of plant viruses. *Mol. Plant-Microbe Interact.* 9, 435-442.

Shams-bakhsh, M., and Symons, R. H. (1997). Barley yellow dwarf virus - PAV RNA does not have a VPg. *Arch. Virol.* 142, 2529-2535.

Shen, P., Kaniewska, M., Smith, C., and Beachy, R. N. (1993). Nucleotide sequence and genomic organisation of rice tungro spherical virus. *Virology* 193, 621-630.

Shintaku, M. H., Carter, S. A., Bao, Y., and Nelson, R. S. (1996). Mapping nucleotides in the 126 kDa protein gene that control the differential symptoms induced by two strains of tobacco mosaic virus. *Virology* 221, 218-225.

Shirako, Y., and Wilson, T. M. A. (1993). Complete nucleotide sequence and organisation of the bipartite RNA genome of soilbrorne wheat mosaic virus. *Virology* 195, 16-32.

Sit, T. L., Leclerc, D., and AbouHaidar, M. G. (1994). The minimal 5` sequence for *in vitro* initiation of papaya mosaic potexvirus assembly. *Virology* 199, 238-242.

Skuzeski, J. M., Nichols, L. M., Gesteland, R. F., and Atkins, J. F. (1991). The signal for a leaky UAG stop codon in several plant viruses includes the 2 downstream codons. *J. Mol. Biol.* 218, 365-373.

Sleat, D. E., and Palukaitis, P. (1990). Site-directed mutagenesis of a plant viral satellite RNA changes its phenotype from ameliorative to necrogenic. *Proc. Natl. Acad. Sci. U.S.A.* 87, 2946-2950.

Smirnyagina, E., Hsu, Y.-H., Chua, N. , and Ahlquist, P. (1994). Secondary mutations in the brome mosaic virus RNA 3 intercistronic region partially suppress a defect in coat protein m RNA transcription. *Virology* 198, 427-436.

Sriskanda, V. S., Pruss, G., Ge, X., and Vance, V. B. (1996). An eight nucleodite sequence in the potato virus X 3` untranslated region is required for both host protein binding and viral multiplication. *J. Virol.* 70, 5266-5271.

Suzuki, M., Kuwata, S., Masuta, C., and Takanami, Y. (1995). Point mutations in the coat protein of cucumber mosaic virus affect symptom expression and virion accumulation in tobacco. *J. Gen. Virol.* 76, 1791-1799.

Suzuki, N. (1995). Molecular analysis of the rice dwarf virus genome. *Semin. Virol.* 6, 89-95

Suzuki, N., Kusano, T., Matsuura, Y., and Omura, T. (1996). Novel NTP binding property of rice dwarf phytoreovirus minor core potein P5. *Virology* 219, 471-474

Takamatsu, N., Watanabe, Y., Meshi, T., and Okada , Y. (1990). Mutational analysis of the pseudoknot region in the 3` non-coding region of tobacco mosaic virus RNA. *J. Virol.* 64, 3686-3693.

Taliansky, M. E., and Robinson, D. J. (1997). *Trans*-acting untranslated elements of groundnut rosette virus satellite RNA are involved in symptom production. *J. Gen. Virol.* 78, 1277-1285.

Taliansky, M. E., Robinson, D. J., and Murant, A. F. (1996). Complete nucleotide sequence and organization of the RNA genome of groundnut rosette umbravirus. *J. Gen. Virol.* 77, 2335-2345.

Tamada, T., Schmitt, C., Saito, M., Guilley, H., Richards, K., and Jonard, G. (1996). High resolution analysis of the readthrough domain of beet necrotic yellow vein virus readthrough protein: A KTER motif is important for efficient transmission of the virus by *Polymyxa betae. J. Gen. Virol.* 77, 1359-1367.

Ten Dam, E. B., Pleij, C. W. A., and Bosch, L. (1990). RNA pseudoknots : Translational frameshifting and readthrough on viral RNAs. *Virus Genes* 4, 121-136.

Thole, V., and Hull, R. (1998) . Rice tungro spherical virus polyprotein processing : Identification of a virus-encoded protease and mutational analysis of putative cleavage sites. *Virology* 247, 106-114.

Thomas, A. A. M., ter Haar, E., Wellink, J., and Voorma, H. O. (1991). Cowpea mosaic virus middle component RNA contains a sequence that allows internal binding of ribosomes and that requires eukaryotic initiation factor 4F for optimal translation. *J. Virol.* 65, 2953 -2959.

Tomashevskaya, O. L., Solovyev, A. G., Karpova, O. V., Fedorkin, O. N., Rodionova, N. P., Morozov, S. Y., and Atabekov, J. G. (1993). Effects of sequence elements in the potato virus X RNA 5` non-translated aB leader on its translation enhancing activity. *J. Gen. Virol.* 74, 2717-2724.

Toriyama, S., Takahashi, M., Sano, Y., Shimizu, T., and Ishihama, A. (1994). Nucleotide sequence of RNA1, the largest genomic segment of rice stripe virus,the prototype of the tenuvirus. *J. Gen. Virol.* 75, 3569-3579.

Toriyama, S., Kimishima, T., Takahashi, M., Shimizu, T., Minaka, N., and Akutsu, K. (1998). The complete nucleotide sequence of the rice grassy stunt virus genomic and genomic comparisons with viruses of the genus *Tenuivirus. J. Gen. Virol.* 79, 2051-2058.

Traynor, P., Young, B. M., and Ahlquist, P. (1991). Deletion analysis of brome mosaic virus 2a protein: Effects on RNA replication and systemic spread. *J. Virol.* 65, 2807-2815.

Tsai, C.-H., and Dreher, T. W. (1991). Tunnip yellow mosaic virus RNAs with anticodon loop substitutions that result in decreased valylation fail to replicate efficiently.*J. Virol.* 65, 3060 -3067.

Turnbull-Ross, A. D., Mayo, M. A., Reavy, B., and Murant, A. F. (1993). Sequence analysis of the parsnip yellow fleck virus polyprotein: Evidence of affinities with picornaviruses. *J. Gen. Virol.* 74, 555-561.

Upadhyaya, N. M., Yang, M., Kositratana, W., Gosh, A., and Waterhouse, P. M. (1995). Molecular analysis of rice ragged stunt oryzavirus segement 9 and sequence conservation among isolates from Thailand and India. *Arch. Virol.* 140, 1945-1956.

Upadhyaya, N. M., Ramm, K., Gellatly, J. A., Li, Z., Kositratana, W., and Waterhouse, P. M. (1997). Rice ragged stunt oryzavirus genome segments S7 and S10 nonstructural proteins of M-r 68025 (pns7) and M-r 32364 (pns10). *Arch. Virol.* 142, 1719-1726.

Upadhyaya, N. M., Ramm, K., Gellatly, J. A., Li, Z., Kositratana, W., and Waterhouse, P. M. (1998). Rice ragged stunt oryzavirus genome segement S4 could encode an RNA-dependent RNA polymerase and a second protein of unknown function. *Arch. Virol.* 143, 1815-1822.

Uyeda, I., Kimura, I., and Shikata, E. (1995). Charaterization of genome structure and establishment of vector cell lines for plant reoviruses. *Adv. Virus Res.* 45, 249-279.

Valle, R. P. C., Drugeon, G., Devignes-Morch, M.-D., Legocki, A. B., and Haenni, A.-L. (1992). Codon context effect in virus translational readthrough: A study *in vitro* of the determinants of TMV and Mo-MuLV amber suppression *FEBS Lett.* 306, 133 -139.

van den Heuvel, J. F., Bruyère, A., Hogenhout, S. A., Ziegler-Graff, V., Brault, V., Verbeek, M., van der Wilk, F., and Richards, K. (1997). The N-terminal region of the luteovirus readthrough domain determines virus binding to *Buchnera* GroEL and is essential for virus persistence in the aphid. *J. Virol.* 71, 7258-7265.

van der Vossen, E. A. G., and Bol, J. F. (1996). Analysis of *Cis*-acting elements in the leader sequence of alfalfa mosaic virus RNA 3. *Virology* 220, 539-543.

van der Vossen, E. A. G., Neeleman, L., and Bol, J. F. (1993). Role of the 5` leader sequence of alfalfa mosaic virus RNA 3 in replication and translation of the viral RNA. *Nucleic Acids Res.* 21, 1361-1367.

van der Vossen, E. A. G., Neeleman, L., and Bol, J. F. (1994). Early and late functions of alfalfa mosaic virus coat protein can be mutated separately. *Virology* 202, 891-903.

van der Vossen, E. A. G., Notenboom, T., and Bol, J. F. (1995). Characterization of sequences controlling the synthesis of alfalfa mosaic virus subgenomic RNA *in vivo.Virology* 212, 663 - 672.

van der Wel, N. N., Goldbach, R., and van Lent, J. W. (1998). The movement protein and coat protein of alfalfa mosaic virus accumulate in structurally modified plasmodesmata. *Virology* 244, 322 -329.

van der Wilk, F., Verbeek, M., Dullemans, A. M., and van den Heuvel, J. F. (1997). The genome-linked protein of potato leafroll virus is located downstream of the putative protease domain of ORF 1 product. *Virology* 234, 300-303.

van Loon, L. C. (1987). Disease induction by plant viruses. *Adv. Virus Res.* 33, 205-255

van Rossum, C. M., Neeleman, L., and Bol, J. F. (1997a). Comparison of the role of 5` terminal sequences of alfalfa mosaic virus RNAs 1, 2 and 3 in viral RNA replication. *Virology* 235, 333 -341.

van Rossum, C. M., Brederode, F. T., Neeleman, L., and Bol, J. F. (1997b). Functional equivalence of common and unique sequences in the 3` untranslated regions of alfalfa mosaic virus RNAs 1, 2 and 3. *J. Virol.* 71, 3811-3816.

van Rossum, C. M., Reusken, C. B. E. M., Brederode, F. T., and Bol, J. F (1997c). The 3` untranslated region of alfalfa mosaic virus RNA 3 contains a core promoter for minus strand RNA synthesis and an enhancer element. *J. Gen. Virol.* 78, 3045 -3049.

Vaquero, C., Turner, A. P., Demangeat, G., Sanz, A., Serra, M. T., Roberts, K., and García-Luque, I. (1994). The 3a protein from cucumber mosaic virus increases the gating capacity of plasmodesmata in transgenic tobacco plants. *J. Gen. Virol.* 75, 3193 -3197.

Vaquero, C., Liao, Y. C., Nahring, J., and Fischer, R. (1997). Mapping of the RNA-binding domain of the cucumber mosaic viurs movement protein. *J. Gen. Virol.* 78, 2095 -2099.

Verchot, J., Herndon, K. L., and Carrington, J. C. (1992). Mutational analysis of the tobacco etch Potyviral 35 kDa proteinase: Identification of essential residues and requirements for autoproteolysis *Virology* 190, 298-306.

Verver, J., Le Gall, O., van Kammen, A., and Wellink, J. (1991). The sequence between nucleotides 161and 512 of cowpea mosaic virus M RNA is able to support internal initation of translation *in vitro J. Gen. Virol.* 72, 2339-2345.

Wang, J., and Simon, A. E. (1997). Analysis of the two subgenomic RNA promoters for turnip crinkle virus *in vivo* and *in vitro*. *Virology* 232, 174-186.

Wang, J. Y., Chay, C., Gildow, F. E., and Gray, S. M. (1995). Readthrough protein associated with virions of barley yellow dwarf luteovirus and its potential role in regulating the efficiency of aphid transmission. *Virology* 206, 954-962.

Wang, S., Browning, K. S., and Miller, W. A. (1997). A viral sequence in the 3` untranslated region mimics a 5` cap in stimulating translation of uncapped mRNA. *EMBO J.* 16, 4107-4116.

Wei, N., and Morris, T. J. (1991). Interactions between viral coat protein and a specific binding region on turnip crinkle virus RNA. *J. Mol. Biol.* 222, 437- 443.

Wei, N., Heaton, L. A., Morris, T. J., and Harrison, S. C. (1990). Structure and assembly of turnip crinkle virus. VII. Identification of coat protein binding sites on the RNA. *J. Mol. Biol.* 214, 85-95.

Weiland, J. J., and Dreher, T. W. (1993). *Cis*-preferential replication of the turnip yellow mosaic virus RNA genome. *Proc. Natl. Acad. Sci. U.S.A.* 90, 6095-6099.

Weiland, J. J., and Edwards, M. C. (1994). Evidence that the αa gene of barley stripe mosaic virus encodes determinants of pathogenicity to oat (*Avena sativa*). *Virology* 201, 116-126.

Wellink, J., and van Kammen, A. (1989). Cell-to-cell transport of cowpea mosaic virus requires both the 58k/48k proteins and the capsid proteins. *J. Gen. Virol.* 70, 2279-2286.

Wetzel, T., Dietzgen, R. G., and Dale, J. L. (1994). Genomic organization of lettuce necrotic yellows rhabdovirus. *Virology* 200, 401- 412.

Xiong, Z., Kim, K. H., Kendall, T. L., and Lommel, S. A. (1993). Synthesis of the putative red clover necrotic mosaic virus RNA polymerase by ribosomal frameshifting *in vitro*. *Virology* 193, 213-221.

Xu, H., Li, Y., Mao, Z., Li, Y., Wu, Z., Qu, L., An, C., Ming, X., Schiemann, J., Casper, R., and Chen, Z. (1998). Rice dwarf phytoreovirus segment S11 encodes a nucleic acid binding protein. *Virology* 240, 267-272.

Xu, Z., Anzola, J. V., Nalin, C. M., and Nuss, D. L. (1989). The 3` terminal sequence of a wound tumor virus transcript can influence conformational and functional properties associated with the 5` terminus. *Virology* 170, 511-522.

Zaccomer, B., Haenni, A.-L., and Macaya, G. (1995). The remarkable variety of plant RNA virus genomes. *J. Gen. Virol.* 76, 231-247.

Zavriev, S. K., Kanyuka K. V., and Levay K. E. (1991). The genome organization of potato virus M RNA. *J. Gen. Virol.* 72, 9-14.

Zavriev, S. K., Hickey, C. M., and Lommel, S. A. (1996). Mapping of the red clover necrotic mosaic virus subgenomic RNA. *Virology* 216, 407-410.

Zelenina, D. A., Kulaeva, O. I., Smirnyagina, E. V., Solovyev, A. G., Miroshnichenko, N. A., Fedorkin, O.N., Rodionova, N. P., Morozov, S.Y., and Atabekov, J. G. (1992). Translation enhancing properties of the 5`-leader of potato virus X genomic RNA. *FEBS Lett.* 296, 267-270.

Zuidema, D., and Jaspars, E. M. J. (1984). Comparative investigations on the coat protein binding sites of the genomic RNAs of alfalfa mosaic and tobacco streak viruses. *Virology* 135, 43-52.

2 GENOME OF DNA VIRUSES

Thomas Frischmuth

INTRODUCTION

The genome of plant DNA viruses can be either single-stranded or double-stranded. Family *Geminiviridae* (Mayo and Pringle, 1998; geminiviruses) and the proposed genus *Nanovirus* (Pringle, 1998; Nanoviruses) contain single-stranded DNA (ssDNA) genome while family *Caulimoviridae* (Mayo and Pringle, 1998) contains double-stranded DNA (dsDNA) genome. Geminiviruses *are* small plant viruses with monopartite (one DNA molecule) or bipartite (two DNA molecules) circular ssDNA genomes of 2.5-3.0 kb encapsided in twinned (geminate) particles of 20 by 30 nm diameter (Harrison, 1985; Frischmuth and Stanley, 1993). The geminiviruses contain three genera of *Mastrevirus, Curtovirus,* and *Begomovirus* (Mayo and Pringle, 1998) and are persistently transmitted by insect vectors.

Nanoviruses are isometric plant viruses of about 20 nm diameter, and contain a multicomponent circular ssDNA genome with each component of about 1kb. They have five members (Pringle, 1998) namely banana bunchy top virus (BBTV; Harding *et al.*, 1993), subterranean clover stunt virus (SCSV; Chu *et al.*, 1993), faba bean necrotic yellows virus (FBNYV; Katul *et al.*, 1993) and milk vetch dwarf virus (MDV; Sano *et al.*, 1993) and a tentative member coconut foliar decay virus (CFDV; Rohde *et al.*, 1990). CFDV has a monopartite genome of 1.3kb (Rohde *et al.* 1990) and is transmitted by *Myndus taffini* planthopper while other nanoviruses have aphid vectors.

Caulimoviridae contains two genera, isometric *Caulimovirus (*caulimoviruses) and bacilliform *Badnavirus* (badnaviruses) (Mayo and Pringle, 1998). Caulimo- as well as badnaviruses are plant pararetroviruses because their replication cycle resembles that of animal retroviruses (Hull, 1992; Rothnie *et al.,* 1994). In contrast to caulimoviruses, badnaviruses infect a wide range of economically important crop plants such as rice, cassava, sugarcane etc.

The caulimoviruses (Hull *et al.* 1987; Rothnie *et al.*, 1994) (type member cauliflower mosaic virus, CaMV) contain circular dsDNA of 7.8-8.2 kb size with one or more site-specific discontinuities in each strand and genome is encapsided in a nonenveloped spherical particle of about 45-50 nm diameter. The badnaviruses (Medberry *et al.*, 1990; Hay *et al.*, 1991; Qu *et al.*, 1991) (bacilliform DNA viruses) have a circular dsDNA genome of 7.5-8 kb encapsided in bacilliform virions of 120-150 x 30 nm size. Most caulimoviruses have aphid vectors while badnaviruses are transmitted semi-persistently by mealybugs or leafhoppers.

GENOME ORGANIZATION

Family *Geminivirdae*

The overall genomic structure is very similar in all the three geminivirus genera (Fig. 1). Open reading frames (ORFs) are distributed on the virion-sense and complementary-sense strands whereby the AUG initiation codons are orientated towards an intergenic region (LIR, IR, CR) and the stop codons to a sequence region rich in polyadenylation signal sequences (Fig. 1). The intergenic region contains sequences for viral replication and gene expression (Lazarowitz, 1992 and Timmermans *et al.*, 1994).

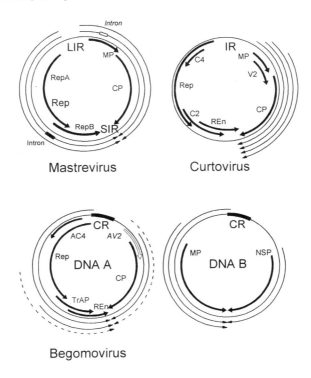

Figure 1. Genome organization in geminiviruses. The diagram shows the consensus genome organization of double stranded replicative forms of geminiviruses. Intergenic regions (IR, LIR, SIR) are indicated. The part of the IR which is almost sequence identical between DNA A and B of begomoviruses is indicated by a black box (CR). Open reading frames are indicated by black arrows within the circle and where a gene's function is known, the name of the gene product is indicated. CP: coat protein; MP: movement protein; NSP: nuclear shuttle protein; Rep: replication associated protein; TrAP: transcriptional activator; REn: replication enhancer protein. ORF *AV2* is only present in Old World begomoviruses (open arrow). Arrows outside the circle represent transcripts, whereby transcripts indicated by dotted arrows are not produced by all begomoviruses (arrow heads 3' ends). Intron sequences of mastreviruses are indicated by a black and open box.

Geminiviruses with monopartite genomes transmitted by leafhopper vectors to monocotylednous and dicotyledonous plants (Morris *et al.*, 1992; Liu *et al.*, 1997) belong to the genus *Mastrevirus*, of which maize streak virus (MSV) is the type species (Fig. 1). Mastreviruses have four ORFs distributed between the virion-sense and complementary-sense strands, whereby replication protein B (RepB) of most mastreviruses does not contain an AUG initiation codon (Fig. 1; Mullineaux *et al.*, 1984; MacDowell *et al.*, 1985; Chatani *et al.*,1991; Briddon *et al.*, 1992; Morris *et al.*,1992; Hughes *et al.*,1993; Liu *et al.*, 1997). Two features of the *Mastrevirus* genome set them apart from the genome organization of the other two genera, the presence of a second non-coding sequence region, the small intergenic region (SIR), and an intron sequence in the replication-associated gene (*Rep*) (Fig. 1).

Viruses of *Curtovirus* genus have a monopartite genome, infect dicotyledonous plants and are transmitted by leafhopper or treehopper vectors (Stanley *et al.* 1986; Klute *et al.*, 1996; Briddon *et al.*, 1996). Beet curly top virus (Fig. I; BCTV; Stanley *et al.*, 1986) is the type member. Two other members are horseradish curly top virus (HrCTV; Klute *et al.*, 1996) and tomato pseudo-curly top virus (TPCTV; Briddon *et al.*, 1996). Their genome organization differs from that of the BCTV, HrCTV lacks the replication enhancer protein (REn) and TPCTV the movement protein (MP) gene (Fig. 1).

Except the mediterranean isolates of tomato yellow leaf curl virus (TYLCV; Navot *et al.*, 1991) and tomato leaf curl virus from Australia (TLCV; Dry *et al.*, 1993), all viruses of the *Begomovirus* genus have a bipartite genome (Fig. 1; DNA A and DNA B), infect dicotyledonous plants and are transmitted by the whitefly *Bemisia tabaci*. The type member is bean golden mosaic virus from Puerto Rico (BGMV; Howarth *et al.*, 1985). The *AV2* gene is only present in Old World mono- as well as bipartite begomoviruses (Fig. 1). Approximately 180 nucleotides of the IR of bipartite begomoviruses are almost identical between DNA A and B (CR: common region).

Genus *Nanovirus*

So far six distinguishable components have been isolated from genome of BBTV- (Burns *et al.*, 1995) and FBNYV (Katul *et al.*, 1997) infected plants and seven from SCSV (Boevink *et al.*, 1995). Each component encodes, in principle, for a single gene in virion-sense orientation (Fig. 2a). Due to transcript analysis of BBTV DNA 1, the presence of a second ORF with a coding capacity of 5kD was postulated. However, no product or function has been associated with this putative ORF. Some components like FBNYV DNA 2 encode for putative duplicates of the *Rep* gene normally encoded by DNA 1 while other components like BBTV DNA 2 have no coding capacity at all (Fig. 2a). The function of such of the components in the pathogenicity of the nanoviruses is not known as yet. All of the components possess a non-coding intergenic region of which, in general, two sequence parts have sequence homologies (Fig. 2a; CR-M: major common region and CR-SL: stem-loop common region). Both common regions harbour sequences which are involved in the viral replication.

In contrast, only a single DNA component with a coding capacity for six ORFs in virion- and complementary-sense orientations was isolated from the tentative member CFDV infected plants (Fig. 2b).

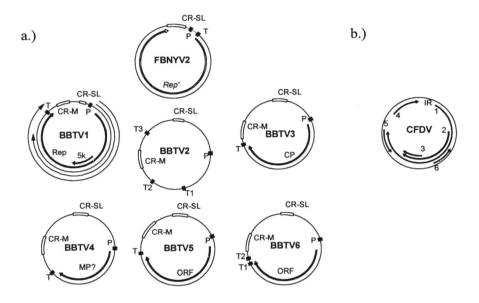

Figure 2. Genome organization of nanoviruses. The genome organization of banana bunchy top virus (BBTV) and faba bean necrotic yellows virus (FBNYV) component 2 are shown in a.) and of coconut foliar decay virus (CFDV) in b.). The location of the stem-loop common region (CR-SL) and major common region (CR-M) are indicated by open boxes and putative promoter (P) and termination (T) sequences by black boxes. Position and orientation of ORFs are indicated by black arrows and for FBNYV DNA 2 by an open arrow. Where a gene's function is known, the name of the gene product is indicated (see figure legend 1). Arrows outside the circle of BBTV1 represents transcripts detected in infected plants (arrow heads 3' ends).

Family *Caulimoviridae*

Genus Caulimovirus

The caulimoviruses have circular dsDNA genomes with one or more site-specific discontinuities in each strand of the DNA (Fig. 3). Only one interruption generally occurs in the minus-sense strand. The interruptions are associated with small nonbase-paired tails which protrude over the double helix. Ribonucleotides are found covalently attached to the 5' termini at the interruptions. The genetic organization reveals seven ORFs encoded by the minus-sense strand and none by the plus-sense strand (Fig. 3). The coding regions are interrupted by an IR of approximately 600bp, with various regulatory elements such as the 35S promoter.

From this promoter a transcript, spanning the entire genome plus an approximately 180 nucleotide long duplication, is produced. A second smaller intergenic region occurs upstream of gene VI, containing the so called 19S promoter. From this promoter a transcript spanning the gene VI is produced (Fig. 3) (Rothnie *et al.*, 1994).

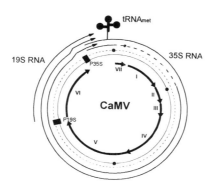

Figure 3. Genome organization of caulimoviruses. The genome organization of the type-member CaMV is shown. The double circle represents the genomic DNA (dotted circle: minus-sense strand and continuos circle: plus-sense strand) with the discontinuities shown as spots. ORFs are indicated by black arrows within the double circle. The position of the two promoter sequences (P35S and P19S) are indicated by black boxes. Outer arcs and circles showing the genomic location of the two major transcripts (35S RNA and 19S RNA) and spliced transcripts whereby the dotted line indicates the positions of intron sequences (arrow heads 3' ends). The position where the replication initiator methionine tRNA binds in the pregenomic 35S RNA is indicated.

Genus Badnavirus

Badnaviruses possess a circular dsDNA ranging in size from 7489bp [commelina yellow mottle virus (CoYMV); Medberry *et al.*, 1990] to 8158bp [cassava vein mosaic virus (CVMV); Calvert *et al.*, 1995]. The genome organization of badnaviruses is very similar to those of caulimoviruses . Like caulimoviruses both strands have site specific discontinuities and only the minus-sense strand encodes for genes . The coding capacity varies from three ORFs (I-III) of CoYMV, four (I-IV) of rice tungro bacilliform virus (RTBV; Hay *et al.*, 1991) to five (I-V) of CVMV (Fig. 4). ORF I of RTBV lacks an AUG initiation codon and the translation is initiated at an AUU codon (Hay *et al.*, 1991). The coding regions are interrupted by an IR of approximately 700bp which contains various regulatory elements such as a promoter similar to the 35S promoter of CaMV (Fig. 4). Like in CaMV, this promoter produces a transcript spanning the entire viral genome (Fig. 4).

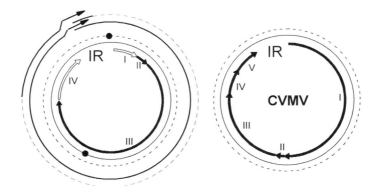

Figure 4. Genome organization of badnaviruses. The double circle represents the genomic DNA (dotted circle: minus-sense strand and continuos circle: plus-sense strand) with the discontinuities shown as spots. ORFs are indicated by black or open arrows within the double circle and the position of the intergenic region (IR) is indicated. Outer circles showing the genomic location of the major transcript as well as a spliced transcript whereby the dotted line indicates the positions of the intron sequence (arrow heads 3' ends). On the left side the general genome organization of badnaviruses is shown and on the right site of cassava vein mosaic virus (CVMV).

GENE FUNCTIONS

Family *Geminiviridae*

Rep is the only protein absolutely necessary for viral replication (Elmer *et al.*, 1988; Briddon *et al.*, 1989; Etessami *et al.*, 1991; Sung and Coutts, 1995; Liu *et al.*, 1998). Rep of curtoviruses and begomoviruses is produced by a single ORF whereas splicing of the complementary sense transcript is essential for production of functional RepA-RepB fusion protein (Rep) of *Mastrevirus* (Fig. 1; Accotto *et al.*, 1989; Schalk *et al.*, 1989). Rep plays a key role in geminivirus replication and also functions as a transcriptional repressor by blocking host-mediated activation of its promoter (Eagle *et al.*, 1994; Hong and Stanley, 1995). Rep confers virus-specific recognition of its origin of replication and initiates plus-strand DNA synthesis (Fontes *et al.*, 1992; Lazarowitz *et al.*, 1992; Heyraud *et al.*,1993; Fontes *et al.*, 1994a,b; Orozco and Hanley-Bowdoin, 1996; Suárez-López and Gutiérrez, 1997; Sanz-Burgos and Gutiérrez, 1998). Biochemical studies established that Rep has cleavage, ligation and ATPase enzymatic activities (Laufs *et al.*, 1995; Hanson *et al.*, 1995; Desbiez *et al.*, 1995; Hoogstraten *et al.*, 1996; Orozco *et al.*, 1997). For recognition of the viral replication origin, oligomerization of Rep is required while homodimers of Rep function as transcriptional repressors. Interaction of Rep oligomeres with REn enhances viral replication (Settlage *et al.*, 1996; Orozco *et al.*, 1997). Beside these virus-specific functions, Rep interacts with Rb-like plant proteins and induces expression of host DNA synthesis proteins in terminally differentiated cells (Xie *et al.*, 1995; Nagar *et al.*, 1995; Ach *et al.*, 1997). DNA binding, cleavage, ligation, Rep-Rep interaction and Rb binding activities are mediated by overlapping protein domains in the amino-terminus of begomovirus Rep protein (Fig. 5).

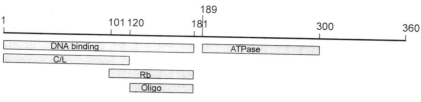

Figure 5. Linear map of begomovirus Rep protein. Grey boxes indicate the position of functional domains. DNA binding: DNA binding domain. C/L: cleavage and ligation domain. Rb: Rb binding. Oligo: Rep-Rep interaction. Numbers correspond to the amino acids position of Rep.

Transcriptional activator protein (TrAP) gene of begomoviruses is required for expression of CP and nuclear shuttle protein (NSP) (Sunter *et al.*, 1990; Sunter and Bisaro 1991, 1992; Haley *et al.*, 1992). This explains the pleiotropic nature of TrAP mutants. These mutants affect infectivity by preventing systemic viral movement in plants, produce no CP, and accumulate reduced amounts of ssDNA in transient assays (Hanley-Bowdoin *et al.*, 1989; Hayes and Buck, 1989; Sunter *et al.*, 1990; Etessami *et al.*, 1991). TrAP stimulates CP expression at the level of transcription by two alternative mechanisms in different tissues: it activates the CP in mesophyll cells, and acts to derepress the promoter in phloem tissue (Sunter and Bisaro, 1997). This small zinc-binding phosphoprotein contains a transcriptional activation domain and has DNA binding activity (Sung and Coutts, 1996; Noris *et al.*, 1996). Because TrAP DNA binding activity is none-specific and strongly prefers ssDNA over dsDNA, the mechanism of how it interacts with its responsive promoter sequences is still unclear.

Replication enhancer protein (REn) mutant-infected plants develop mild symptoms and viral DNA is greatly reduced in these plants (Sunter *et al.*, 1990; Etessami *et al.*, 1991; Morris *et al.*, 1991; Stanley *et al.*, 1992; Hormuzdi and Bisaro 1995). REn functions as a replication enhancer via interaction with Rep and by that modulates the function of Rep, replication verses repression (Settlage *et al.*, 1996).

CP is the most abundant protein in infected tissues (Morris-Krsinich *et al.*, 1985; Townsend *et al.*, 1985). The CP of bipartite begomoviruses is not essential for viral replication, systemic spread or symptom development (Stanley and Townsend, 1986; Gardiner *et al.*, 1988). In contrast, the CP of mastreviruses and curtoviruses is required for viral movement (Briddon *et al.*, 1989; Boulton *et al.*, 1989; Woolston *et al.*, 1989; Lazarowitz *et al.*, 1989). The CP protects viral DNA during transmission and determines vector specificity (Briddon *et al.*, 1990; Azzam *et al.*, 1994; Höfer *et al.*, 1997).

MP, besides the CP of mastreviruses, (Fig. 1) is involved in viral movement. MP gene product is located in secondary plasmodesmata (Dickinson *et al.*, 1996). The

BCTV gene product of virion-sense ORF V3 seems to be involved in viral movement (Frischmuth *et al.*, 1993). However, this ORF is not present in TPCTV, a second member of the *Curtovirus* genus.

Little is known about movement of monopartite begomoviruses in plants except of one report that ORF AC4 seems to be involved in movement of TYLCV (Jupin *et al.*, 1994). Mutational analysis showed that both genes encoded by DNA B of bipartite begomoviruses are involved in viral movement (Fig. 1; Brough *et al.*, 1988; Etessami *et al.*, 1988). The two proteins have different functions. NSP, also a movement protein, moves the viral DNA in and out of the nucleus (Noueiry *et al.*, 1994; Pascal *et al.*, 1994; Sanderfoot *et al.*, 1996). In the cytoplasm NSP interacts with MP and is translocated to the cell periphery (Sanderfoot and Lazarowitz, 1995). At the cytoplasmic membrane, MP forms endoplasmic reticulum-derived tubules, which extend through cell wall to the next cell (Ward *et al.*, 1997). Other data suggest that MP is able to increase the molecular size exclusion limit (SEL) of plasmodesmata (Noueiry *et al.*, 1994). Therefore, two hypotheses for movement of bipartite begomoviruses are currently discussed. Either viral DNA moves to the next cell by endoplasmic reticulum-derived tubules or via MP modified plasmodesmata.

Role of AC4 protein (AC4) in viral pathogenicity is not clear. The AC4 protein of curtoviruses is involved in symptom development and ectopic expression of AC4 in transgenic plants which confirmed that this gene is sufficient to initiate cell division (Stanley and Latham, 1992; Latham *et al.*, 1997). Expression of the TLCV *AC4* gene, a monopartite member of the begomovirus group, produced virus-like symptoms in transgenic plants (Krake *et al.*, 1998) and AC4 of TYLCV seems to be involved in viral movement (Jupin *et al.*, 1994). The *AC4* gene of bipartite begomoviruses had some influence on *Rep* expression in protoplast studies but mutations in this gene showed no influence on viral pathogenicity in plants (Gröning *et al.*, 1994).

Genus *Nanovirus*

Little is known about gene products of nanoviruses and their functions. Genomic component 5 of FBNYV encodes the CP and has counterpart on component 3 of BBTV and component 5 of SCSV (Fig. 2; Katul *et al.*, 1997, Wanitchakorn *et al.*, 1997). The second characterized gene product is Rep on BBTV genomic component 1 (Fig. 2; Harding *et al.*, 1993). The Rep protein has cleavage and ligation activities similar to those of the geminivirus Rep protein (Hafner *et al.*, 1997). Functions of other ORFs are not known except that the gene product of component 4 might be involved in viral movement (Katul *et al.*, 1997).

Family *Caulimoviridae*

Genus *Caulimovirus*

The functions of ORF I to VII of type member CaMV are summarized in Table 1. ORF I encodes MP (Thomas *et al.*, 1993) and cell-to-cell spread occurs via ectopic plasmodesmata modified by the tubular extensions (Linstead *et al.*, 1988).

Similar tubular structures extending from surface of protoplasts derived from infected plants are composed, at least in part, of MP (Perbal *et al.*, 1993). The gene II product is involved in transmission of CaMV by aphids (Rothnie *et al.*, 1994). The protein accumulates in electron-lucent inclusion bodies in infected cells (Espinoza *et al.*, 1991) and has probably two functional domains, one interacting with virus particle and the second with specific sites in aphid stylets (Blanc *et al.*, 1993). The ORF III product has sequence-nonspecific nucleic acid binding activity and is required for viral infectivity and may function in virus assembly (Dautel *et al.*, 1994; Jacquot *et al.*, 1998). ORF IV encodes for the CP of CaMV. During assembly of particles, CP interacts with inclusion body protein and is modified by proteolytic cleavage after assembly (Himmelbach *et al.*, 1996). The gene V product is a multifunctional protein with reverse transcriptase (RT), RNaseH and protease activities and has the central function in viral replication (Fig. 6; Rothnie *et al.*, 1994). The gene VI product is also a multifunctional protein (Table 1). It is the matrix protein for the major inclusion body and is absolutely necessary for translation of genes from polycistronic 35S RNA (Fig. 3, Rothnie *et al.*, 1994).

Table 1. Functions of cauliflower mosaic virus genes

ORF	Functions
I	Movement
II	Aphid-transmission factor (ATF)
III	Infectivity, virus assembly
IV	Capsid protein
V	Reverse transcriptase, RNaseH, Protease
VI	Major virus inclusion body protein, symptom production, host range, translational *trans*-activator (TAV), virus assembly
VII	Unknown

Furthermore, the gene VI product is involved in virus assembly (Himmelbach *et al.*, 1996), host range determination and severity of symptoms (Schoelz *et al.*, 1986). Transgenic plants expressing gene VI show symptom-like phenotypes (Rothnie *et al.*, 1994; Cecchini *et al.*, 1997). The function and role of ORF VII remains obscure. It is not required for replication and its product is not detected in infected plants (Rothnie *et al.*, 1994).

Genus Badnavirus

Not much is known about gene functions of badnaviruses. From sequence comparisons of badnaviruses with CaMV, some enzymatic functions can be associated to certain ORFs (Fig. 6). Most of these functions are located on multifunctional proteins. Although the insect vectors involved in transmission of some badnaviruses are known, viral functions have not been identified. In case of

RTBV, transmission by leafhoppers requires coinfection with rice tungro spherical virus (RTSV; Jones *et al.*, 1991).

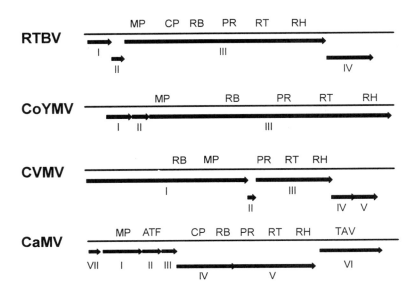

Figure 6. Linear representation of RTBV, CoYMV, CVMV and CaMV genomes. ORFs are indicated by black arrows and where a gene's function is known, the name is given. MP: movement protein; CP: capsid protein; RB: RNA binding domain; PR: protease motif; RT: reverse transcriptase motif; RH: RNaseH domain, ATF: aphid transmission factor; TAV: translational *trans*-activator.

GENE EXPRESSION

Family *Geminiviridae* and Genus *Nanovirus*

The virion- as well as complementary-sense strands of geminiviruses contain ORFs (Fig. 1) whereas in nanoviruses, with the exception of CFDV, only the virion-sense strand has coding capacity (Fig. 2). Accordingly transcripts were mapped in virion- and complementary-sense orientation for geminiviruses (Townsend *et al.*, 1985; Accotto *et al.*, 1989; Sunter *et al.*, 1989; Frischmuth *et al.*, 1991; Frischmuth *et al.*, 1993; Mullineaux *et al.*, 1993) and only in virion-sense for nanoviruses (Beetham *et al.*, 1997). Splicing of complementary-sense transcripts occurs in mastreviruses but not in curto- and begomoviruses (Morris-Krsinich *et al.*, 1985; Mullineaux *et al.*, 1990; Accotto *et al.*, 1989; Dekker *et al.*, 1991, Morris *et al.*, 1992). Splicing of virion-sense genes has been observed in MSV and digitaria streak virus (DSV)-infected plants but not in wheat dwarf virus (WDV) and tobacco yellow dwarf virus (TYDV), although the intron sequence within MP is present in both viruses (Fig. 1;

Wright *et al.*, 1997). Therefore, it remains unclear whether splicing of virion-sense genes is a common feature of mastrevirus for MP and CP expression.

Fine mapping of transcripts confirmed that transcription initiation occurs from predicted promoter sequences and termination at the termination signal sequences (Figs. 1 and 2). The translation of overlapping genes of begomovirus complementary-sense genes of DNA A occurs from polycistronic mRNAs (Fig. 1; Frischmuth *et al.*, 1991; Mullineaux *et al.*, 1993), whereas transcripts with different 5' ends were mapped for the overlapping virion-sense genes of BCTV (Fig. 1; Frischmuth *et al.*, 1993), indicating distinct expression strategies for overlapping genes in geminiviruses. So far no model has been developed how the translation of overlapping genes from these polycistronic mRNAs is performed.

For transcription of virion-sense genes of DNA A and B (CP and NSP) the virus gene product TrAP is required. This means that the complementary-sense gene product TrAP must be produced before the virion-sense genes. Therefore, the following model for gene regulation of geminiviruses has been proposed: In the first step of infection complementary-sense gene products Rep, TrAP and REn are produced (early genes), whereby the transcription is host-mediated activated, then Rep and REn interact and transcription of complementary-sense genes is shut off and replication enhanced. In the third step CP and NSP (late genes) promoters are activated by TrAP and encapsidation and spread of viral DNA beginns.

Whether a similar model, early and late gene activation, applies also to nanoviruses is not clear because much is not known about their gene functions.

Family *Caulimoviridae*

Like retroviruses, pararetroviruses produce a full-length, terminally redundant transcript (Figs. 3 and 4). This pregenomic RNA is the template for reverse transcription and translation. Caulimoviruses have a second promoter, specific for transcription of gene VI (Fig. 3). This promoter gives rise to the 19S RNA transcript, which is colinear with 3' end of pregenomic transcript (Figs. 3 and 7).

The transcript leader sequence of caulimo- and badnaviruses pregenomic RNAs are unusually long; they are predicted to form extensive secondary structure and contain several short ORFs (Fig. 7; Fütterer *et al.*, 1994; Pooggin *et al.*, 1998). In plants infected with CaMV or RTBV, the polycistronic pregenomic RNA is able to express all viral proteins except gene VI product of CaMV which is translated from the 19S RNA (Rothnie *et al.*, 1994; Fütterer *et al.*, 1997).

How is the translation of successive genes of the polycistronic RNA controlled? Two rather unusual mechanisms have been proposed in caulimoviruses to play a role in allowing translation of 35S RNA of CaMV. The first mechanism is termed "Ribosome Shunting" (Fütterer *et al.*, 1993; Dominguez *et al.*, 1998) for which extensive secondary structure of the 5' leader of the 35S RNA is essential (Dominguez *et al.*, 1998). Mutational analysis, abolishing the formation of stem-loop structures, showed that the expression is drastically reduced (Dominguez *et al.*, 1998). The model for the ribosome shunting suggests that the scanning ribosome is transferred to the AUG of the first gene (gene VII) by bypassing the AUG codons of the short ORFs in the leader at the bottom of the stem-loop structures (Dominguez *et al.*, 1998). The translation of downstream ORFs of ORF VII depends on TAV (ORF VI). TAV functions as a translational *trans*-activator protein in such way, that TAV allows ribosomes that have translated one ORF to remain competent to translate further downstream ORFs, or to become initiation competent once more (Rothnie *et al.*, 1994).

Beside the genomic-length RNA and 19S RNA, spliced RNA products have been detected in infected plants (Figs. 3, 4 and 7; Kiss-László and Hohn, 1996). In CaMV the role of spliced RNAs is not clear because the full-length genomic RNA is capable to produce all proteins. However, mutations in splice donors within ORF I, preventing formation of spliced RNA species in which the ORF I and II are fused (Fig. 7), renders the virus non-infectious.

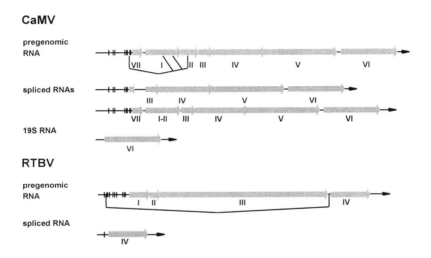

Figure 7. The RNAs of plant pararetroviruses cauliflower mosaic virus (CaMV) and rice tungro bacilliform virus (RTBV). Pregenomic, spliced and 19S RNA transcripts are shown. Location of genes are indicated in gray arrows. Short ORFs preceding the first ORF are indicated as black boxes.

For the badnavirus RTBV, a different model for translation of the polycistronic genomic RNA has been suggested. ORFs I, II and III are translated from pregenomic RNA by leaky scanning; some ribosomes initiate at the weak AUU initiation codon of ORF I and the AUG codon in the unfavorable context of ORF II, whereas others continue scanning until they reach the strong start codon of ORF III (Fütterer *et al.*, 1996, Fütterer *et al.*, 1997). ORF IV is translated from a spliced transcript (Fig. 7). ORFs I, II and III are part of an intron and the removal of this intron results in the in-frame fusion of a small ORF in the transcript leader sequence to ORF IV allowing the ribosomes to initiate translation of ORF IV (Fig. 7; Fütterer *et al.*, 1994; Kiss-László and Hohn, 1996).

REFERENCES

Accotto, G. P., Donson, J., and Mullineaux, P. M. (1989). Mapping of *Digitaria* streak virus transcripts reveals different RNA species from the same transcription unit. *EMBO J.* 8, 1033 -1039.

Ach, R. A., Durfee, T., Miller, A. B., Taranto, P., Hanley-Bowdoin, L., Zambryski, P. C., and Gruissem, W. (1997). RRB1 and RRB2 encode maize retinoblastoma-related proteins that interact with a plant D-type cyclin and geminivirus replication protein. *Mol. Cell Biol.* 17, 5077-5086.

Azzam, O., Frazer, J., de la Rosa, D., Beaver, J. S., Ahlquist, P., and Maxwell, D. P. (1994). Whitefly transmission and efficient ssDNA accumulation of bean golden mosaic geminivirus require functional coat protein. *Virology* 204, 289 -296.

Beetham, P. R., Hafner, G. J., Harding, R. M., and Dale, J. L. (1997). Two mRNAs are transcribed from banana bunchy top virus DNA-1. *J. Gen. Virol.* 78, 229 - 236.

Blanc, S., Cerutti, M., Usmany, M., Vlak, J. M., and Hull, R. (1993). Biological activity of cauliflower mosaic virus aphid-transmission factor expressed in a heterologous system. *Virology* 192, 643 -50.

Boevink, P., Chu, P. W., and Keese, P. (1995). Sequence of subterranean clover stunt virus DNA: affinities with the geminiviruses. *Virology* 207, 354 -361.

Boulton, M. I., Steinkellner, H., Donson, J., Markham, P. G., King, D. I., and Davies, J. W. (1989). Mutational analysis of the virion-sense genes of maize streak virus. *J. Gen. Virol.* 70, 2309 - 2323.

Briddon, R. W., Bedford, I. D., Tsai, J. H., and Markham, P. G. (1996). Analysis of the nucleotide sequence of the treehopper-transmitted geminivirus, tomato pseudo-curly top virus, suggests a recombinant origin. *Virology* 219, 387-394.

Briddon, R. W., Lunness, P., Chamberlin, L. C., Pinner, M. S., Brundish, H., and Markham, P. G. (1992). The nucleotide sequence of an infectious insect-transmissible clone of the geminivirus *Panicum* streak virus. *J. Gen. Virol.* 73, 1041-1047.

Briddon, R. W., Pinner, M. S., Stanley, J., and Markham, P. G. (1990). Geminivirus coat protein gene replacement alters insect specificity. *Virology* 177, 85 -94.

Briddon, R. W., Watts, J., Markham, P. G., and Stanley, J. (1989). The coat protein of beet curly top virus is essential for infectivity. *Virology* 172, 628 - 633.

Brough, C. L., Hayes, R. J., Morgan, A. J., Coutts, R. H. A., and Buck, K. W. (1988). Effects of mutagenesis *in vitro* on the ability of cloned tomato golden mosaic virus DNA to infect *Nicotiana benthamiana* plants. *J. Gen. Virol.* 69, 503 -514.

Burns, T. M., Harding, R. M., and Dale, J. L. (1995). The genome organization of banana bunchy top virus: analysis of six ssDNA components. *J. Gen. Virol.* 76, 1471-1482.

Calvert, L. A., Ospina, M. D., and Shepherd, R. J. (1995). Characterization of cassava vein mosaic virus: A distinct plant pararetrovirus. *J. Gen. Virol.* 76, 1271-1278.

Cecchini, E., Gong, Z., Geri, C., Covey, S. N., and Milner, J. J. (1997). Transgenic Arabidopsis lines expressing gene VI from cauliflower mosaic virus variants exhibit a range of symptom-like phenotypes and accumulate inclusion bodies. *Mol. Plant-Microbe Interact.* 10, 1094 -1101.

Chatani, M., Matsumoto, Y., Mizuta, H., Ikegami, M., Boulton, M. I., and Davies, J. W. (1991). The nucleotide sequence and genome structure of the geminivirus miscanthus streak virus. *J. Gen. Virol.* 72, 2325 - 2331.

Chu, P. W., Keese, P., Qiu, B. S., Waterhouse, P. M., and Gerlach, W. L. (1993). Putative full-length clones of the genomic DNA segments of subterranean clover stunt virus and identification of the segment coding for the viral coat protein. *Virus Res.* 27, 161-171.

Covey, S. N., Turner, D. S., Lucy, A. P., and Saunders, K. (1990). Host regulation of the cauliflower mosaic virus multiplication cycle. *Proc. Natl. Acad. Sci. U S A* 87, 1633 -1637.

Dautel, S., Guidasci, T., Pique, M., Mougeot, J. L., Lebeurier, G., Yot, P., and Mesnard, J. M. (1994). The full-length product of cauliflower mosaic virus open reading frame III is associated with the viral particle. *Virology* 202, 1043 -1045.

Dekker, E. L., Woolston, C. J., Xue, Y. B., Cox, B., and Mullineaux, P. M. (1991). Transcript mapping reveals different expression strategies for the bicistronic RNAs of the geminivirus wheat dwarf virus. *Nucleic Acids Res.* 19, 4075 - 4081.

Desbiez, C., David, C., Mettouchi, A., Laufs, J., and Gronenborn, B. (1995). Rep protein of tomato yellow leaf curl geminivirus has an ATPase activity required for viral DNA replication [published erratum appears in *Proc. Natl. Acad. Sci. U S A* 1995 Nov 21, 92 (24), 11322]. *Proc. Natl. Acad. Sci. U S A* 92, 5640 -5644.

Dickinson, V. J., Halder, J., and Woolston, C. J. (1996). The product of maize streak virus ORF V1 is associated with secondary plasmodesmata and is first detected with the onset of viral lesions. *Virology* 220, 51-59.

Dominguez, D. I., Ryabova, L. A., Pooggin, M. M., Schmidt-Puchta, W., Futterer, J., and Hohn, T. (1998). Ribosome shunting in cauliflower mosaic virus. Identification of an essential and sufficient structural element. *J. Biol. Chem.* 273, 3669 -3678.

Dry, I. B., Rigden, J. E., Krake, L. R., Mullineaux, P. M., and Rezaian, M. A. (1993). Nucleotide sequence and genome organization of tomato leaf curl geminivirus. *J. Gen. Virol.* 74, 147-151.

Eagle, P. A., Orozco, B. M., and Hanley-Bowdoin, L. (1994). A DNA sequence required for geminivirus replication also mediates transcriptional regulation. *Plant Cell* 6, 1157-1170.

Elmer, J. S., Brand, L., Sunter, G., Gardiner, W. E., Bisaro, D. M., and Rogers, S. G. (1988). Genetic analysis of the tomato golden mosaic virus. II. The product of the AL1 coding sequence is required for replication. *Nucleic Acids Res.* 16, 7043 -7060.

Engel, M., Fernández, O., Jeske, H., and Frischmuth, T. (1998). Molecular characterization of a new whitefly-transmissible bipartite geminivirus infecting tomato in Panama. *J. Gen. Virol.* 79, 2313 - 2317.

Espinoza, A. M., Medina, V., Hull, R., and Markham, P. G. (1991). Cauliflower mosaic virus gene II product forms distinct inclusion bodies in infected plant cells. *Virology* 185, 337-344.

Etessami, P., Callis, R., Ellwood, S., and Stanley, J. (1988). Delimitation of essential genes of cassava latent virus DNA 2. *Nucleic Acids Res.* 16, 4811- 4829.

Etessami, P., Saunders, K., Watts, J., and Stanley, J. (1991). Mutational analysis of complementary-sense genes of African cassava mosaic virus DNA A. *J. Gen. Virol.* 72, 1005 - 1012.

Fontes, E. P., Eagle, P. A., Sipe, P. S., Luckow, V. A., and Hanley-Bowdoin, L. (1994a). Interaction between a geminivirus replication protein and origin DNA is essential for viral replication. *J. Biol. Chem.* 269, 8459 - 8465.

Fontes, E. P., Gladfelter, H. J., Schaffer, R. L., Petty, I. T., and Hanley-Bowdoin, L. (1994b). Geminivirus replication origins have a modular organization. *Plant Cell* 6, 405 - 416.

Fontes, E. P., Luckow, V. A., and Hanley-Bowdoin, L. (1992). A geminivirus replication protein is a sequence-specific DNA binding protein. *Plant Cell* 4, 597 - 608.

Frischmuth, S., Frischmuth, T., and Jeske, H. (1991). Transcript mapping of Abutilon mosaic virus, a geminivirus. *Virology* 185, 596 - 604.

Frischmuth, S., Frischmuth, T., Latham, J. R., and Stanley, J. (1993). Transcriptional analysis of the virion-sense genes of the geminivirus beet curly top virus. *Virology* 197, 312 -319.

Frischmuth, T., and Stanley, J. (1993). Strategies for the control of geminivirus diseases. *Semin. Virol.* 4, 329 -337.

Fütterer, J., Kiss-Laszlo, Z., and Hohn, T. (1993). Nonlinear ribosome migration on cauliflower mosaic virus 35S RNA. *Cell* 73, 789 - 802.

Fütterer, J., Potrykus, I., Bao, Y., Li, L., Burns, T. M., Hull, R., and Hohn, T. (1996). Position-dependent ATT initiation during plant pararetrovirus rice tungro bacilliform virus translation. *J Virol* 70, 2999-3010.

Fütterer, J., Potrykus, I., Valles Brau, M. P., Dasgupta, I., Hull, R., and Hohn, T. (1994). Splicing in a plant pararetrovirus. *Virology* 198, 663 - 670.

Fütterer, J., Rothnie, H. M., Hohn, T., and Potrykus, I. (1997). Rice tungro bacilliform virus open reading frames II and III are translated from polycistronic pregenomic RNA by leaky scanning. *J. Virol.* 71, 7984-7989.

Gardiner, W. E., Sunter, G., Brand, L., Elmer, J. S., Rogers, S. G., and Bisaro, D. M. (1988). Genetic analysis of tomato golden mosaic virus: the coat protein is not required for systemic spread or symptom development. *EMBO J.* 7, 899-904.

Gröning, B. R., Hayes, R. J., and Buck, K. W. (1994). Simultaneous regulation of tomato golden mosaic virus coat protein and AL1 gene expression: expression of the AL4 gene may contribute to suppression of the AL1 gene. *J. Gen. Virol.* 75, 721-726.

Hafner, G. J., Harding, R. M., and Dale, J. L. (1997). A DNA primer associated with banana bunchy top virus. *J. Gen. Virol.* 78, 479 - 486.

Hafner, G. J., Stafford, M. R., Wolter, L. C., Harding, R. M., and Dale, J. L. (1997). Nicking and joining activity of banana bunchy top virus replication protein in vitro. *J. Gen. Virol.* 78, 1795 -1799.

Haley, A., Zhan, X., Richardson, K., Head, K., and Morris, B. (1992). Regulation of the activities of African cassava mosaic virus promoters by the AC1, AC2, and AC3 gene products. *Virology* 188, 905 -909.

Hanley-Bowdoin, L., Elmer, J. S., and Rogers, S. G. (1989). Functional expression of the leftward open reading frames of the A component of tomato golden mosaic virus in transgenic tobacco plants. *Plant Cell* 1, 1057-1067.

Hanson, S. F., Hoogstraten, R. A., Ahlquist, P., Gilbertson, R. L., Russell, D. R., and Maxwell, D. P. (1995). Mutational analysis of a putative NTP-binding domain in the replication-associated protein (AC1) of bean golden mosaic geminivirus. *Virology* 211, 1-9.

Harding, R. M., Burns, T. M., Hafner, G., Dietzgen, R. G., and Dale, J. L. (1993). Nucleotide sequence of one component of the banana bunchy top virus genome contains a putative replicase gene. *J. Gen. Virol.* 74, 323 -328.

Harrison, B. D. (1985). Advances in geminivirus research. *Annu. Rev. Phytopathol.* 23, 55 - 82.

Hay, J. M., Jones, M. C., Blakebrough, M. L., Dasgupta, I., Davies, J. W., and Hull, R. (1991). An analysis of the sequence of an infectious clone of rice tungro bacilliform virus, a plant pararetrovirus. *Nucleic Acids Res.* 17, 9993 -10012.

Hayes, R. J., and Buck, K. W. (1989). Replication of tomato golden mosaic virus DNA B in transgenic plants expressing open reading frames (ORFs) of DNA A: Requirement of ORF AL2 for production of single-stranded DNA. *Nucleic Acids Res.* 17, 10213 -10222.

Heyraud, F., Matzeit, V., Kammann, M., Schaefer, S., Schell, J., and Gronenborn, B. (1993). Identification of the initiation sequence for viral-strand DNA synthesis of wheat dwarf virus. *EMBO J.* 12, 4445 - 4452.

Himmelbach, A., Chapdelaine, Y., and Hohn, T. (1996). Interaction between cauliflower mosaic virus inclusion body protein and capsid protein: implications for viral assembly. *Virology* 217, 147-157.

Höfer, P., Bedford, I. D., Markham, P. G., Jeske, H., and Frischmuth, T. (1997). Coat protein gene replacement results in whitefly transmission of an insect nontransmissible geminivirus isolate. *Virology* 236, 288 -295.

Hong, Y., and Stanley, J. (1995). Regulation of African cassava mosaic virus complementary-sense gene expression by N-terminal sequences of the replication-associated protein AC1. *J. Gen. Virol.* 76, 2415-2422.

Hoogstraten, R. A., Hanson, S. F., and Maxwell, D. P. (1996). Mutational analysis of the putative nicking motif in the replication-associated protein (AC1) of bean golden mosaic geminivirus. *Mol. Plant-Microbe Interact.* 9, 594 -599.

Hormuzdi, S. G., and Bisaro, D. M. (1995). Genetic analysis of beet curly top virus: examination of the roles of L2 and L3 genes in viral pathogenesis. *Virology* 206, 1044 -1054.

Howarth, A. J., Caton, J., Bossert, M., and Goodman, R. M. (1985). Nucleotide sequence of bean golden mosaic virus and a model for gene regulation in geminiviruses. *Proc. Natl. Acad. Sci. USA* 82, 3572 -3576.

Hughes, F. L., Rybicki, E. P., and Kirby, R. (1993). Complete nucleotide sequence of sugarcane streak Monogeminivirus. *Arch. Virol.* 132, 171-182.

Hull, R. (1992). Genome organization of retroviruses and retroelements: Evolutionary considerations and implications. *Semin. Virol.* 3, 373 -382.

Hull, R., Covey, S. N., and Maule, A. J. (1987). Structure and replication of caulimovirus genomes. *J. Cell Sci. (Suppl)* 7, 213 -229.

Jacquot, E., Geldreich, A., Keller, M., and Yot, P. (1998). Mapping regions of the cauliflower mosaic virus ORF III product required for infectivity. *Virology* 242, 395 - 402.

Jones, M. C., Gough, K., Dasgupta, I., Rao, B. L., Cliffe, J., Qu, R., Shen, P., Kaniewska, M., Blakebrough, M., Davies, J. W., and et al. (1991). Rice tungro disease is caused by an RNA and a DNA virus. *J. Gen. Virol.* 72, 757-761.

Jupin, I., De Kouchkovsky, F., Jouanneau, F., and Gronenborn, B. (1994). Movement of tomato yellow leaf curl geminivirus (TYLCV): involvement of the protein encoded by ORF C4. *Virology* 204, 82-90.

Katul, L., Maiss, E., Morozov, S. Y., and Vetten, H. J. (1997). Analysis of six DNA components of the faba bean necrotic yellows virus genome and their structural affinity to related plant virus genomes. *Virology* 233, 247-259.

Katul, L., Vetten, H. J., and Mauss, E. (1993). Characterisation and serology of virus-like particles associated with faba bean necrotic yellows. *Ann. Appl. Biol.* 123, 629 - 647.

Kiss-László, Z., and Hohn, T. (1996). Pararetro- and retrovirus RNA: Splicing and the control of nuclear export. *Trends in Microbiology* 4, 480 - 485.

Klute, K. A., Nadler, S. A., and Stenger, D. C. (1996). Horseradish curly top virus is a distinct subgroup II geminivirus species with rep and C4 genes derived from a subgroup III ancestor. *J. Gen. Virol.* 77, 1369 -1378.

Krake, L. R., Rezaian, M. A., and Dry, I. B. (1998). Expression of the tomato leaf curl geminivirus C4 gene produces viruslike symptoms in transgenic plants. *Mol. Plant-Microbe Interact.* 11, 413-417.

Latham, J. R., Saunders, K., Pinner, M. S., and Stanley, J. (1997). Induction of plant cell division by beet curly top virus gene C4. *Plant J.* 11, 1273 - 1283.

Laufs, J., Traut, W., Heyraud, F., Matzeit, V., Rogers, S. G., Schell, J., and Gronenborn, B. (1995). *In vitro* cleavage and joining at the viral origin of replication by the replication initiator protein of tomato yellow leaf curl virus. *Proc. Natl. Acad. Sci. U S A* 92, 3879 -3883.

Lazarowitz, S. G. (1992). Geminiviruses: genome structure and gene function. *Crit. Rev. Plant Sci.* 11, 327-349.

Lazarowitz, S. G., Pinder, A. J., Damsteegt, V. D., and Rogers, S. G. (1989). Maize streak virus genes essential for systemic spread and symptom development. *EMBO J.* 8, 1023 -1032.

Lazarowitz, S. G., Wu, L. C., Rogers, S. G., and Elmer, J. S. (1992). Sequence-specific interaction with the viral AL1 protein identifies a geminivirus DNA replication origin. *Plant Cell* 4, 799 - 809.

Linstead, P. J., Hills, G. J., Plaskitt, K. A., Wilson, I. G., Harker, C. L., and Maule, A. J. (1988). The subcellular location of the gene I product of cauliflower mosaic virus is consistent with a function associated with virus spread. *J. Gen. Virol.* 69, 1809 -1818.

Liu, L., Davies, J. W., and Stanley, J. (1998). Mutational analysis of bean yellow dwarf, a geminivirus of the genus *Mastrevirus* that is adapted to dicotyledonous plants. *J. Gen. Virol.* 79, 2265 -2274.

Liu, L., van Tonder, T., Pietersen, G., Davies, J. W., and Stanley, J. (1997). Molecular characterization of a subgroup I geminivirus from a legume in South Africa. *J. Gen. Virol.* 78, 2113 -2117.

MacDowell, S. W., Macdonald, H., Hamilton, W. D. O., Coutts, R. H. A., and Buck, K. W. (1985). The nucleotide sequence of cloned wheat dwarf virus DNA. *EMBO J.* 4, 2173-2180.

Mayo, M. P., and Pringle, C. R. (1998). Virus taxonomy – 1997. *J. Gen. Virol.* 79, 649-657.

Medberry, S. L., Lockhart, B. E. L., and Olszewski, N. E. (1990). Properties of Commelina yellow mottle virus's complete DNA sequence, genomic discontinuities and transcript suggest that it is a pararetrovirus. *Nucleic Acids Res.* 18, 5505 -5513.

Morris, B., Richardson, K., Eddy, P., Zhan, X. C., Haley, A., and Gardner, R. (1991). Mutagenesis of the AC3 open reading frame of African cassava mosaic virus DNA A reduces DNA B replication and ameliorates disease symptoms. *J. Gen. Virol.* 72, 1205 -1213.

Morris, B. A., Richardson, K. A., Haley, A., Zhan, X., and Thomas, J. E. (1992). The nucleotide sequence of the infectious cloned DNA component of tobacco yellow dwarf virus reveals features of geminiviruses infecting monocotyledonous plants. *Virology* 187, 633 - 642.

Morris-Krsinich, B. A. M., Mullineaux, P. M., Donson, J., Boulton, M. I., Markham, P. G., Short, M. N., and Davies, J. W. (1985). Biderectional transcription of maize streak virus DNA and idenfication of the coat protein gene. *Nucleic Acids Res.* 13, 7237-7256.

Mullineaux, P. M., Donson, J., Morris-Krsinich, B. A., Boulton, M. I., and Davies, J. W. (1984). The nucleotide sequence of maize streak virus DNA. *EMBO J.* 3, 3063 -3068.

Mullineaux, P. M., Guerineau, F., and Accotto, G. P. (1990). Processing of complementary sense RNAs of Digitaria streak virus in its host and in transgenic tobacco. *Nucleic Acids Res.* 18, 7259 - 7265.

Mullineaux, P. M., Rigden, J. E., Dry, I. B., Krake, L. R., and Rezaian, M. A. (1993). Mapping of the polycistronic RNAs of tomato leaf curl geminivirus. *Virology* 193, 414 - 423.

Nagar, S., Pedersen, T. J., Carrick, K. M., Hanley-Bowdoin, L., and Robertson, D. (1995). A geminivirus induces expression of a host DNA synthesis protein in terminally differentiated plant cells. *Plant Cell* 7, 705 -719.

Navot, N., Pichersky, E., Zeidan, M., Zamir, D., and Czosnek, H. (1991). Tomato yellow leaf curl virus: A whitefly-transmitted geminivirus with a single genomic component. *Virology* 185, 151-161.

Noris, E., Jupin, I., Accotto, G. P., and Gronenborn, B. (1996). DNA-binding activity of the C2 protein of tomato yellow leaf curl geminivirus. *Virology* 217, 607 - 612.

Noueiry, A. O., Lucas, W. J., and Gilbertson, R. L. (1994). Two proteins of a plant DNA virus coordinate nuclear and plasmodesmal transport. *Cell* 76, 925 -932.

Orozco, B. M., and Hanley-Bowdoin, L. (1996). A DNA structure is required for geminivirus replication origin function. *J. Virol.* 70, 148 -158.

Orozco, B. M., and Hanley-Bowdoin, L. (1998). Conserved sequence and structural motifs contribute to the DNA binding and cleavage activities of a geminivirus replication protein. *J Biol. Chem.* 273, 24448 -24456.

Orozco, B. M., Miller, A. B., Settlage, S. B., and Hanley-Bowdoin, L. (1997). Functional domains of a geminivirus replication protein. *J. Biol. Chem.* 272, 9840 -9846.

Palmer, K. E., and Rybicki, E. P. (1997). The use of geminiviruses in biotechnology and plant molecular biology, with particular focus on mastreviruses. *Plant Sci.* 129, 115 -130.

Pascal, E., Sanderfoot, A. A., Ward, B. M., Medville, R., Turgeon, R., and Lazarowitz, S. G. (1994). The geminivirus BR1 movement protein binds single-stranded DNA and localizes to the cell nucleus. *Plant Cell* 6, 995 -1006.

Perbal, M. C., Thomas, C. L., and Maule, A. J. (1993). Cauliflower mosaic virus gene I product (P1) forms tubular structures which extend from the surface of infected protoplasts. *Virology* 195, 281-285.

Pooggin, M. M., Hohn, T., and Futterer, J. (1998). Forced evolution reveals the importance of short open reading frame A and secondary structure in the cauliflower mosaic virus 35S RNA leader. *J. Virol.* 72, 4157 - 4169.

Pringle, C. R. (1998). Virus taxonomy - San Diego 1998. *Arch Virol.* 143, 1449-1459.

Qu, R. D., Bhattacharyya, M., Laco, G. S., De Kochko, A., Rao, B. L., Kaniewska, M. B., Elmer, J. S., Rochester, D. E., Smith, C. E., and Beachy, R. N. (1991). Characterization of the genome of rice tungro bacilliform virus: Comparison with Commelina yellow mottle virus and caulimoviruses [published erratum appears in *Virology* 1992, Feb; 186 (2), 798]. *Virology* 185, 354 - 364.

Rohde, W., Randles, J. W., Langridge, P., and Hanold, D. (1990). Nucleotide sequence of a circular single-stranded DNA associated with coconut foliar decay virus. *Virology* 176, 648 -51.

Rothnie, H. M., Chapdelaine, Y., and Hohn, T. (1994). Pararetroviruses and retroviruses: A comparative review of viral structure and gene expression strategies. *Adv. Virus Res.* 44, 1- 67.

Sanderfoot, A. A., Ingham, D. J., and Lazarowitz, S. G. (1996). A viral movement protein as a nuclear shuttle. The geminivirus BR1 movement protein contains domains essential for interaction with BL1 and nuclear localization. *Plant Physiol.* 110, 23 -33.

Sanderfoot, A. A., and Lazarowitz, S. G. (1995). Cooperation in viral movement: The geminivirus BL1 movement protein interacts with BR1 and redirects it from the nucleus to the cell periphery. *Plant Cell* 7, 1185 -1194.

Sano, Y., Isogai, M., Satoh, S., and Kojima, M. (1993). Small virus-like particles containing single-stranded DNAs associated with milk vetch dwarf disease in Japan. "6th International Congress on Plant Pathology", Montreal, Abstract No. 17.1.27.

Sanz-Burgos, A. P., and Gutierrez, C. (1998). Organization of the *cis*-acting element required for wheat dwarf geminivirus DNA replication and visualization of rep protein-DNA complex. *Virology* 243, 119-129.

Saunders, K., Lucy, A., and Stanley, J. (1991). DNA forms of the geminivirus African cassava mosaic virus consistent with a rolling circle mechanism of replication. *Nucleic Acids Res.* 19, 2325 -2330.

Saunders, K., Lucy, A., and Stanley, J. (1992). RNA-primed complementary-sense DNA synthesis of the geminivirus African cassava mosaic virus. *Nucleic Acids Res.* 20, 6311- 6315.

Schalk, H. J., Matzeit, V., Schiller, B., Schell, J., and Gronenborn, B. (1989). Wheat dwarf virus, a geminivirus of graminaceous plants needs splicing for replication. *EMBO J.* 8, 359 -364.

Schoelz, J., Shepherd, R. J., and Daubert, S. (1986). Region VI of cauliflower mosaic virus encodes a host range determinant. *Mol. Cell Biol.* 6, 2632 -2637.

Settlage, S. B., Miller, A. B., and Hanley-Bowdoin, L. (1996). Interactions between geminivirus replication proteins. *J. Virol.* 70, 6790 - 6795.

Stanley, J. (1995). Analysis of African cassava mosaic virus recombinants suggests strand nicking occurs within the conserved nonanucleotide motif during the initiation of rolling circle DNA replication. *Virology* 206, 707 -712.

Stanley, J., and Latham, J. R. (1992). A symptom variant of beet curly top geminivirus produced by mutation of open reading frame C4. *Virology* 190, 506 - 609.

Stanley, J., Latham, J. R., Pinner, M. S., Bedford, I., and Markham, P. G. (1992). Mutational analysis of the monopartite geminivirus beet curly top virus. *Virology* 191, 396 - 405.

Stanley, J., Markham, P. G., Callis, R. J., and Pinner, M. S. (1986). The nucleotide sequence of an infectious clone of the geminivirus beet curly top virus. *EMBO J.* 5, 1761-1767.

Stanley, J., and Townsend, R. (1986). Infectious mutants of cassava latent virus generated *in vivo* from intact recombinant DNA clones containing single copies of the genome. *Nucleic Acids Res.* 14, 5981-5998.

Suarez-Lopez, P., and Gutierrez, C. (1997). DNA replication of wheat dwarf geminivirus vectors: effects of origin structure and size. *Virology* 227, 389 -399.

Sung, Y. K., and Coutts, R. H. (1995). Mutational analysis of potato yellow mosaic geminivirus. *J. Gen. Virol.* 76, 1773-1780.

Sung, Y. K., and Coutts, R. H. (1996). Potato yellow mosaic geminivirus AC2 protein is a sequence non-specific DNA binding protein. *FEBS Lett.* 383, 51-54.

Sunter, G., and Bisaro, D. M. (1991). Transactivation in a geminivirus: AL2 gene product is needed for coat protein expression. *Virology* 180, 416 - 419.

Sunter, G., and Bisaro, D. M. (1992). Transactivation of geminivirus AR1 and BR1 gene expression by the viral AL2 gene product occurs at the level of transcription. *Plant Cell* 4, 1321-1331.

Sunter, G., and Bisaro, D. M. (1997). Regulation of a geminivirus coat protein promoter by AL2 protein (TrAP): evidence for activation and derepression mechanisms. *Virology* 232, 269 -280.

Sunter, G., Gardiner, W. E., and Bisaro, D. M. (1989). Identification of tomato golden mosaic virus-specific RNAs in infected plants. *Virology* 170, 243 - 250.

Sunter, G., Hartitz, M. D., Hormuzdi, S. G., Brough, C. L., and Bisaro, D. M. (1990). Genetic analysis of tomato golden mosaic virus: ORF AL2 is required for coat protein accumulation while ORF AL3 is necessary for efficient DNA replication. *Virology* 179, 69 -77.

Thomas, C. L., Perbal, C., and Maule, A. J. (1993). A mutation of cauliflower mosaic virus gene I interferes with virus movement but not virus replication. *Virology* 192, 415 - 421.

Timmermans, M. C. P., O. P. Das, and J. Messing (1994). Geminiviruses and their uses as extrachromosomal replicons. *Annu. Rev. Plant Physiol. Plant Mol. Biol.* 45, 79 -112.

Townsend, R., Stanley, J., Curson, S. J., and Short, M. N. (1985). Major polyadenylated transcripts of cassava latent virus and location of the gene encoding the coat protein. *EMBO J.* 4, 33 -37.

Wanitchakorn, R., Harding, R. M., and Dale, J. L. (1997). Banana bunchy top virus DNA-3 encodes the Viral coat protein. *Arch. Virol.* 142, 1673 -1680.

Ward, B. M., Medville, R., Lazarowitz, S. G., and Turgeon, R. (1997). The geminivirus BL1 movement protein is associated with endoplasmic reticulum-derived tubules in developing phloem cells. *J. Virol.* 71, 3726 -3733.

Woolston, C. J., Reynolds, H. V., Stacey, N. J., and Mullineaux, P. M. (1989). Replication of wheat dwarf virus DNA in protoplasts and analysis of coat protein mutants in protoplasts and plants. *Nucleic Acids Res.* 17, 6029 - 6041.

Wright, E. A., Heckel, T., Groenendijk, J., Davies, J. W., and Boulton, M. I. (1997). Splicing features in maize streak virus virion- and complementary-sense gene expression [In Process Citation]. *Plant J.* 12, 1285 -1297.

Xie, Q., Suarez-Lopez, P., and Gutierrez, C. (1995). Identification and analysis of a retinoblastoma binding motif in the replication protein of a plant DNA virus: Requirement for efficient viral DNA replication. *EMBO J.* 14, 4073 - 4082.

3 GENOME ORGANIZATION IN RNA VIRUSES

Sergey Morozov and Andrey Solovyev

INTRODUCTION

Molecular taxonomy of plant viruses is now nearing completion (Mayo and Pringle, 1998; Pringle, 1998) primarily because of the evolutionary relationships that emerged from comparative studies on nucleotide and protein sequences and RNA-dependent RNA polymerase gene of various plant viruses (Koonin, 1991; Koonin and Dolja 1993; Zanotto et al., 1996). Four distinct supergroups of plus-sense RNA plant viruses are now recognized. Based on similarity of the basic genome features between different groups of animal and plant viruses, two distinct supergroups of plus-RNA plant viruses initially delimited were plant "picorna-like" and plant "alpha-like" supergroups. The two more supergroups of plus-RNA plant viruses recognized recently are the 'sobemo-like" and "carmo-like" supergroups.

Plant "picorna-like" viruses have virion RNA with 5'-linked protein (VPg), and their genome encodes a polyprotein containing a conserved replication module, which is similar in arrangement of the encoded proteins and their sequences to that of picornaviruses and includes helicase, VPg, chymotrypsin-like protease, and RNA-dependent RNA polymerase (RdRp) (Goldbach, 1987; Koonin and Dolja, 1993; Gorbalenya and Koonin, 1993; Buck, 1996; Kadare and Haenni, 1997; Ryan and Flint, 1997). The families of filamentous and spherical viruses with monopartite or divided genomes of the "picorna-like" plant viruses are *Comoviridae*, *Potyviridae,* and *Sequiviridae* (Mayo and Pringle, 1998).

Genomic RNAs of plant "alpha-like" viruses are capped and code for a conserved set of the replicative protein domains including RdRp, helicase, and methyl/guanylyltransferase (Goldbach, 1987; Gorbalenya et al., 1989a, b; Koonin and Dolja, 1993; Rozanov et al., 1992; Buck, 1996). "Alpha-like" viruses are subdivided into several families, of which only some have official status (Mayo and Pringle, 1998; Pringle, 1998): tentative family *Tubiviridae* have seven rod-like genera (Torrance and Mayo, 1997); family *Closteroviridae* with two genera; family *Bromoviridae*, combining tripartite isometric viruses belonging to five genera and closely related genus *Idaeovirus;* and genera *Potexvirus, Carlavirus, Allexvirus,* and *Foveavirus* constituting the family *Potexviridae* (Song et al., 1998; Martelli and Jelkmann, 1998; Pringle, 1998), and also genera *Capillovirus, Trichovirus, Vitivirus, Marafivirus*, and *Tymovirus* (Mayo and Pringle, 1998); these nine genera of viruses with different particle morphologies and genome organization were combined into a proposed order *"Tymovirales"* (Koonin and Dolja, 1993).

"Sobemo-like" supergroup includes plant viruses whose genomes code for the VPg, similar to "picorna-like" viruses. However, they do not code for the helicase and code for polyproteins where the replicative domains are in the order 'protease-VPg-RdRp', not 'VPg-protease-RdRp' as in the "picorna-like" viruses (Koonin and Dolja, 1993; van der Wilk *et al.*, 1997a; K.Makinen, personal communication). This supergroup includes genus *Sobemovirus* and proposed genera *Polerovirus* (subgroup 2 luteoviruses) and *Enamovirus* (Koonin and Dolja, 1993; Mayo and Ziegler-Graff, 1996; Miller and Rasochova, 1997; Pringle, 1998).

"Carmo-like" viruses, have rather small RNA genomes (capped or uncapped) coding for only one replicative domain, RdRp (Koonin and Dolja, 1993). This supergroup includes genera *Luteovirus* (subgroup 1 luteoviruses), *Umbravirus,* and family *Tombusviridae* (Taliansky *et al.*, 1996; Gibbs *et al.*, 1996; Mayo and Ziegler-Graff, 1996; Miller and Rasochova, 1997; Mayo and Pringle, 1998; Pringle, 1998).

Monopartite negative-sense RNA plant viruses, along with some animal viruses, belong to the genera *Cytorhabdovirus* and *Nucleorhabdovirus* of the order *Mononegavirales* (the family *Rhabdoviridae*). Multipartite minus-stranded RNA plant viruses are classified into the genus *Tospovirus* (family *Bunyaviridae*) and the closely related genus *Tenuivirus* (Ramirez and Haenni, 1994; Jackson *et al.*, 1998). The genomes of all plant minus RNA viruses include three principal gene modules (gene blocks): block 1 coding for the nucleocapsid protein and a phosphoprotein; block 2 coding for the envelope membrane proteins; block 3 coding for the polymerase (Tordo *et al.*, 1992). The key difference in the mRNA synthesis in multipartite and monopartite plant minus-stranded RNA viruses is the following: in monopartite viruses, viral polymerase catalyses both the polyadenylation and capping (methyl/guanylyltransferase reaction), while in segmented viruses, mRNAs are not polyadenylated and their capping requires the cap-snatching mechanism when mRNA transcription is primed by short fragments generated from the 5'-termini of host mRNAs (Tordo *et al.*, 1992; Ramirez and Haenni, 1994).

The double-stranded RNA plant viruses belong to the genera *Phytoreovirus*, *Fijivirus*, and *Oryzavirus* (family *Reoviridae*), and genera *Alphacryptovirus* and *Betacryptovirus* (family *Partitiviridae*) (Mayo and Pringle, 1998). Plant viruses of families *Reoviridae* and *Partitiviridae* are strikingly different not only in their genome sizes (the *Reoviridae* members have RNA genomes of 26-28 kb in 10-12 RNA segments, while the *Partitiviridae* genomes are of 3-4 kb in 2-3 RNA segments), but also in their basic replication mechanisms, namely, conservative *versus* semiconservative synthesis, respectively (Boccardo *et al.*, 1987; Xie *et al.*, 1993; Uyeda *et al.*, 1995; Buck, 1996).

The aim of this review is the comprehensive description of the virus genome maps of all characterized groups of plant RNA viruses where genomic nucleotide sequences are available (Mayo and Pringle, 1998; Pringle, 1998).

GENOME MAPS OF POSITIVE-STRANDED RNA VIRUSES

Picorna-like Supergroup

Family Potyviridae

Flexous filamentous particles of potyviruses are composed of one type of coat protein (CP) and RNA molecules with VPg at the 5'-end and poly(A) track at the 3'-end (Riechmann *et al.,* 1992). Representatives of the best studied genus of the family *Potyvirus* (type member, potato virus Y, PVY) have monopartite genomes of 10-11 kb coding for a single polyprotein of 3000-3400 amino acids, which is cleaved by virus-coded proteinases to give the individual viral proteins. The order of the final products is: (from N- to C-terminus of the polyprotein): P1, HC-Pro (helper component-proteinase), P3, $6K_1$, CI (cylindrical inclusion protein), $6K_2$, VPg part of NIa (nuclear inclusion protein), protease part of NIa, NIb (large nuclear inclusion protein), and CP (Riechmann *et al.,* 1992; Dougherty and Semler, 1993) (Fig. 1).

PVY (*Potyvirus*)

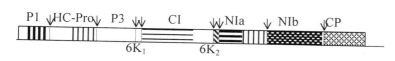

RGMV (*Rymovirus*)

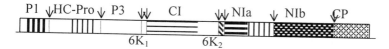

BaYMV (*Bymovirus*)

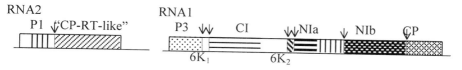

Figure 1. Genome organization in the genera of the family *Potyviridae*. PVY, potato virus Y: 9703 nts [M95491]. RGMV, ryegrass mosaic virus: 9542 nts [AF035818]. BaYMV, barley yellow mosaic virus: RNA1, 7632 nts [D01091]; RNA2, 3585 nts [D01092]. The following shading is used in this and subsequent figures to denote the functional domains:

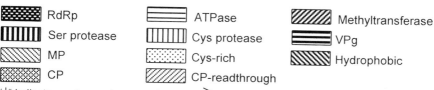

"ψ" indicates protease cleavage site;   "ℑ" shows translational frameshift;
"→" indicates translational read-through.

P1 protein has at least two activities: *cis*-cleavage of P1-(HC-Pro) precursor performed by the chymotrypsin-like serine protease domain of P1, and activation of viral RNA replication which probably involves the P1-specific RNA-binding activity (Koonin and Dolja, 1993; Verchot and Carrington, 1995; Ryan and Flint, 1997). HC-Pro is a multifunctional protein. Its papain-like cysteine protease domain located in the C-terminal portion of HC-Pro cleaves the polyprotein *in cis* at the boundary between HC-Pro and P3; HC domain participates in aphid transmission of the virus, probably playing role of a "bridge" attaching virions within insect food canal; and HC-Pro central region promotes genome amplification and the virus spread through the plant tissues. Potyvirus movement is likely associated with ability of HC-Pro to modify the size exclusion limit of plasmodesmata (Carrington *et al.*, 1996; Maia *et al.*, 1996; Kasschau *et al.*, 1997; Blanc *et al.*, 1997; Rojas *et al.*, 1997; Ryan and Flint, 1997).

The functions of P3 and $6K_1$ proteins are still unclear, though they contain some of the pathogenicity determinants of potyviruses (Riechmann *et al.*, 1992). CI protein is necessary for genome replication. It contains ATPase-helicase domain of the superfamily 2 and has RNA helicase activity *in vitro*, is localized in the vicinity of and inside the plasmodesmata, and is essential for virus cell-to-cell movement (Koonin and Dolja, 1993; Rodriguez-Cerezo *et al.*, 1997; Kadare and Haenni, 1997; Roberts *et al.*, 1998). $6K_2$ protein associates with large vesicules derived from endoplasmic reticulum *via* interaction of its central hydrophobic domain with the membranes. Presumably, $6K_2$ targets replication complexes (primarily NIa protein) to membranous sites of replication (Schaad *et al.*, 1997).

NIa (VPg-protease) is an essential replication protein. Its N-terminal part represents a VPg with the molecular mass of 24-27 kDa, which is covalently linked to the 5'-end of viral RNA through the tyrosine residue. NIa proteinase located in the C-terminal portion of NIa is a chymotrypsin-like cysteine protease which cleaves at seven specific sites in the extended region of the polyprotein including $P3-6K_1-CI-6K_2-VPg-NIa(Pro)-NIb-CP$ (Fig. 1). The cleavage proceeds in several stages and includes both *cis*- and *trans*-reactions. Additionally, NIa protease domain specifically interacts with NIb (RdRp), thus participating in formation of replication complex (see Riechmann *et al.*, 1992; Dougherty and Semler, 1993; Murphy *et al.*, 1996; Schaad *et al.*, 1996; Li *et al.*, 1997; Ryan and Flint, 1997).

NIb protein is the potyviral RdRp which belongs to supergroup 1. The ability of NIb to polymerize RNA chains was demonstrated *in vitro*. Recruiting of the NIb protein for the genome replication occurs *via* its specific interactions with NIa protease, coat protein, and, possibly, VPg (see Koonin and Dolja, 1993; Hong and Hunt, 1996; Li *et al.*, 1997).

Potyviral CP, in addition to the encapsidation of genomic RNA, performs at least two more functions. The conserved tripeptide DAG located at the N-terminus is a determinant of aphid transmission, and mutations of the CP affect both the internal regions (that caused defects in assembly) and the terminal regions (that gave assembly-competent virus) resulting in significant inhibition of the virus cell-to-cell spread which suggests a role of the CP in virus transport function (Riechmann *et al.*, 1992; Atreya *et al.*, 1995; Dolja *et al.*, 1995; Rojas *et al.*, 1997; Wu and Shaw, 1998).

Among the *cis*-elements, several conserved sequence motifs in the untranslated regions (UTRs) and the 3'-terminal portion of the potyvirus CP gene are involved in recognition of the replication complex during plus- and minus-strand RNA synthesis (Haldemann-Cahill *et al.*, 1998). Potyviruses seem to contain origin of assembly near the 5' terminus of the viral RNA (Wu and Shaw, 1998). Efficient translation of potyviral RNA requires synergistic interaction between the 5'-UTR and poly(A), that provides a cap-independent translational enhancer (for review, see Gallie, 1996).

The genome organization in viruses of the *Rymovirus* genus (transmitted by eriophyid gall mites) and some other monopartite potyviruses of the proposed genera *Ipomovirus*, *Macluravirus*, and *Tritimovirus* are principally similar to that of PVY-like *Potyviridae* representatives (Gotz and Maiss, 1995; Salm *et al.*, 1996; Badge *et al.*, 1996; Schubert *et al.*, 1996; Colinet *et al.*, 1998; Stenger *et al.*, 1998; Pringle, 1998). However, viruses of the genus *Bymovirus* (the type member, barley yellow mosaic virus) transmitted by *Polymyxa graminis* have bipartite genomes encapsidated in two slightly flexous, rod-shaped particles of different sizes. Large particles possess RNA1 coding for the counterparts of the *Potyvirus* proteins $6K_1$-CI-$6K_2$-VPg-NIa(Pro)-NIb-CP. This block is preceeded by a zinc-finger protein gene which is unique for bymoviruses. The smaller RNA2 codes for two proteins of 28 kDa (a counterpart of HC-Pro papain-like protease) and 70 kDa (a fungal transmission factor related to benyvirus CP readthrough product) (Kashiwasaki *et al.*, 1990; 1991; Koonin and Dolja, 1993; Dessens *et al.*, 1995; Namba *et al.*, 1998).

Family Sequiviridae

This family includes two genera of monopartite plant viruses, *Sequivirus* and *Waikavirus*, with genomic RNAs of 10-12 kb that possibly have a VPg. Spherical virions of *Sequiviridae* are composed of three types of CP molecules. The *Sequiviridae* genomes are expressed as large polyproteins of more than 3000 residues (Fig. 2). Although the details of their genome structure and expression are still unknown, some of the polyprotein cleavage products have been delimited and

PYFV (*Sequivirus*)

RTSV (*Waikavirus*)

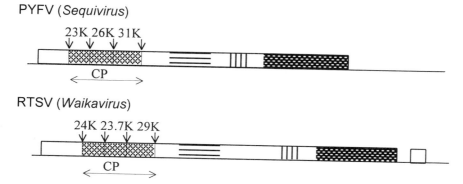

Figure 2. Genome organization in genera of family *Sequiviridae*. PYFV, parsnip yellow fleck virus: 9871 nts [D14066]. RTSV, rice tungro spherical virus: 12226 nts [M95497]. For shading legend see Figure 1.

their functions predicted. The sequences of the three CPs are within the N-terminal one-third of the polyprotein (but not the most N-terminal region). The central part of polyprotein contains conserved domain of a helicase of superfamily 3 and a large polyprotein area of unknown function, while the C-terminal one-third of *Sequiviridae* polyprotein represents the typical picornavirus-like replication module, VPg-cystein chymotrypsin-like protease-RdRp of supergroup 1 (Fig. 2) (Turnbull-Ross *et al.*, 1993; Koonin and Dolja, 1993; Hull, 1996; Reddick *et al.*, 1997). There are two main differences between the genera *Sequivirus* and *Waikavirus*. First, poly(A) tail is found at the 3'-end of the *Waikavirus* genomic RNA (Reddick *et al.*, 1997), whereas the 3' end of the *Sequivirus* genomic RNA lacks poly(A) and folds into a stem-loop structure (Turnbull-Ross *et al.*, 1992). Second, genomic RNA of waikaviruses has an additional short open reading frame (ORF) of 200-240 nucleotides downstream of the large polyprotein ORF (Fig. 2), which could be expressed using the small subgenomic RNAs (sgRNAs) found in the infected cells (Shen *et al.*, 1993; Hull, 1996; Reddick *et al.*, 1997).

Family Comoviridae

The components of bipartite genomes of this family (represented by genera *Comovirus*, *Nepovirus*, and *Fabavirus*) are encapsidated separately into spherical particles composed of one or two types of the CP molecules. Genomic RNAs are polyadenylated and possess VPg. Translation of both genomic segments results in polyproteins which give rise to the mature proteins after processing by virus-coded protease (Fig. 3).

CPMV (*Comovirus*)

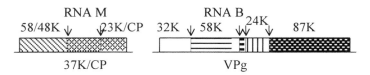

TBRV (*Nepovirus*)

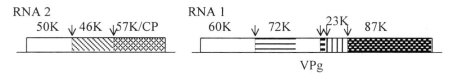

Figure 3. Genome organization in the genera of the family *Comoviridae*. CPMV, cowpea mosaic virus: B RNA, 5889 nts [X00206]; M RNA, 3481 nts [X00729]. TBRV, tomato black ring virus: RNA1, 7362 nts [D00322]; RNA2, 4662 nts [X04062]. For shading legend see Figure 1.

In CPMV and other viruses of the family, all replicative functions are perfomed by RNA1-coded proteins. The C-terminal part of the RNA1-specific polyprotein represents picornavirus-like replication module: helicase of superfamily 3-VPg-cystein chymotrypsin-like protease-RdRp of supergroup 1 (Fig. 3). The N-terminal segment of the RNA1-specific polyprotein is probably a specific viral protease co-factor required for *trans*-cleavage of the RNA2-coded polyprotein (Dougherty and Semler, 1993; Koonin and Dolja, 1993; Rott *et al.*, 1995; Shanks *et al.*, 1996; Zalloua *et al.*, 1996; Ryan and Flint, 1997; Kadare and Haenni, 1997).

RNA2 of the *Comoviridae* members specifies encapsidation, cell-to-cell movement, and long-distance spread of the virus. The RNA2 polyprotein translation in the genus *Comovirus* initiates at two alternative sites (Chapter 4). The CP synthesis and virion formation are essential for the movement, which occurs in virion form through special tubules modifying plasmodesmal structures. These tubules are composed of the RNA2-coded movement protein (MP), which also displays activities of plasmodesmal targeting and modification, and interaction with virions (Chapter 7).

Among the *cis*-elements, several conserved sequence motifs involving stem-loop structures in the 5'- and 3'-UTRs of comovirus RNAs are thought to be involved in replication complex recognition during plus- and minus-strand RNA synthesis (Buck, 1996). A 5'-terminal translated region located downstream of the first translational start codon of CPMV RNA2 contains internal ribosome binding site (IRES) that is responsible in part for bypassing the first initiation translation codon to synthesize the MP-CP precursor (Verver *et al.*, 1991; Gallie, 1996).

Sobemo-like Supergroup

It includes spherical plant viruses of the genera *Sobemovirus* and *Polerovirus* (subgroup 2 luteoviruses). Their genomes have VPg and contain no poly(A). A sobemo-like RNA component is also found in the virus of symbiotic nature belonging to the genus *Enamovirus* (Mayo and Pringle, 1998; Pringle, 1998).

Genus Sobemovirus

The size of monopartite sobemovirus genomes ranges from 4 to 5 kb. Genomic RNAs are non-polyadenylated and contain a VPg at the 5'-end. The genomes of typical sobemoviruses, southern bean mosaic virus (SBMV, the type member), rice yellow mottle virus (RYMV), and cocksfoot mottle virus (CfMV), contain four ORFs that, however, are arranged differently (Fig. 4). ORFs 2 and 3 of CfMV give rise to the fusion protein due to ribosomal frameshifting, and the resulting product has similar structure to polyprotein of SBMV and RYMV (Ngon A Yassi *et al.*, 1994; Futterer and Hohn, 1996; Makinen *et al.*, 1995a, b; Ryabov *et al.*, 1996).

ORF1-encoded protein of different sobemoviruses has no significant sequence similarity. Although deletions of the ORF1 region affect replication, ORF1 protein *per se* is not required for RNA synthesis. This protein seems to be involved in virus cell-to-cell and long-distance movement (Makinen *et al.*, 1995a; Bonneau *et al.*, 1998; Fowler *et al.*, 1998).

Depending on the genome organization, sobemoviruses express the RdRp of supergroup 1 as the C-terminal part of either the large ORF2-encoded polyprotein (as in SBMV) or ORF2-ORF3 fusion protein (as in CfMV) (Fig. 4). The N-terminal part of polyproteins contains long stretches of hydrophobic amino acids suggesting a membrane association for this protein domains. The hydrophobic region is followed by serine chymotrypsin-like protease domain. Compared to those in most picorna-like viruses, protease and VPg domain are transposed in sobemovirus polyproteins (Fig. 4) (Gorbalenya *et al.*, 1988; Dougherty and Semler, 1993; Koonin and Dolja, 1993; Ngon A Yassi *et al.*, 1994; Makinen *et al.*, 1995a, b; Ryabov *et al.*, 1996; Sivakumaran and Hacker, 1998; K. Makinen, personal communication).

SBMV (*Sobemovirus*)

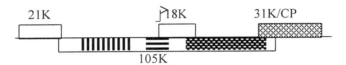

CfMV (*Sobemovirus*)

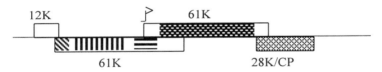

PLRV (*Polerovirus*)

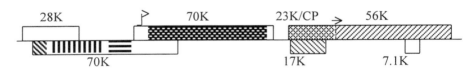

Figure 4. Genome organization in the genera of sobemo-like supergroup. SBMV, southern bean mosaic virus: 4194 nts [M23021]. CfMV, cocksfoot mottle virus: 4082 nts [Z48630]. PLRV, potato leafroll virus: 5882 nts [Y07496]. For shading legend see Figure 1.

The CP gene occupies the 3'-terminal position in the genome, synthesis of CP protein requires formation of a sgRNA that is encapsidated, and the CP (and probably virus particles) is involved in virus spread through the infected plants (Fuentes and Hamilton, 1993; Ryabov *et al.*, 1996; Hacker and Sivakumaran, 1997; Fowler *et al.*, 1998).

Among the *cis*-elements, extended stem-loop structures are found in the 5'- and 3'-terminal regions of sobemovirus genomes, which may act as signals for replicase recognition (Ngon A Yassi *et al.*, 1994; Ryabov *et al.*, 1996). The viral sgRNA start is in the vicinity of (Ryabov *et al.*, 1996) or included in (Hacker and Sivakumaran, 1997) a conserved hairpin structure found in all sequenced sobemovirus genomes (S. Morozov, unpublished). It can be proposed that this structure is an essential element of subgenomic promoter.

Genus Polerovirus (Subgroup 2 luteoviruses)

It was previously regarded as one of the subgroups of spherical phloem-limited luteoviruses. Apparently, the known luteoviruses arose by recombinational combining of luteovirus-specific 3'-terminal genome part with sobemovirus-like or carmovirus-like 5'-terminal genome portions coding for replicative proteins (Koonin and Dolja, 1993; Mayo and Ziegler-Graff, 1996; Miller and Rasochova, 1997). The typical sobemovirus replication module found in genomes of subgroup 2 luteoviruses helped include these viruses into the "sobemo-like" supergroup as the genus *Polerovirus* with potato leafroll virus (PLRV) as the type member.

The VPg-linked genome of PLRV contains seven ORFs, which are divided by a long intergenic region into two clusters of overlapping genes, the 5'-terminal sobemovirus-like replicative module and the 3'-terminal CP-coding gene module common for both subgroups of luteoviruses (Miller and Rasochova, 1997; Mayo and Ziegler-Graff, 1996; Ashoub *et al.*, 1998) (Fig. 4). Description of the polypeptides encoded by the poleroviruses (named as in PLRV) is given below.

ORF0-encoded protein (28K) is poorly conserved in poleroviruses and is not involved in virus replication. Its expression determines pathogenesis phenotype by a yet unknown mechanism (van der Wilk *et al.*, 1997b; Ashoub *et al.*, 1998).

The polerovirus genome region coding for ORF1-encoded protein (70K) and ORF1/ORF2 fusion product is arranged similarly to the sobemovirus ORF2-ORF3 region. The N-terminal segment of the ORF1 protein includes highly hydrophobic stretches of amino acids, which could be regarded as functional analogs of the potyvirus $6K_2$. The hydrophobic region is followed by serine chymotrypsin-like protease domain. The C-terminal part of the PLRV ORF1 protein contains the VPg domain. ORF2 overlaps ORF1 and is translated through a -1 frameshift as ORF1/ORF2 polyprotein, thereby expressing domain of RdRp of supergroup 1 (Rohde *et al.*, 1994; Futterer and Hohn, 1996; Makinen *et al.*, 1995a, b; Ryabov *et al.*, 1996; Mayo and Ziegler-Graff, 1996; van der Wilk *et al.*, 1997a).

The major CP is encoded by ORF3, whereas suppression of amber stop codon in CP gene forms ORF3-ORF5 readthrough protein, which is a minor component of viral capsids. Both proteins do not influence viral replication. The CP readthrough domain is responsible for persisting virus particles in vector hemolymph by acting as a tag that binds extracellular GroEL protein of a bacterial endosymbiont of the aphid. Major and minor CPs are expressed through formation of specific sgRNA1 (Rohde *et al.*, 1994; Bruyere *et al.*, 1997; van den Heuvel *et al.*, 1997; Ashoub *et al.*, 1998). ORF4-encoded protein (17K) is an MP specialized in the cell-to-cell virus transport within vascular tissue. ORF4 is nested into ORF3 and expressed from the same sgRNA1 (Sokolova *et al.*, 1997; Schmitz *et al.*, 1997). Recently, additional ORF6-coded protein of unknown function was predicted to be expressed from the PLRV-specific sgRNA2 (Ashoub *et al.*, 1998).

Genus Enamovirus

Pea enation mosaic virus (PEMV) is an example of the symbiotic partnership between two autonomously replicating RNAs. PEMV RNA1 (presently referred to as 'pea enation mosaic virus-1') (Fig. 5) is non-polyadenylated, has a VPg, and displays obvious similarity in gene arrangement and encoded proteins to the genomes of sobemo-like viruses. In contrast, the genomic RNA2 ('pea enation mosaic virus-2') codes for the RdRp related to those of carmoviruses (Demler *et al.*, 1993; 1994; Koonin and Dolja, 1993; Wobus *et al.*, 1998).

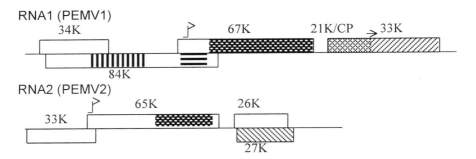

Figure 5. Genome organization of pea enation mosaic virus: PEMV1 (the genus *Enamovirus*), 5706 nts [L04573]; PEMV2 (the genus *Umbravirus*), 4253 nts [U03563, S53233]. For shading legend see Figure 1.

PEMV RNA1 contains five ORFs structurally and functionally corresponding to the counterparts in the PLRV polerovirus genome (Fig. 5). The function of PEMV ORF0-encoded protein (34K) is unknown. ORF0 significantly overlaps the replicative module which comprises ORF1 and ORF2. They code for sobemo-like pattern of replicative proteins, i.e. serine chymotrypsin-like protease-VPg-RdRp of supergroup 1, representing the parts of the ORF1 polyprotein and ORF1/ORF2 frameshifting product. PEMV CP gene (ORF3) is expressed from specific sgRNA and can be translated as ORF3-ORF5 read-through protein, which is a minor virion component (Fig. 5). Both the CP and its read-through product are dispensable for virus movement through infected plants. However, they are required for formation of virions (containing both PEMV RNAs 1 and 2) capable of circulative aphid transmission. Like luteoviruses, PEMV CP read-through domain contains determinants for binding the GroEL protein of a bacterial endosymbiont that probably protects virus particles from degradation in aphid hemolymph. PEMV RNA1 does not code for PLRV ORF4 protein analog (phloem-specific MP) and is incapable of moving cell-to-cell or long-distance on its own. Movement of RNA1 in plants depends on PEMV RNA2 (Demler *et al.*, 1993; 1994; 1997; Koonin and Dolja, 1993; van den Heuvel *et al.*, 1997; Skaf *et al.*, 1997; Wobus *et al.*, 1998).

PEMV RNA2 resembles the genomic RNA of umbraviruses, and its genetic organization is described later.

Carmo-like Supergroup

In it, viral genomes encode RdRp of supergroup 2, but not other replication modules conserved in viruses of other supergroups. Genomic RNAs are non-polyadenylated and may contain a cap-structure (Koonin and Dolja, 1993; Mayo and Pringle, 1998).

Genus Umbravirus

The umbraviruses do not code for CP, are unable to form virus particles, and depend on a helper virus, a luteovirus as a rule, in encapsidation and transmission by aphids. Their genome has four ORFs (Fig. 6). ORF1 codes for a protein (37K) of unknown function. ORF2 codes for RdRp of supergroup 2 which is expressed as ORF1/ORF2 frameshifting fusion. Formation of a sgRNA is required for translation of the significantly overlapping ORF3 and ORF4. ORF4 protein (29K) is a MP related to those in viruses of family *Bromoviridae* and capable of unspecific cell-to-cell transport of viral RNAs. ORF3 protein can form cytoplasmic and nuclear membrane-associated inclusions and is probably involved in phloem-specific long-distance movement (Gibbs *et al.*, 1996; Taliansky *et al.*, 1996; Ryabov *et al.*, 1998; M. Taliansky, personal communication).

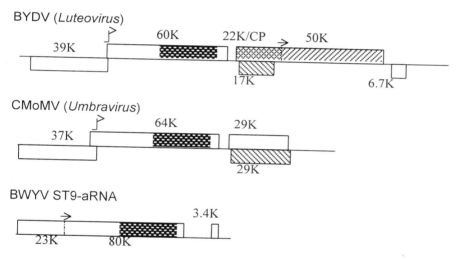

Figure 6. Genome organization in the genera of carmo-like supergroup. BYDV, barley yellow dwarf virus (BYDV-PAV): 5677 nts (X 07653). CMoMV, carrot mottle mimic virus: 4201 nts (U57305). BWYV ST9-aRNA, beet western yellows virus ST9-associated RNA: 4201 nts (L04281). For shading legend see Figure 1.

PEMV RNA2 ('pea enation mosaic virus-2', recently proposed member of genus *Umbravirus*) contains four ORFs (Fig. 5) and provides functions essential for viral replication and movement in plants. RdRp of supergroup 2 is encoded by ORF2 and is expressed as ORF1/ORF2 frameshift fusion. Overlapping ORFs 3 and 4 are expressed from a sgRNA. ORF4 protein (27K) is proposed to be a MP by analogy with umbravirus ORF4 protein. The function of ORF3 protein (26K) remains unknown. PEMV RNA2 lacks genes responsible for encapsidation and aphid transmission, and completely depends on sobemovirus-like RNA1 in these functions (Demler *et al.*, 1993; Taliansky *et al.*, 1996; Ryabov *et al.*, 1998; Pringle, 1998).

Beet WesternYellows Virus ST9-associated RNA

This self-replicating RNA together with carrot red leaf virus-associated RNA represent another example of a partnership between sobemovirus- and carmovirus-like genomes. Despite the extremely small genomes (2.8 kb), replication of these luteovirus-associated RNAs does not depend on a helper virus. However, they are encapsidated and transmitted with the aid of luteovirus virion proteins. Two ORFs are translated directly from genomic RNA (Fig. 6) giving rise to ORF1-encoded 23-kDa protein and ORF1/ORF2 translational readthrough protein. The ORF2 part of this fusion represents RdRp of supergroup 2. Very small ORF3-coded protein (ca. 3-4K) is translated from the respective 3'-coterminal sgRNA (Chin *et al.*, 1993; Sanger *et al.*, 1994; Watson *et al.*, 1997).

Genus Luteovirus (Subgroup 1 luteoviruses)

Monopartite RNA genomes in viruses of this genus are of 5.6-5.8 kb in length, contain neither cap-structure nor VPg, and are non-polyadenylated (Shams-bakhsh and Symons, 1997; Miller and Rasochova, 1997) (Fig. 6). Genomic RNAs of barley yellow dwarf luteovirus (BYDV, strain PAV) and closely related viruses have six ORFs, and only five of them are present in more distantly related soybean dwarf virus (SDV) (Rathjen *et al.*, 1994; Miller and Rasochova, 1997).

The function of ORF1-encoded protein (39K) is obscure; however, deletions in this gene destroy the ability of genomic RNA to replicate (Mohan *et al.*, 1995; Miller and Rasochova, 1997). ORF2-encoded protein is expressed only as ORF1/ORF2 frameshift fusion and includes RdRp domain of supergroup 2 (Brault and Miller, 1992; Koonin and Dolja, 1993; Rohde *et al.*, 1994; Futterer and Hohn, 1996; Miller and Rasochova, 1997). ORF3 codes for CP, and ORF5 is expressed as ORF3-ORF5 readthrough protein, which is a minor CP and is involved in aphid transmission. Translation of CP and CP-readthrough requires the synthesis of sgRNA1 (Dinesh-Kumar *et al.*, 1992; Cheng *et al.*, 1994; Rathjen *et al.*, 1994; Chay *et al.*, 1996; Miller and Rasochova, 1997). ORF4 nested into ORF3 is expressed from the same sgRNA1 and codes for a MP (17K) (Mohan *et al.*, 1995; Chay *et al.*, 1996; Miller and Rasochova, 1997). Small ORF6 can be translated *in vitro* from sgRNA2 and its product (6.7K) influences genome replication in protoplasts; however, the function of this protein remains unknown (Mohan *et al.*, 1995; Miller and Rasochova, 1997).

Among *cis*-elements, mutagenesis of sgRNA1 promoter of BYDV-PAV showed that activity of this promoter is influenced by two stem-loop structures (Koev *et al.*, 1998). Genomic RNA and sgRNA1 have complex cap-independent translational enhancer composed of 5'-UTR of genomic RNA (or of sgRNA1) and an element located between ORF5 and ORF6 (Gallie, 1996; Miller and Rasochova, 1997).

Family Tombusviridae

This family includes five approved genera, namely, *Tombusvirus*, *Carmovirus*,

Machlomovirus, *Necrovirus*, and *Dianthovirus*, and the recently proposed genera *Panicovirus*, *Aureusvirus*, and *Avenavirus* (Russo *et al.*, 1994; Mayo and Pringle, 1998; Pringle, 1998). All representatives are rather small spherical viruses with capped and non-polyadenylated genomes.

The viruses of the *Tombusvirus* and *Aureusvirus* genera have monopartite genomes of 4.4-4.8 kb containing five ORFs (Russo *et al.*, 1994; Rubino and Russo, 1997; Miller *et al.*, 1997; Pringle, 1998) (Fig. 7). Oat chlorotic stunt virus (the proposed genus *Avenavirus*) contains only four ORFs and can be regarded as a member of *Tombusviridae* (Boonham *et al.*, 1995; Pringle, 1998).

ORF1-encoded protein and ORF1-ORF2 read-through fusion protein are directly translated from the genomic RNA of tombusviruses (Fig. 7). Both proteins are involved in replication of genomic RNA and can function *in trans*. ORF2 codes for RdRp of supergroup 2, whereas ORF1-encoded protein is proposed to be involved in the formation of vesicular bodies from the membranes of mitochondria or peroxisomes. These vesicles are believed to be the sites of virus RNA replication (Kollar and Burgyan, 1994; Russo *et al.*, 1994; Scholthof *et al.*, 1995a; Burgyan *et al.*, 1996; Molinari *et al.*, 1998; Oster *et al.*, 1998).

CP is encoded by ORF3 and translated from sgRNA1. The CP is dispensable for the cell-to-cell and, probably, long-distance transport, but is a determinant of fungal transmission (Scholthof *et al.*, 1993; Campbell, 1996; Robbins *et al.*, 1997; Rubino and Russo, 1997). ORF4- and ORF5-encoded proteins are translated from viral sgRNA2. ORF4 codes for a 22-27 kDa MP of TMV 30K subgroup of single-gene coded movement proteins. ORF5-encoded protein promotes long-distance transport in a host-dependent manner and is responsible for symptom severity (Mushegian and Koonin, 1993; Scholthof *et al.*, 1995b; Johnston and Rochon, 1996; Rubino and Russo, 1997).

Among the *cis*-elements, three sequences essential for tombusvirus replication are found in genome: 5'- and 3'-termini and 5'-proximal region inside ORF2. Stem-loop structures located in these regions probably represent replicase recognition sites (Finnen and Rochon, 1993; Chang *et al.*, 1995; Havelda and Burgyan, 1995; Buck, 1996). Additional *cis*-element, pX gene, is the host-dependent determinant of replication (Scholthof and Jackson, 1997; Molinari *et al.*, 1998). The core *cis*-elements involved in sgRNA synthesis are delimited by deletion analysis (Johnston and Rochon, 1995).

The viruses of the *Carmovirus*, *Machlomovirus*, *Panicovirus*, and *Necrovirus* genera have monopartite genomes of 3.6-4.3 kb containig five principal ORFs. However, some minor variations in gene arrangement occur between different viruses so that one or two additional ORFs can exist (Grieco *et al.*, 1996; Lot *et al.*, 1996; Molnar *et al.*, 1997; Weng and Xiong, 1997; Ciuffreda *et al.*, 1998; Turina *et al.*, 1998) (Fig. 7).

ORF1-encoded protein has usually a molecular mass of 22-29 kDa; however, in genera *Machlomovirus* and *Panicovirus* this protein is significantly larger (48-50kDa). ORF1 protein is essential for replication and could be involved in vesicular formation on mitochondrial membranes. Leaky terminator of ORF1 makes possible the synthesis of ORF1/ORF2 translational read-through product; ORF2 codes for the RdRp of supergroup 2 (Hacker *et al.*, 1992; Koonin and Dolja, 1993; Futterer and Hohn, 1996; Molnar *et al.*, 1997; Ciuffreda *et al.*, 1998).

Genome Organization in RNA Viruses

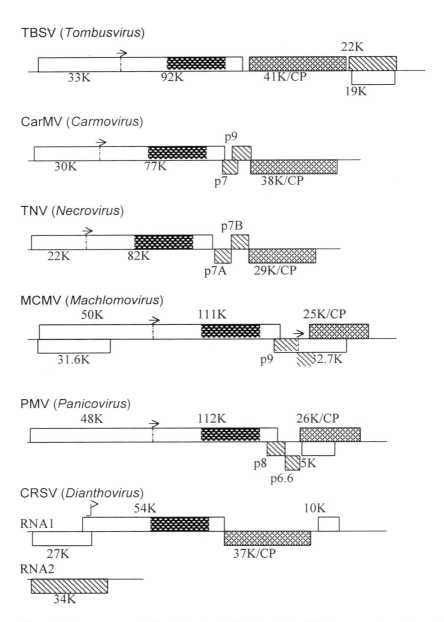

Figure 7. Genome organization in the family *Tombusviridae*. The type members of each genus are shown. TBSV, tomato bushy stunt virus: 4776 nts [M21958, M31019]. CarMV, carnation mottle virus: 4003 nts [X02986]. TNV, tobacco necrosis virus: 3762 nts [D00942]. MCMV, maize chlorotic mottle virus: 4437 nts [X14736]. PMV, panicum mosaic virus: 4326 nts [U55002]. CRSV, carnation ringspot virus: RNA1, 3756 nts [L18870]; RNA2, 1394 nts [M88589]. For shading legend see Figure 1.

ORF3 and ORF4-coded proteins of 6-10 kDa (Fig. 7) are expressed from sgRNA1. Both of them are required for virus cell-to-cell and long distance movement. First gene in this conserved movement gene module (ORF3) codes for the hydrophylic protein with nucleic-acid binding activity, whereas the second gene (ORF4) encodes a polypeptide with an obvious hydrophobic domain that is possibly involved in membrane binding (Hacker *et al.*, 1992; Offei *et al.*, 1995; Molnar *et al.*, 1997; Weng and Xoing, 1997; Wobbe *et al.*, 1998; Turina *et al.*, 1998).

CP is encoded by ORF5 and expressed from sgRNA2. CP is dispensable for virus cell-to-cell movement but essential for long-distance transport and eliciting plant hypersensitive reaction (Hacker *et al.*, 1992; Molnar *et al.*, 1997).

Cis-elements involved in replication, sgRNA synthesis, and encapsidation are mapped in turnip crinkle carmovirus (TCV) genome. The sequences required for the initiation of plus-strand and minus-strand RNA synthesis include discrete 5' and 3'-terminal regions with characteristic secondary structure and linear signals (Song and Simon, 1995; Buck, 1996; Guan *et al.*, 1997). Promoters for the synthesis of sgRNA1 and sgRNA2 contain stem-loop structures located just upstream of the transcription start sites (Wang and Simon, 1997). The virus assembly origin was located in the 3'-terminal portion of TCV coat protein gene (Qu and Morris, 1997).

In contrast to other genera of family *Tombusviridae*, the genus *Dianthovirus* includes viruses with bipartite genomes, consisting of RNAs of 3.8 and 1.4 kb (Fig. 7). RNA1 can replicate in protoplasts in the absence of RNA2 and is translated into ORF1 protein of unknown function and ORF1/ORF2 frameshift fusion protein. ORF2 contains RdRp domain of supergroup 2.

ORF3 codes for the CP, which is involved in host-dependent long-distance virus transport but dispensable for cell-to-cell movement. The CP is expressed from a 3'-coterminal sgRNA, and promoter of this sgRNA is a complex genetic element consisting of a core *cis*-element located upstream of sgRNA transcription start in RNA1 and *trans*-acting element found in RNA2. Base pairing between these specific sequences in genomic RNAs 1 and 2 is required for promoter activation (Xiong and Lommel, 1989; Xiong *et al.*, 1993; Kim and Lommel, 1994; Zavriev *et al.*, 1996; Sit *et al.*, 1998).

RNA2 of dianthoviruses is monocistronic and codes for a MP. The MP has multidomain organization and following activities are described for the dianthovirus MP. It possesses the nucleic-acid binding properties. It also increases the size exclusion limit of plasmodesmata, and transports itself and the ribonucleoproteins (RNPs) through such plasmodesmata. This protein has distinct signals for targeting to the plasmodesmata and for the host-specific phloem entry of the viruses (Xiong *et al.*, 1993; Giesman-Cookmeyer and Lommel, 1993; Wang *et al.*, 1998).

Alpha-like Supergroup

Its viruses have two main peculiarities: their genomes are capped, and code for a conserved set of replicative protein domains which include RdRp of supergroup 3, helicase of superfamily 1, and methyl/guanylyltransferase (Goldbach, 1987; Koonin and Dolja, 1993; Buck, 1996). Three lineages of RdRp were delimited within supergroup 3, namely, tobamo, rubi, and tymo lineages (Koonin and Dolja, 1993).

Family Closteroviridae

This family includes filamentous viruses with capped and non-polyadenylated genomes, which are either monopartite (genus *Closterovirus*) or bipartite (genus *Crinivirus*). Among plant positive-strand RNA viruses, closteroviruses have the largest genomes, with the size ranging from 15.5 kb in beet yellows virus (BYV) to 19.3 kb in citrus tristeza virus (CTV). The genomes contain 9-12 ORFs (Fig. 8), of

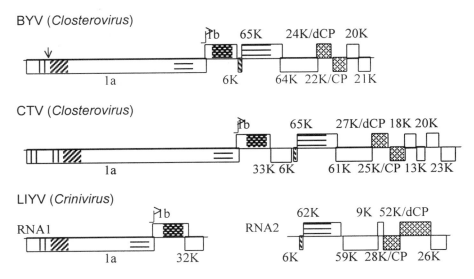

Figure 8. Genome organization of closteroviruses. BYV, beet yellows virus: 15480 nts [X73476]. CTV, citrus tristeza virus : 19296 nts [U16304, U02547, L 20760]. LIYV, lettuce infectious yellows virus: RNA1, 8118 nts [U15440]; RNA2, 7193 nts [U15441, U05242]. "dCP" indicates diverged duplicate of CP gene. For shading legend see Figure 1.

which seven are conserved and correspond to two gene modules represented in BYV, the type closterovirus, by ORFs 1a/1b and ORFs 2, 3, 4, 5 and 6. ORFs 1a/1b are expressed from the genomic RNA, whereas expression of other genes in monopartite closterovirus genomes requires sgRNAs (at least five in BYV and nine in CTV) (Agranovsky *et al.,* 1994; Dolja *et al.,* 1994; Klaassen *et al.,* 1995; Karasev *et al.,* 1995; 1997; Agranovsky, 1996; Jelkmann *et al.,* 1997; Zhu *et al.,* 1998). A description of the closterovirus proteins with reference to BYV is outlined below.

ORF1a- and ORF1b-coded proteins are necessary and sufficient for RNA replication. At least three conserved protein domains were identified in ORF1a that is translated directly from genomic RNA (RNA1 in the case of the *Crinivirus* genomes). The N-terminal portion contains the domain of papain-like proteinase that can be duplicated in some closteroviruses. The protease domain is cleaved from the rest part of polyprotein autocatalytically, and the resulting leader protein has at least one activity as an enhancer of RNA replication and transcription. The domains of methyl/guanylyltransferase and helicase of superfamily 1 in the ORF1a product are separated by a long non-conserved spacer. Expression of ORF1b that codes for

RdRp of supergroup 3 (tobamo lineage) occurs by a translational frameshift resulting in ORF1a/1b fusion protein (Koonin and Dolja, 1993; Agranovsky *et al.*, 1994; Dolja *et al.*, 1994; Agranovsky, 1996; Jelkmann *et al.*, 1997; Zhu *et al.*, 1998; Peremyslov *et al.*, 1998).

ORF2 codes for a small hydrophobic protein of 5-7 kDa. It includes a stretch of nonpolar amino acids which are predicted to act as membrane-binding tag (Dolja *et al.*, 1994; Agranovsky *et al.*, 1996; Karasev *et al.*, 1996; Jelkmann *et al.*, 1997; Zhu *et al.*, 1998). ORF3-coded protein (65K) is homologous to the HSP70 family of cell heat shock proteins. It is highly conserved between different closteroviruses and has an ATPase activity *in vitro*. However, the 65K protein is unable to bind peptides and, therefore, is functionally different from the orthodox HSP70 proteins. The BYV 65K protein is able to interact with microtubules *in vitro* and probably participates in the virus movement through infected plants (Agranovsky *et al.*, 1997; 1998; Karasev *et al.*, 1992; 1995; 1996; Zhu *et al.*, 1998). ORF4 is well conserved among closteroviruses; however, no functions have been proposed for its product (Dolja *et al.*, 1994; Karasev *et al.*, 1995; 1996; Agranovsky, 1996; Jelkmann *et al.*, 1997).

ORF5 and ORF6 represent a result of gene duplication. ORF6 codes for the major CP, whereas ORF5 encodes a diverged duplicate of CP. In some closteroviruses, these genes are transposed, i.e. the CP gene is located closer to the 5' terminus of genomic RNA. There is a correlation between the degree of relatedness of the closterovirus CPs and their vector specificity that might suggest involvement of CP in vector transmission. Diverged duplicate of CP represents a minor CP that encapsidates a small part of the genomic RNA, thus giving rise to the unique "rattlesnake" structure of the filamentous virions (Boyko *et al.*, 1992; Klaassen *et al.*, 1994; Agranovsky *et al.*, 1995; Agranovsky, 1996; Karasev *et al.*, 1996; Jelkmann *et al.*, 1997; Zhu *et al.*, 1998). ORF7 encodes a protein (20K), which is unique for BYV. ORF8-encoded 20-22 kDa protein demonstrates marginal similarity between several aphid-transmissible closteroviruses and acts as an enhancer of replication in BYV (Karasev *et al.*, 1996; Zhu *et al.*, 1998; Peremyslov *et al.*, 1998).

Cis-acting element is predicted to be present at the 3' terminus of the closterovirus genomes. This element involves stem-loop structure where it could serve as the recognition signal for the replication proteins (Karasev *et al.*, 1996; Zhu *et al.*, 1998).

Family Bromoviridae

It includes five genera with tripartite genomes (*Bromovirus*, *Cucumovirus*, *Alfamovirus*, *Ilarvirus*, and *Oleavirus*) and genus *Idaeovirus* with bipartite genome. The genomic RNAs in viruses of *Bromoviridae* are capped and non-polyadenylated, and code for closely related proteins (Ahlquist, 1992; Ziegler *et al.*, 1993; Palukaitis *et al.*, 1992; Grieco *et al.*, 1996; Mayo and Pringle, 1998).

RNAs 1 and 2 encode the 1a and 2a proteins, respectively (Fig. 9). In *Idaeovirus* raspberry bushy dwarf virus (RBDV), these proteins are represented by a single protein encoded by RNA1. The 1a protein and 1a-like region in the RNA1-encoded

protein of RBDV have two conserved domains, N-terminal methyl/
guanylyltransferase domain and C-terminal domain of helicase of superfamily 1,
which are separated by a spacer region. Domain of RdRp of supergroup 3 (tobamo
lineage) is located in the 2a protein or the C-terminal part of RBDV RNA1-coded
protein (Fig. 9). Apart from the main enzymatic functions, these replicative proteins
have some additional activities: replication of bromoviruses requires specific
interactions between the helicase domain of the 1a protein and the N-terminal
region of the 2a protein, and both 1a and 2a proteins are involved in virus spread in
the infected plants (Ahlquist, 1992; Ziegler *et al.,* 1993; Palukaitis *et al.,* 1992;
Rozanov *et al.,* 1992; O'Reilly *et al.,* 1995; Grieco *et al.,* 1996; Buck, 1996; Ge *et
al.,* 1997; Wintermantel *et al.,* 1997; Ishikawa *et al.,* 1997).

In viruses of the genera *Cucumovirus* and *Ilarvirus,* but not other genera of
Bromoviridae, RNA2 encodes an additional 2b protein (Fig. 9) that is most likely
expressed from a sgRNA. The 2b protein of cucumoviruses is involved in host-
specific long-distance virus movement and symptom determination (Shi *et al.,*
1997; Xin *et al.,* 1998).

RNA3 in viruses of the family *Bromoviridae* (RNA2 in genus *Idaeovirus*) is
bicistronic. The 5'-proximal gene is translated directly from genomic RNA and
codes for 3a MP. The 3a MPs in *Bromoviridae* bind single-stranded RNA and GTP
in vitro, change size-exclusion limit of plasmodesmata (promote the cell-to-cell
movement of large fluorescent dextran molecules), traffic themself and RNA from
cell-to-cell upon microinjection experiments, and associate with cell wall (close to
or inside plasmodesmata). In addition, alfamovirus, oleavirus and bromovirus MPs
can form tubular structures of which role in spread of these particular viruses is
still obscure. Apart from its role in virus movement, the 3a MP is also an important
determinant of host specificity and symptom expression (Vaquero *et al.,* 1994,
1997; De Jong *et al.,* 1995; Ding *et al.,* 1995; Mise and Ahlquist, 1995; Kaplan *et
al.,* 1995, 1997; Li and Palukaitis, 1996; Schmitz and Rao, 1996; Carrington *et al.,*
1996; Canto *et al.,* 1997; Itaya *et al.,* 1997; Kasteel *et al.,* 1997; van der Wel *et al.,*
1998).

The CP gene in the viruses of the family *Bromoviridae* is located in the 3'-
terminal region of RNA3 (RNA2 in *Idaeovirus*) and expressed from a sgRNA.
Besides the role in genome encapsidation and virion transmission by vectors, the
CP in viruses of this family is involved in cell-to-cell and long-distance spread.
However, the significance of the full-length CP and the formation of virions for
such spread varies substantially between different viruses (Ahlquist, 1992;
Palukaitis *et al.,* 1992; van der Vossen *et al.,* 1994; Taliansky and Garcia-Arenal,
1995; Schmitz and Rao, 1996; Rao, 1997; Schneider *et al.,* 1997). In alfamovirus
and ilarviruses, the CP is also involved in several steps of genome replication
including a very early function in the initiation of infection (phenomenon of
'genome activation') and a role in the assymetric accumulation of viral plus-RNA
(van der Vossen *et al.,* 1994; De Graaff *et al.,* 1995).

Among the *cis*-elements, the 3'-UTRs of genomic RNAs in brome mosaic virus
(the genus *Bromovirus*) and cucumber mosaic virus (the genus *Cucumovirus*) can be
folded into a specific secondary structure that mimics tRNA and can perform tRNA-
like functions including the aminoacylation and adenylation. This genome region
contains a promoter for the minus-strand genomic RNA synthesis (Ahlquist, 1992;

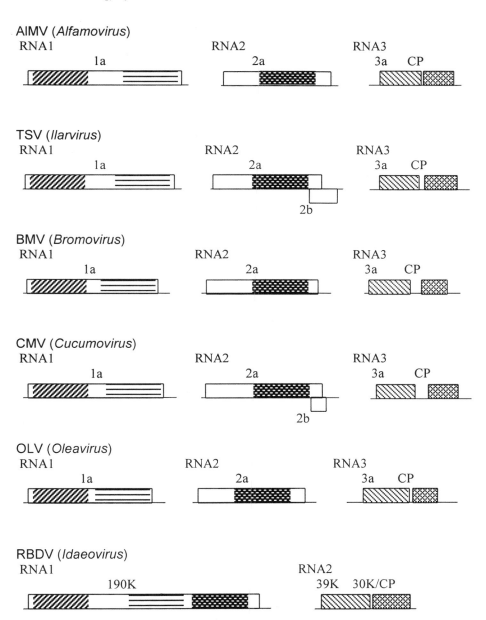

Figure 9. Genome organization in the family *Bromoviridae*. AIMV, alfalfa mosaic virus: RNA1, 3644 nts [L00163, J02000]; RNA2, 2593 nts, [K02702, J02002]; RNA3, 2037 nts [K02703]. TSV, tobacco streak virus: RNA1, 3491 nts [U80934]; RNA2, 2926 nts [U75538]; RNA3, 2205 nts [X00435]. BMV, brome mosaic virus: RNA1, 3234 nts [K02706]; RNA2, 2865 nts [K02707]; RNA3, 2111 nts [J02042, J02043]. CMV, cucumber mosaic virus: RNA1, 3389 nts [X02733]; RNA2, 3035 nts [X00985]; RNA3, 2193 nts [J02059]. OLV, olive latent virus 2: RNA1, 3126 nts [X94346]; RNA2, 2734 nts [X94347]; RNA3, 2438 nts [X76993]. RBDV, raspberry bushy dwarf virus: RNA1, 5449 nts [S51557]; RNA2, 2231 nts [S55890]. For shading legend see Figure 1.

Palukaitis *et al.*, 1992; Duggal *et al.*, 1994; Buck, 1996). The genomes of *Bromovirus* and *Cucumovirus* contain the 3'-tRNA-like structures accepting tyrosine. Virus RNAs in other four genera have 3'-terminal regions with a series of stem-loop structures important for interaction with the CP and RdRp as in alfamo- and ilarviruses (Grieco *et al.*, 1996; Buck, 1996; van Rossum *et al.*, 1997a). Studies on plus-strand RNA synthesis in viruses of family *Bromoviridae* suggest that two sequence elements (ICR1 and ICR2) at the 5'-terminal region of genomic RNAs bind to putative cell transcription factor associated with DNA-dependent RNA-polymerase III and play an important role in formation of an initiation complex for plus-RNA synthesis (Pogue and Hall, 1992; Pogue *et al.*, 1992; Duggal *et al.*, 1994; Buck, 1996; van Rossum *et al.*, 1997b). *Cis*-elements involved in virus RNA replication are also located in the intercistronic region between 3a and CP genes, in which ICR2-like motif is present in addition to the 5'-terminus of the genomic RNA (Pacha *et al.*, 1990; Boccard and Baulcombe, 1993; Buck, 1996). This intercistronic region contains the elements required for subgenomic promoter activity (Smirnyagina *et al.*, 1994; Duggal *et al.*, 1994; Buck, 1996; Siegel *et al.*, 1997).

The elements enhancing BMV RNA translation in a cap-structure-dependent manner are mapped to the pseudoknotted region upstream of the tRNA-like structure (Gallie and Kobayashi, 1994).

Family "Tubiviridae"

This tentative family (Torrance and Mayo, 1997) consists of mono-, bi-, and tripartite rod-like viruses belonging to genera *Tobamovirus*, *Tobravirus*, *Furovirus*, *Hordeivirus*, *Pomovirus*, *Pecluvirus*, and *Benyvirus* (Fig. 10).

Tobacco mosaic virus (TMV) of genus *Tobamovirus* is the type member. Tobamoviruses have monopartite, capped, and non-polyadenylated genomes of 6.3-6.6 kb with the 3'- tRNA-like structure (Fig. 10). Three subgroups of tobamoviruses are distiguished; however, genetic organization is common for all viruses of the genus, and their genomes encode four major proteins (Dawson and Lehto, 1990; Dawson, 1992; Dorokhov *et al.*, 1994; Chng *et al.*, 1996; Lartey *et al.*, 1996; Zaitlin, 1998).

The 126K protein is a major translation product of TMV genomic RNA. The N-terminal part of this protein contains methyl/guanylyltransferase domain, whereas the C-terminal domain represents helicase of superfamily 1. The 183K protein is synthesized by translational readthrough of the 126K ORF leaky termination codon. The C-terminal region of 183K protein contains domain of RdRp of supergroup 3. The 126K and 183K proteins, together with at least one host protein, form an active membrane-bound replication complex at the endoplasmic reticulum of tobamovirus-infected cells. In addition, 126K protein influences tobamovirus long-distance movement through vasculature and induction of host hypersensitive response (Dawson and Lehto, 1990; Dawson, 1992; Koonin and Dolja, 1993; Futterer and Hohn, 1996; Buck, 1996; Chng *et al.*, 1996; Derrick *et al.*, 1997; Deom *et al.*, 1997; Heinlein *et al.*, 1998; Zaitlin, 1998).

The 30K protein is a tobamovirus MP. It is translated from a non-capped sgRNA at early stages of infection, and several activities are associated with it. It binds

GTP and single-stranded RNA *in vitro,* and can increase the plasmodesmata size exclusion limit and transports itself and RNA molecules through plasmodesmata to neighboring cells. In TMV-infected cells, 30K MP is localized to plasmodesmata and forms complexes with cortical endoplasmic reticulum and components of cytoskeleton, tubulin and actin. The RNA-binding activity of 30K protein is associated with the formation of untranslatable RNP capable of translocation from cell-to-cell (Dawson and Lehto, 1990; Dawson, 1992; Padgett *et al.,* 1996; Carrington *et al.,* 1996; Karpova *et al.,* 1997; Oparka *et al.,* 1997; Heinlein *et al.,* 1998; Zaitlin, 1998).

The TMV CP is encoded by the 3'-proximal gene expressed from specific sgRNA. The CP is essential for long-distance movement through phloem and contains a specific determinant that is recognized by host components and elicits the hypersensitive response. (Okada, 1986; Dawson, 1992; Hilf and Dawson, 1993; Wang *et al.,* 1997; Zaitlin, 1998).

Additional putative gene (ORF-X) overlaps the 3' end of the MP gene and the 5' end of the CP gene in tobamoviruses of subgroup 1a. The respective short protein of 4-5-kDa is able to associate with eukaryoitic translation elongation factor EF-1α. Interestingly, EF-1α binds plant virus genomic RNAs bearing tRNA-like structures. However, ORF-X protein could have only accessory significance (if any), since the ORF-X-encoding genome region (the 3'-terminal part of the 30K protein gene) is not directly involved in replication and movement functions (Morozov *et al.,* 1993; Duggal *et al.,* 1994; Fedorkin *et al.,* 1995; Fenczik *et al.,* 1995; Goodwin and Dreher, 1998).

Among the *cis*-elements, the 3'-terminal tRNA-like structure of tobamoviruses can be aminoacylated in most cases with histidine (Giege *et al.,* 1993). The sequences affecting tobamovirus RNA replication are mapped to the 5'- and 3'-UTRs including the tRNA-like structure and an upstream pseudoknotted region (Takamatsu *et al.,* 1990; 1991; Ogawa *et al.,* 1992; Duggal *et al.,* 1994). Subgenomic promoters are chsaracterized upstream of the 30K and CP genes (Lehto *et al.,* 1990; Dawson and Lehto, 1990; Duggal *et al.,* 1994). The 5'- and 3'-terminal sequences of the tobamovirus genomes affect the efficiency of RNA translation. The leader sequence of TMV RNA (ω-sequence) enhances translation of the viral genome and foreign genes. Pseudoknotted region upstream of the TMV tRNA-like structure acts synergistically with cap-structure to facilitate genomic RNA translation (Gallie, 1996). Internal ribosome entry site is located upstream of the crucifer tobamovirus CP gene suggesting a possibility of CP expression directly from genomic RNA (Ivanov *et al.,* 1997). In TMV and related tobamoviruses, the origin of virion assembly (Oa) is located in MP gene, while tobamoviruses of another group have the Oa in the CP gene (Okada, 1986; Chng *et al.,* 1996).

The genomes of tobraviruses of genus *Tobravirus* are bipartite and consist of capped and non-polyadenylated RNAs encapsidated separately in rod-like particles. RNA1 is about 6.8 kb and RNA2 ranges from 1.8 to 4.5 kb in different virus isolates (Fig. 10). The 3'-terminal region is conserved between both genomic RNAs and contains tRNA-reminiscent pseudoknotted structure (MacFarlane *et al.,* 1989; Uhde *et al.,* 1998; Sudarshana and Berger, 1998).

RNA1 codes for four proteins of molecular masses 134-kDa, 194-kDa, 29-kDa, and 16-kDa in tobacco rattle virus (TRV, the type member of the genus) (Hamilton *et al.,* 1987). The 134-kDa protein is analogous to the TMV 126K protein and contains domains of methyl/guanylyltransferase and helicase of superfamily 1 at the N- and C-termini, respectively. Translational readthrough of the 134-kDa ORF UGA terminator (resulting in decoding stop codon as Trp or Cys) gives rise to the 194-kDa protein with the domain of RdRp of supergroup 3 at the C-terminus. Tobravirus MP (29-kDa) encoded by ORF3 and expressed from a sgRNA has significant sequence similarity to the tobamoviral MPs. Additional shorter sgRNA synthesized from RNA1 codes for the Cys-rich zinc-finger 16-kDa protein that is distantly related to the hordeivirus Cys-rich γb-protein involved in regulation of virus RNA synthesis. RNA1 is capable of replication and movement in plants in the absence of RNA2 (Hamilton *et al.,* 1987; MacFarlane *et al.,* 1989; Ziegler-Graff *et al.,* 1991; Mushegian and Koonin, 1993; Urban and Beier, 1995; Savenkov *et al.,* 1998; Sudarshana and Berger, 1998).

RNA2 of tobraviruses does not have messenger activity, and the 5'-proximal CP gene is translated from a sgRNA. The corresponding sg promoter is highly conserved in tobraviruses (Goulden *et al.,* 1990; Uhde *et al.,* 1998). The RNA1-specific genes can be located in the long 3'-terminal region of RNA2 because of identity between genomic RNAs. Additionally, RNA2 of several tobravirus strains codes for two or three specific ORFs. In TRV strains capable of vector transmission, an ORF located downstream of the CP gene (for example, ORF3 in pea early browning tobravirus RNA2) codes for moderately conserved 27-37-kDa protein that is required for virus transmission by root-feeding nematodes (Goulden *et al.,* 1990; Hernandez *et al.,* 1997; Schmitt *et al.,* 1998; Uhde *et al.,* 1998; Sudarshana and Berger, 1998).

The regions containing *cis*-elements responsible for the initiatin of RNA replication are mapped in TRV RNA2 (Angenent *et al.,* 1989).

Hordeivirus (genus *Hordeivirus*) genome consist of three RNAs which are encapsidated by a single 22-kDa CP to give rod-shaped particles. The hordeivirus RNAs are designated as α, β and γ and are of 3.7-3.9 kb, 3.1-3.6 kb, and 2.6-3.2 kb, respectively. Each RNA has a cap at the 5'-end and a highly conserved tRNA-like structure at the 3' end. In barley stripe mosaic virus (BSMV, the type hordeivirus), this structure can be charged with tyrosine. In the BSMV and lychnis ringspot hordeivirus genomes, a poly(A) sequence separates the coding region from the tRNA-like structure; however, this sequence is absent from the genome of poa semilatent hordeivirus. The BSMV genome (ND18 strain) encodes eight proteins (Jackson *et al.,* 1991; Agranovsky *et al.,* 1992; Donald *et al.,* 1995; Solovyev *et al.,* 1996; Savenkov *et al.,* 1998).

BSMV RNAα has a single ORF from which the replicase subunit (130 kDa αa protein) is translated. This protein contains domains of methyl/guanylyltransferase and helicase of superfamily 1 (Fig. 10). The region just upstream of the helicase domain in αa protein influences cell-to-cell movement in a host-dependent manner (Gustafson *et al.,* 1989; Gorbalenya *et al.,* 1989a; Koonin and Dolja, 1993; Weiland and Edwards, 1994; Donald *et al.,* 1995; Savenkov *et al.,* 1998).

RNAγ is bicistronic and encodes the 87-kDa γa protein (RdRp of supergroup 3). The Cys-rich 17-kDa γb protein expressed from sgRNA has an RNA-binding activity and affects regulation of virus gene expression and RNA replication. Additionally, the γb protein is a pathogenicity determinant and a seed transmission

TMV (*Tobamovirus*)

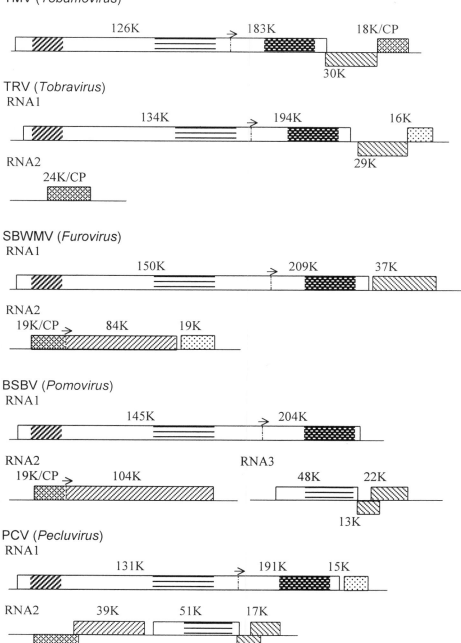

TRV (*Tobravirus*)
RNA1

RNA2
24K/CP

SBWMV (*Furovirus*)
RNA1

RNA2
19K/CP   84K   19K

BSBV (*Pomovirus*)
RNA1

RNA2   RNA3
19K/CP   104K   48K   22K
                                    13K

PCV (*Pecluvirus*)
RNA1

RNA2   39K   51K   17K
23K/CP   14K

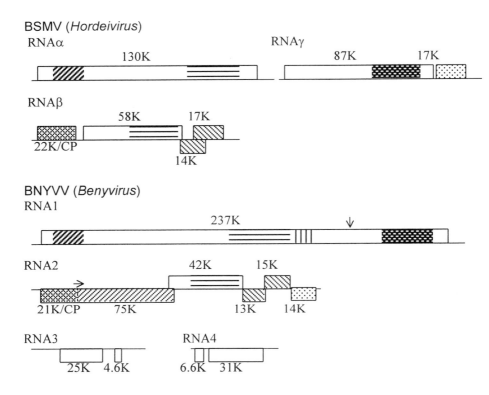

Figure 10. Genome organization in the genera of the proposed family *Tubiviridae*. TMV, tobacco mosaic virus: 6395 nts [J02415]. TRV, tobacco rattle virus: RNA1, 6791 nts [D00155] (strain SYM); RNA2, 1905 nts [X03686] (strain PSG). SBWMV, soil-borne wheat mosaic virus: RNA1, 7099 nts [L07937]; RNA2, 3593 nts [L07938]. BSBV, beet soil-borne virus: RNA1, 5834 nts [Z97873]; RNA2, 3454 nts [U64512]; RNA3, 3006 nts [Z66493]. PCV, peanut clump virus: RNA1, 5897 nts [X78602]; RNA2, 4504 nts [L07269]. BSMV, barley stripe mosaic virus: RNAa, 3783 nts [U35768]; RNAb, 3237 nts [U35772]; RNAg, 3168 nts [U13918]. BNYVV, beet necrotic yellow vein virus: RNA1, 6746 nts [X05147, D00115]; RNA2, 4612 nts [X04197]; RNA3, 1775 nts [M36894]; RNA4, 1370 nts, [M36897]. For shading legend see Figure 1.

factor (Jackson *et al.,* 1991; Donald *et al.,* 1995; Donald and Jackson, 1994; 1996 Savenkov *et al.,* 1998).

RNAβ encodes the CP (βa protein) at the 5'-terminal region and the "triple gen block" (TGB) which encodes three non-structural proteins, 58-kDa βb, 17-kDa β and 14kDa βd; the βd ORF overlaps the other two TGB genes. The βb protein expressed from a larger sgRNA, and the βc and βd proteins are expressed from smaller sgRNA. The βb product contains conserved helicase motifs and binds RN giving rise to non-virion RNP *in planta*. The βb protein binds NTPs and exhil ATPase activity, that is attributed to the helicase motifs 1 and 2. The βc and proteins are small hydrophobic proteins capable of interacting with cell wall membranes. Each of the TGB proteins is required for the viral cell-to-movement. The recent studies on the intra-TGB recombinants between diffc hordeiviruses suggest highly specific functional interactions of the βb protein with the βc and βd proteins (Morozov *et al.,* 1989; Jackson *et al.,* 1991; Donald *et al.,* 1995, 1997; Donald and Jackson, 1996; Zhou and Jackson, 1996a; Solovyev *et al.,* 1996, 1998; Samuilova *et al.,* 1998).

Cis-elements potentially involved in the replication of the hordeivirus genomes and sgRNA synthesis are predicted and studied experimentally (Jackson *et al.,* 1991; Donald *et al.,* 1995; Zhou and Jackson, 1996a, b; Solovyev *et al.,* 1996; Savenkov *et al.,* 1998).

The type virus of the genus *Furovirus,* soil-borne wheat mosaic virus (SBWMV), has bipartite genome. The genomic RNAs are encapsidated separately. SBWMV is transmitted by a plasmodiophoraceous fungus. Virus RNAs are capped and their 3'-ends represent valilated tRNA-like structures which are preceeded by a long pseudoknot domain (Fig. 10). RNA1 codes for the 150-kDa protein which is an analog of the TMV 126-kDA protein, i.e. contains domains of methyl/guanylyl-transferase and helicase of superfamily 1. The 209-kDa protein contains the domain of RdRp of supergroup 3 and is synthesized as a translational read-through of the the 150-kDa protein gene. The 3'-terminal portion of RNA1 encodes the 37-kDa protein expressed from a sgRNA. This protein is similar to the MPs of the TMV 30K MP family showing close relation to the MPs of dianthoviruses (Shirako and Wilson, 1993; Koonin and Dolja, 1993; Mushegian and Koonin, 1993; Torrance and Mayo, 1997; Goodwin and Dreher, 1998). RNA2 has the CP gene which is terminated by a leaky UGA codon and, thus, 84-kDa read-through form of the CP is translated. The 84-kDa protein is believed to be required for fungus transmission: parts of its read-through domain undergo deletions upon prolonged mechanical virus transmission. An additional type of N-terminally extended SBWMV CP can be synthesized after initiation of translation at CUG triplet in the 5'-UTR. The 3'-terminal part of RNA2 contains a gene for the 19-kDa protein that shows a similarity to the Cys-rich proteins of hordeiviruses (Shirako and Wilson, 1993; Chen *et al.,* 1995; Savenkov *et al.,* 1998; Shirako, 1998).

Genomic RNA sequences of genus *Pomovirus* are available for potato mop-top virus (PMTV, the type pomovirus), beet soil-borne virus (BSBV), beet virus Q (BVQ), and broad bean necrosis virus (BBNV). BSBV and BBNV have only minor differences in genome organization. All pomoviruses have tripartite genomes, and the 3'-terminal regions of their genomic RNAs can be folded into tRNA-like structures (valilated in BSBV) which are preceeded by a long pseudoknot domain (Fig. 10). The BSBV RNA1 codes for the 145-kDa protein and its read-through product of 204-kDa. These proteins are the closely related counterparts of the SBWMV 150-kDa and 209-kDa proteins and code for domains of methyl/guanylyl-transferase, helicase of superfamily 1, and RdRp of supergroup 3 (Kashiwazaki *et al.,* 1995; Torrance and Mayo, 1997; Koenig and Loss, 1997; Koenig *et al.,* 1998; Lu *et al.,* 1998). RNA2 of BSBV, BVQ, BBNV and PMTV RNA3 contain the CP gene with a leaky termination codon. The read-through product of BSBV CP gene, the 104-kDa protein, considerably varies in length but is conserved in sequence between different pomoviruses. This protein is found in virus particles where it plays a role of the determinant for virus transmission by a fungal vector. In BVQ, the CP read-through domain is splitted into three ORFs (Kashiwazaki *et al.,* 1995; Torrance and Mayo, 1997; Cowan *et al.,* 1997; Koenig and Loss, 1997; Koenig *et al.*, 1998; Lu *et al.*, 1998). RNA3 of BSBV (RNA2 in PMTV) codes for the TGB proteins of 48-kDa, 13-kDa, and 22-kDa proteins. These pomovirus proteins are thought to be the MPs because of close similarity to the corresponding hordeivirus MPs. In PMTV RNA2, TGB is followed by small gene coding for the 8-kDa Cys-rich protein (Scott *et al.*, 1994; Koenig *et al.*, 1998; Solovyev *et al.*, 1996; Lu *et al.*, 1998).

The genome of peanut clump virus (PCV), the type spieces of genus *Pecluvirus*, is bipartite, and the genomic RNAs have the 3'-terminal tRNA-like structures capable of valylation. RNA1 contains three ORFs. ORF2 is translated by suppression of ORF1 termination codon. ORF1 and ORF2 code for the 131-kDa and 191-kDa proteins, respectively, which have domains of methyl/guanylyltransferase, helicase of superfamily 1, and RdRp of supergroup 3 (Fig. 10). ORF3 codes for the 15-kDa Cys-rich protein which is similar to the hordeivirus Cys-rich proteins (Herzog *et al.*, 1994; Miller *et al.*, 1996; Torrance and Mayo, 1997; Goodwin and Dreher, 1998; Savenkov *et al.*, 1998). RNA2 has five ORFs. ORF1 is the CP gene which is followed by ORF2 coding for the 39-kDa protein of unknown function. The TGB occupies the 3'-terminal part of RNA2. The amino acid sequences of the TGB proteins in PCV are similar to those in pomo- and hordeiviruses (Manohar *et al.*, 1993; Solovyev *et al.*, 1996).

The genome of beet necrotic yellow vein benyvirus (BNYVV) (genus *Benyvirus*) consists of four capped and polyadenylated RNAs, while some Japanese isolates contain a fifth component. RNA1 and RNA2 are necessary and sufficient for infection of some hosts, whereas RNA3 and RNA4 are present in field isolates of BNYVV and influence symptom phenotype and host-dependent vascular movement (Richards and Tamada, 1992; Jupin *et al.*, 1992; Lauber *et al.*, 1998). BNYVV RNA1 carries a single ORF which encodes the 237-kDa protein. Principally, this protein is similar to the replicative readthrough proteins encoded by tobamo-, tobra-, furo-, pomo-, and pecluviruses, i.e. contains domains of methyl/guanylyltransferase, helicase of superfamily 1, and RdRp of supergroup 3. However, unlike other members of *Tubiviridae*, the BNYVV RdRp belongs to "rubi" lineage and not to "tobamo" lineage. Moreover, this 237-kDa polyprotein contains domain of a papain-like protease located just downstream of the helicase domain. As a result, BNYVV polyprotein can be cleaved autocatalytically into the 150-kDa protein containing first two replicative domains and the 66-kDa RdRp protein (Richards and Tamada, 1992; Koonin and Dolja, 1993; Rozanov *et al.*, 1995; Hehn *et al.*, 1997).

RNA2 contains six ORFs. The 5' proximal ORF encodes the 21-kDa CP which is not required for cell-to-cell movement but can influence long-distance transport. The CP gene terminates with leaky amber codon, and its readthrough gives rise to the 75-kDa protein, which is a minor virion component required for efficient virion assembly and for transmission of BNYVV by soil-borne fungus *Polymyxa betae* (Schmitt *et al.*, 1992; Richards and Tamada, 1992; Tamada *et al.*, 1996).

The TGB proteins encoded by ORFs 2, 3, and 4 in RNA2 BNYVV are structurally similar to the respective hordeivirus proteins, although their amino acid sequences are only distantly related. The first gene in the TGB encodes the 42-kDa protein which is expressed from special sgRNA. This protein contains conserved helicase motifs important for cell-to-cell movement, associates with membrane-rich fraction *in vivo*, and efficiently binds RNA *in vitro*. The two smaller TGB proteins of 13-kDa and 15-kDa are expressed concurrently from a smaller sgRNA. These TGB products are small hydrophobic proteins associated with membranes. Highly specific interactions between BNYVV TGB proteins are proposed to be essential for their function in cell-to-cell movement (Morozov *et al.*, 1989; Gilmer *et al.*, 1992a; Richards and Tamada, 1992; Bleykasten *et al.*, 1996; Lauber *et al.*, 1998). Another sgRNA is synthesized from the 3'-most part of BNYVV RNA2 and encodes the small Nys-rich 14-kDa protein, which is most likely a functional counterpart of the hordeivirus γb protein and involved in regulation of RNA and CP synthesis (Gilmer *et al.*, 1992a; Hehn *et al.*, 1995).

Cis-elements involved in viral plus- and minus-strand RNA synthesis were mapped to a set of stem-loop structures in genomic RNAs 3 and 4 (Richards and Tamada, 1992; Gilmer *et al.*, 1992b, 1993; Lauber *et al.*, 1997).

Order "Tymovirales"

This order is delimited by a sequence similarity between replicative protein domains of diverse groups of filamentous and spherical plant viruses (Rozanov *et al.*, 1990; Ding *et al.*, 1990; Morozov *et al.*, 1990; Koonin and Dolja, 1993).

Tymoviruses (genus *Tymovirus*) have isometric virions and monopartite genomes of 6.0-6.7 kb. The genomic RNA is capped, has valilated tRNA-like structure, and codes for the three ORFs (Fig. 11) (Ding *et al.*, 1990; Giege *et al.*, 1993; Gibbs, 1994; Ranjith-Kumar *et al.*, 1998). RNA is translated into the replicative protein of 194-210 kDa and the MP of 48.5-82 kDa. The replicative protein possesses domains of methyl/guanylyltransferase, helicase of superfamily 1 and RdRp of supergroup 3. This protein exhibits ATPase and GTPase activities *in vitro*, and its N- and C-terminal domains participate in RNA binding. Domain of papain-like protease is located upstream of the helicase domain. In the case of turnip yellow mosaic virus (TYMV, the type tymovirus), the 206 kDa polyprotein underdgoes autocatalytic cleavage, yielding products of 141 kDa and 66 kDa. The latter protein contains RdRp domain (Bransom *et al.*, 1996; Rozanov *et al.*, 1992; 1995; Koonin and Dolja, 1993; Gibbs, 1994; Bransom and Dreher, 1994; Dreher and Weiland, 1994; Kadare *et al.*, 1995; Deiman *et al.*, 1997).

Tymovirus MP gene is nested almost completely in the 206-kDa ORF, and the encoded 69-kDa protein is larger than other known MPs and has a very high proline content. However, the TYMV 69-kDa MP is capable of association with membranes that is typical for MPs (Bozarth *et al.*, 1992; Mushegian and Koonin, 1993; Ranjith-Kumar *et al.*, 1998). Tymovirus CP is translated from a sgRNA. The CP is dispensable for RNA replication but required for long-distance spread and probably for cell-to-cell movement (Bransom *et al.*, 1995; Hellendoorn *et al.*, 1997; Hayden *et al.*, 1998).

Important *cis*-element involved in the synthesis of viral minus-strand RNA is a tRNA-like structure; however, its competence for aminoacylation has only minor effect on replication. The secondary and tertiary structures (including specific pseudoknot region) located in the tymoviral 3'-UTR are the most important for recognition by viral replicase (Giege *et al.*, 1993; Goodwin *et al.*, 1997; Singh and

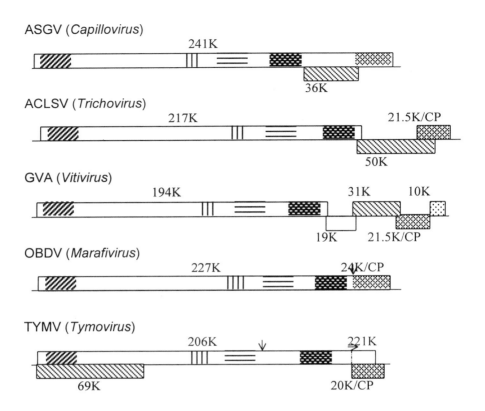

Figure 11. Genome organization in the genera of the proposed order *Tymovirales*. ASGV, apple stem grooving virus: 6496 nts [D14995, 547260]. ACLSV, apple chlorotic leaf spot virus: 7552 nts [D14996]. GVA, grapevine virus A: 7349 nts [X75433]. OBDV, oat blue dwarf virus: 6509 nts [U87832]. TYMV, turnip yellow mosaic virus: 6319 nts [AF035403]. For shading legend see Figure 1.

Dreher, 1997; Deiman *et al.*, 1997; 1998; Ranjith-Kumar *et al.*, 1998; Kolk *et al.*, 1998). Mechanism of the CP sgRNA synthesis has been studied in tymoviruses. The proposed subgenomic promoter includes highly conserved sequence block of 16 nucleotides located approximately ten bases upstream of sgRNA start site (Gargouri *et al.*, 1989; Ding *et al.*, 1990; Gibbs, 1994). The 5'-UTR in genomic RNA of tymoviruses contains conserved hairpins with protonatable internal loops which could have a role in genome RNA encapsidation (Hellendoorn *et al.*, 1997).

The only sequenced member of genus *Marafivirus*, oat blue dwarf virus (OBDV), is a spherical monopartite virus with a capped and polyadenylated genome. The genomic RNA contains a single large ORF coding for the 227-kDa polyprotein (Fig. 11), which has an obvious sequence similarity to the polyprotein of tymoviruses and contains the same set of domains, namely, methyl/guanylyltransferase, papain-like protease, helicase of superfamily 1, and RdRp of supergroup 3. Additionally, the extreme C-terminal part of the 227-kDa protein contains virus CP sequence. The marafivirus CP can be expressed as a result of protease processing of the polyprotein, or alternatively from a specific sgRNA. The putative sgRNA promoter in OBDV is strikingly similar to that in tymoviruses (Edwards *et al.*, 1997).

The genomes of capilloviruses (genus *Capillovirus*) are monopartite. The polyadenylated genomic RNA of 6.5-7.4 kb is encapsidated in filamentous virions. Capilloviruses as well as marafiviruses differ from other alpha-like viruses in coding the CP as a part of the polyprotein which contains also replicative domains (Fig. 11). This polyprotein of 240-260 kDa is encoded by ORF1, and its conserved domains are methyl/guanylyl-transferase, papain-like protease, helicase of superfamily 1, and RdRp of supergroup 3. Their sequences are closly related to those in tymo- and marafiviruses. ORF2 is nested in the 3'-terminal region of ORF1 and codes for a putative MP of the TMV 30K family (Yoshikawa *et al.*, 1993; Rozanov *et al.*, 1995; Jelkmann, 1995; Ohira *et al.*, 1995).

Very flexuous filamentous particles of trichoviruses (genus *Trichovirus*) consist of a single 22-27-kDa CP, and the capped and polyadenylated genomic RNA of 7.5-7.6 kb (Fig. 11). ORF1 is translated from genomic RNA into the 220-kDa polyprotein which contains a set of protein domains typical for tymoviruses: methyl/guanylyltransferase, papain-like protease, helicase of superfamily 1, and RdRp of supergroup 3. ORF2 encodes the MP of the TMV 30K family, which is located to the cell wall and membrane fractions of infected plants. ORF3 codes for the CP. Both 5'-distal genes in trichovirus genomes are most likely expressed from the respective sgRNAs (Sato *et al.*, 1995; Rozanov *et al.*, 1995; German-Retana *et al.*, 1997; Yoshikawa *et al.*, 1997).

The virions of vitiviruses (genus *Vitivirus*) are composed of the 20.5-26-kDa CP and polyadenylated RNA of 7.4-8.2 kb, resembling trichoviruses in particle morphology and genome properties. However, their spread in plants is limited to the phloem. Genomic RNA of the type vitivirus, grapevine virus A, contains five ORFs (Fig. 11), of which ORF1 is analogous to ORF1 of trichoviruses and codes for the 194-kDa protein with three replicative protein domains as well as the domain of papain-like protease. ORF2 encodes 19-20-kDa protein with no sequence similarity to the known viral proteins. ORF3 encodes the MP of the TMV 30K family. The viral CP is encoded by ORF4. ORF5 encodes a Cys-rich protein with some similarity to the small nucleic acid-binding Cys-rich proteins of carlaviruses (Minafra *et al.*, 1994, 1997; Martelli *et al.*, 1997).

The proposed family *"Potexviridae"* includes the viruses of four genera. *Potexvirus, Carlavirus, Foveavirus,* and *Allexvirus*. All viruses of this family are characterized by capped and polyadenylated RNA genomes with similar gene arrangement and considerable homologies in the encoded proteins (Morozov *et al.*, 1989; Rupasov *et al.*, 1989; Kanyuka *et al.*, 1992; Jelkmann, 1994; Cavileer *et al.*, 1994; Solovyev *et al.*, 1994; Wong *et al.*, 1997; Song *et al.*, 1998; Martelli and Jelkmann, 1998; Meng *et al.*, 1998; Zhang *et al.*, 1998).

The genus *Potexvirus* is most abundant and best studied among *Potexviridae*. The potexvirus genomic RNA of 6.0-6.7 kb encodes five ORFs (Fig. 12). ORF1 of potexviruses is translated from virion RNA into 150-170 kDa protein which contains three replicative domains typical for alpha-like viruses, namely, methyl/guanylyltransferase, helicase of superfamily 1, and RdRp of supergroup 3. Mutations in helicase and RdRp domains disable virus replication. This protein does not have protease domain and acts in unprocessed form. Replicase of potato virus X (PVX, type member of potexviruses) is associated with membranes (Morozov *et al.*, 1989; Rozanov *et al.*, 1992; Koonin and Dolja, 1993; Longstaff *et al.*, 1993; Solovyev *et al.*, 1994; Doronin and Hemenway, 1996; Davenport and Baulcombe, 1997).

Expression of the potexvirus ORFs 2-5 requires formation of two or three sgRNAs which can be encapsidated in some viruses (Morozov *et al.*, 1991; Bancroft *et al.*, 1991; Solovyev *et al.*, 1994). ORFs 2, 3, and 4 represent the TGB coding for the virus MPs. First gene in TGB encodes a protein with the conserved set of helicase motifs. This protein has an ATP/GTPase activity and binds RNA *in vitro*. *In planta*, it moves to plasmodesmata in the absence of other viral proteins, participates in increasing of plasmodesmal size exclusion limit and potentiates cell-to-cell transport of virus particles or virus-like RNPs through plasmodesmata. The smaller TGB proteins encoded by ORFs 2 and 3 have highly hydrophobic amino acid segments resembling membrane-embedded domains. In accordance with their predicted properties, the small TGB proteins associate with membranes *in vitro* and *in vivo*. In the infected plants, they are necessary for the viral RNP movement through plasmodesmata (Morozov *et al.*, 1989, 1997; Beck *et al.*, 1991; Kalinina *et al.*, 1996; Angell *et al.*, 1996; Hefferon *et al.*, 1997; Lough *et al.*, 1998). ORF5 codes for the CP. Like the CPs of other members of *Potexviridae*, it contains highly variable N-terminal segment which is not required for virion assembly. The CP in potexviruses is essential for the TGB-potentiated cell-to-cell and long-distance movement, since the CP forms virions or virion-like RNPs which can pass through plasmodesmata (for references, see Chapman *et al.*, 1992a; Forster *et al.*, 1992; Santa Cruz *et al.*, 1998; Lough *et al.*, 1998).

Apple stem pitting foveavirus (ASPV) and two other related viruses encode five ORFs which are anologous to those in potexviruses. However, the ORF1 protein, the CP, and total genome length are much larger in these viruses than in potexviruses (Fig. 12). Moreover, the ASPV ORF1 protein contains domain of papain-like protease upstream of helicase domain and could undergo a proteolytic processing (Jelkmann, 1994; Rozanov *et al.*, 1995; Martelli and Jelkmann, 1998; Meng *et al.*, 1998; Zhang *et al.*, 1998).

Carlaviruses encode six ORFs (Fig. 12): ORFs 1 to 5 correspond to those of potexviruses. As in *Foveavirus*, papain-like protease domain is located upstream of the helicase domain in the ORF1-encoded product of carlaviruses, and the polyprotein undergoes processing. ORF6 in carlaviruses codes for small Cys-rich nucleic acid-binding protein which is expressed by two alternative mechanisms, by translational frameshifting giving rise to the fusion protein with the CP and by internal initiation (Rupasov *et al.*, 1989; Kanyuka *et al.*, 1992; Rohde *et al.*, 1994; Jelkmann, 1994; Cavileer *et al.*, 1994; Lawrence *et al.*, 1995).

Genomes of allexviruses differ from carlavirus genomes in two features: first, they encode additional gene between the TGB and CP ORFs, which codes for 32-42 kDa protein showing no similarity to the known proteins; second, the third ORF in the allexvirus TGB has no AUG initiation codon and is likely translated from non-AUG initiating codon (Kanyuka *et al.*, 1992; Arshava *et al.*, 1995; Song *et al.*, 1998).

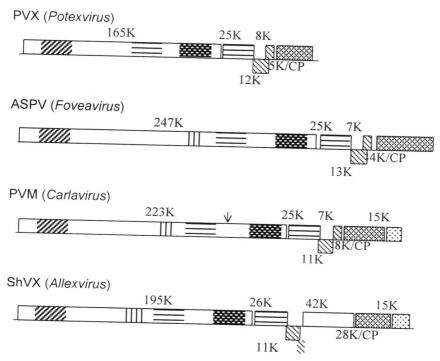

Figure 12. Genome organization in viruses of proposed family *Potexviridae*, order *Tymovirales*. PVX, potato virus X: 6435 nts [X05198, M37458]. ASPV, apple stem pitting virus: 9306 nts [D21829]. PVM, potato virus M: 8535 nts [X53062]. ShVX, shallot virus X: 8832 nts [M97264]. For shading legend see Figure 1.

Cis-elements responsible for sgRNA synthesis are mapped in the PVX genome. A highly conserved nucleotide sequence motif upstream of the TGB and CP gene in potex- and carlaviruses represents a core element of sgRNA promoter. The complementary version of this motif in the 3'-UTRs of potex- and carlavirus RNAs plays a crucial role in initiation of minus-strand RNA synthesis (Solovyev *et al.*, 1994; Sriskanda *et al.*, 1996; Kim and Hemenway, 1997).

The 5'-UTR in potexviruses contains the signals influencing both the virion assembly and the synthesis of the genomic RNA and sgRNAs (Sit *et al.*, 1994; Kim and Hemenway, 1996). Additionally, 5'-UTR of the PVX RNA represents a strong translational enhancer capable of increasing the translation of foreign genes (Tomashevskaya *et al.*, 1993).

GENOME MAPS OF PLANT NEGATIVE-STRANDED RNA VIRUSES

Plant negative-stranded RNA viruses belong to the families *Rhabdoviridae* and *Bunyaviridae*, and the genus *Tenuivirus* closely related to *Bunyaviridae* (Tordo *et al.*, 1992; Ramirez and Haenni, 1994).

Family *Rhabdoviridae*

There are two genera of plant viruses in this family, divided according to the subcellular pattern of their replication, *Cytorhabdovirus* and *Nucleorhabdovirus*. Their negative-stranded RNA genomes, which range from 11 to 14 kb, code for six to seven proteins (Wetzel *et al.,* 1994; Jackson *et al.,* 1998; Fang *et al.,* 1998). The best studied plant rhabdovirus is sonchus yellow net nucleorhabdovirus (SYNV). Transcription of SYNV genomic RNA represents the polar synthesis of monocistronic capped and polyadenylated mRNAs. The RNA polymerase complex slips over the genomic template starting from the 3' terminus between one stop and the following start, eluding the intergenic area, thus, giving rise to the sequential and polar transcription. Proteins required in large amounts are encoded at the 3' end of genomic RNA. Referring to the SYNV genomic map from 3' to 5' terminus, the rhabdovirus proteins have the following functions (Fig. 13)

SYNV (*Nucleorhabdovirus*)

Figure 13. Genome organization in the genus *Nucleorhabdovirus* of the family *Rhabdoviridae*. SYNV, sonchus yellow net virus: 13720 nts [L32603]. Virion strand is shown. For shading legend see Figure 1.

The N protein (nucleocapsid protein) encapsidates genomic minus-strand RNA and is also capable of forming complex with the P protein (phosphoprotein) which is a component of viral nuclear-associated replicase complex. The ORF next to P gene encodes sc4 protein which has no analogs among the virus-specific proteins of animal rhabdoviruses. Its function is not clear, however, sc4 protein is known to be a virion component associated with SYNV envelope. M protein is the matrix protein which, as in animal rhabdoviruses, could be involved in membrane-protein interactions forming the enveloped virions. G protein is a membrane-bound protein which forms the glycoprotein spikes of the rhabdovirus virions. L protein is present in low abundance within the virion, where it specifically associates with P protein. L protein should contain sequence domains related to its replicative functions, namely, RdRp, nucleic acid binding, poly(A) polymerase, and methyl/guanylyltransferase domains (Wetzel *et al.,* 1994; Fang *et al.,* 1994; Wagner and Jackson, 1997; Jackson *et al.,* 1998).

Family *Bunyaviridae*

All plant viruses of the family *Bunyaviridae* belong to the genus *Tospovirus*. The genome of tomato spotted wilt virus (TSWV, the type tospovirus) consists of three RNA segments (L RNA, M RNA, and S RNA) coding for six virus-specific proteins (Fig. 14). Messenger RNAs of bunyaviruses are not polyadenylated and their capping involves the cap-snatching mechanism. The L RNA has single ORF coding for the 331.5-kDa RdRp, and a specific mRNA for its translation is transcribed from the genomic RNA. Tospovirus RdRp is a virion component.

M RNA of tospoviruses has the ambisense coding strategy. This genome component contains an ORF for 33.6-kDa NS_m protein at the 5' terminus, which is the tospoviral MP capable of forming tubular structures protruding from the cell surface and probably also appear in plasmodesmata. Another ORF in M RNA encodes a precursor of two envelope glycoproteins, G1 and G2. S RNA has also an ambisense coding strategy. It codes for the non-structural protein NS_s of unknown function and the nucleocapsid protein N.

TSWV (*Tospovirus*)

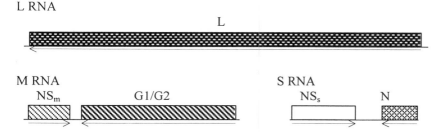

Figure 14. Genome organization in the genus *Tospovirus* of the family *Bunyaviridae*. TSWV, tomato spotted wilt virus: L RNA, 8897 nts [D10066, D01230]; M RNA, 4756 nts [AB010996]; S RNA, 2837 nts [D13926, D00821]. Arrows indicate directions from N- to C-termini of the encoded genes. Virion strands are shown. For shading legend see Figure 1.

The tospovirus genomic RNAs have the conserved terminal sequences at the 5' and 3' termini which can potentially form stable base-paired panhandle structures. Ambisense genomic RNAs (RNA S and RNA M) have an intergenic region with A-U rich inverted repeat participating in termination of mRNA transcription (Adkins *et al.*, 1995; Storms *et al.*, 1995; van Poelwijk *et al.*, 1997; Qiu *et al.*, 1998; Chu and Yeh, 1998; Satyanarayana *et al.*, 1998; Bandla *et al.*, 1998).

Genus *Tenuivirus*

The tenuiviruses represent an unusual type of negative-stranded RNA viruses, since they cannot form enveloped particles and their virions (actually nucleocapsids) have flexous thread-like morphology. Viruses of this genus are distinguished also by their unique genome composition, consisting of 4-6 RNA segments. The 5' and 3'-terminal sequences of about 20 bases are conserved among all RNA segments and are complementary to each other. Viral RNA is therefore able to form an intramolecular secondary structure of the panhandle shape. The virus particles contain the 35-kDa major CP and a minor protein which represents viral RNA polymerase.

Genomic RNA1 of rice stripe virus (RSV, the type tenuivirus) encodes RNA replicase of which RdRp domain is clearly homologous to the L protein of animal phleboviruses (Fig. 15). RNA2 of RSV has an ambisense coding strategy: it codes for the 94-kDa protein with weak similarity to phlebovirus G1/G2 precursor and the protein of 22.8-kDa. RNA3 of RSV is ambisense and codes for the 35-kDa nucleocapsid protein and non-structural protein of 24-kDa. The RNA4 of RSV also uses ambisense coding strategy. It codes for major non-structural protein of 20.5-kDa that accumulates in infected plants to high levels and the 32.5-kDa non-structural protein.

The genome of rice grassy stunt virus contains six genomic RNAs. All six components have an ambisense coding strategy, and the proteins encoded by RNAs 1, 2, 5, and 6 are only distantly related to the RSV proteins, whereas RNAs 3 and 4 are unique (Huiet *et al.*, 1993; Ramirez and Haenni, 1994; Toriyama *et al.*, 1994, 1997, 1998; Estabrook *et al.*, 1996).

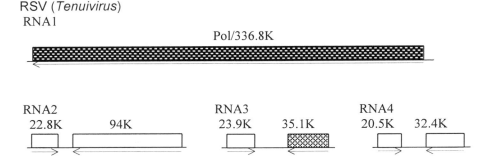

Figure 15. Genome organization in genus *Tenuivirus*. RSV, rice stripe virus: RNA1, 8970 nts [L32603]; RNA2, 3514 nts [D13176]; RNA3, 2504 nts [X53563]; RNA4, 2157 nts [D10979, D01164]. Arrows indicate directions from N- to C-termini of the encoded genes. Virion strands are shown. For shading legend see Figure 1.

GENOME MAPS OF PLANT DOUBLE-STRANDED RNA VIRUSES

Family *Reoviridae*

The plant reoviruses are divided into three genera, *Phytoreovirus*, *Fijivirus*, and *Oryzavirus* according to particle morphology, the number and size of their genomic segments, and vector specificity. The genus *Phytoreovirus* has a genome consisting of 12 dsRNA segments (Table 1)enclosed in virions of 65-70 nm diameter and composed of six proteins. Rice dwarf virus (RDV) is the best studied phytoreovirus. Genomic segment S1 codes for P1 protein having consensus motifs of RdRp in the C-terminal half. RdRp is a minor core component. RDV segment S2 encodes the P2 protein of 127-kDa. This protein is a minor outer capsid protein, is essential for infection of vector cells, and influences transmissibility by vector insects. The P3 protein encoded by segment S3 is a major core protein. Segment S4 codes for the non-structural protein P4 containing nucleic-acid binding zinc-finger motif. Protein P5 encoded by segment S5 is a core component which is assumed to be the mRNA guanylyltransferase. Segment S6 codes for additional non-sructural protein P6, whereas proteins P7 and P8 are components of core and outer capsid, respectively. P7 protein has the ability to bind multiple structural proteins including RdRp (P1), P3, and P8, as well as genomic RNAs during virus assembly. Proteins P9, P10, P11, and P12 are non-structural proteins and their functions are unknown, although P11 is found to be a single-and double-stranded nucleic acid binding protein (Shikata, 1989; Suzuki *et al.*, 1992, 1996; Maruyama *et al.*, 1997; Xu *et al.*, 1998).

Members of the genera *Fijivirus* and *Oryzavirus* contain 10 dsRNA segments. They are transmitted by planthoppers in a persistent manner. Characteristic feature of the fijivirus genomes is the coding of two non-overlapping ORFs in some genome segments (Anzola *et al.*, 1989; Shikata, 1989; Uyeda *et al.*, 1995; Li *et al.*, 1996; Isogai *et al.*, 1998).

Table 1. Genome organization of rice dwarf virus, the genus *Phytoreovirus*

Genome segment	Length, nts	Protein encoded	Accession number
S1	4423	RdRp	D90198
S2	3512	Minor outer capsid protein	U73202
S3	3195	Major core protein	X54620, X17203
S4	2468	NS, nucleic acid binding	X54622, X51432
S5	2570	NS, guanyltransferase	D90033, X16017
S6	1699	NS	M91653, M31298
S7	1696	Minor core protein	D10218
S8	1427	Outer capsid protein	D10219
S9	1305	NS	D10220
S10	1321	NS	D10221
S11	1067	NS, nucleic acid binding	U36568
S12	1066	NS	D90200

NS, nonstructural protein.

Family *Partitiviridae*

It includes two genera of plant viruses, *Alphacryptovirus* and *Betacryptovirus*. The viruses of these genera contain isometric virions of 30-38 nm in diameter composed from coat protein of 50-60 kDa (one or two species) and 2 or 3 molecules of double-stranded RNA where the 3' end of plus strand is polyadenylated. Among sequenced plant cryptovirus genomic RNAs, the genome of beet 3 alphacryptovirus consists of two parts: RNA 1 (1740 bp) codes for the CP and RNA 2 (1600 bp) codes for RdRp. Two monogenic dsRNA components (2238 bp and 2072 bp) are also contained in white clover 2 betacryptovirus (Boccardo *et al.*, 1987; Koonin, 1992; Xie *et al.*, 1993).

CONCLUSIONS

This review illustrates that the genome organization of plant RNA-containing viruses is highly diversified but this does not mean a chaos of all possible combinations of distinct genome elements. Apparently, there are specific rules and restrictions which influence features of newly emerging virus genomes. The majority of these rules is still obscure, although one can speculate about the evolutionary scenario for the putative events that resulted in the present virus genome diversity (Koonin and Dolja, 1993). Nevertheless, some features and tendencies in gene arrangement could be defined. In multipartite negative-stranded RNA genomes, the obvious unique tendency is using the ambisense strategy. On the other hand, despite the extensive gene shuffling in evolution, there are some limitations in the combinations of certain genes and *cis*-elements, which are evident in plant positive-stranded RNA viruses. For example, the 3'-terminal tRNA-like structures capable of aminoacylation can be found only in the genomes of alpha-like plant viruses coding for the RdRp of supergroup 3; the triple block of movement genes is also revealed in the genomes of alpha-like viruses only; the double block of small movement genes is a feature of the *Tombusviridae* genomes coding for the RdRp of supergroup 2. Restictions, which constrain gene shuffling and influence the virus genome structure, are not fully understood, and further analysis of the genome maps in new virus groups and functions of the encoded polypeptides will help in understanding the basic principles of the RNA virus genome design.

ACKNOWLEDGEMENTS

We are grateful to A. Jackson and M. Zaitlin for kindly providing us reviews in advance of publication. We thank A. Agranovsky and N. Vassetzky for helpful comments on the manuscript.

REFERENCES

Adkins, S., Quadt, R., Choi, T. J., Ahlquist, P., and German, T. (1995). An RNA-dependent RNA polymerase activity associated with virions of tomato spotted wilt virus, a plant- and insect-infecting bunyavirus. *Virology* 207, 308-311.

Agranovsky, A. A. (1996). Principles of molecular organization, expression, and evolution of closteroviruses: Over the barriers. *Adv. Virus Res.* 47, 119-158.

Agranovsky, A. A., Karasev, A. V., Novikov, V. K., Lunina, N. A., Loginov, S., and Tyulkina, L. G. (1992). Poa semilatent virus, a hordeivirus having no internal polydisperse poly(A) in the 3' non-coding region of the RNA genome. *J. Gen. Virol.* 73, 2085-2092.

Agranovsky, A. A., Koonin, E. V., Boyko, V. P., Maiss, E., Frotschl, R., Lunina, N. A., and Atabekov, J. G. (1994). Beet yellows closterovirus: Complete genome structure and identification of a leader papain-like thiol protease. *Virology* 198, 311-324.

Agranovsky, A. A., Lesemann, D. E., Maiss, E., Hull, R., and Atabekov, J. G. (1995). "Rattlesnake" structure of a filamentous plant RNA virus built of two capsid proteins. *Proc. Natl. Acad. Sci. USA* 92, 2470-2473.

Agranovsky, A. A., Folimonova, S. Y., Folimonov, A. S., Denisenko, O. N., and Zinovkin, R. A. (1997). The beet yellows closterovirus p65 homologue of HSP70 chaperones has ATPase activity associated with its conserved N-terminal domain but does not interact with unfolded protein chains. *J. Gen. Virol.* 78, 535-542.

Agranovsky, A. A., Folimonov, A. S., Folimonova, S., Morozov, S., Schiemann, J., Lesemann, D., and Atabekov, J. G. (1998). Beet yellows closterovirus HSP70-like protein mediates the cell-to-cell movement of a potexvirus transport-deficient mutant and a hordeivirus- based chimeric virus. *J. Gen.*

Virol. 79, 889-895.

Ahlquist, P. (1992). Bromovirus RNA replication and transcription. *Cur. Opin. Gen. Dev.* 2, 71-76.

Angell, S. M., Davies, C., and Baulcombe, D. C. (1996). Cell-to-cell movement of potato virus X is associated with a change in the size-exclusion limit of plasmodesmata in trichome cells of *Nicotiana clevelandii. Virology* 216, 197-201.

Angenent, G. C., Posthumus, E., and Bol, J. F. (1989). Biological activity of transcripts synthesized in vitro from full-length and mutated DNA copies of tobacco rattle virus RNA 2. *Virology* 173, 68-76.

Anzola, J. V., Dall, D. J., Xu, Z. K., and Nuss, D. L. (1989). Complete nucleotide sequence of wound tumor virus genomic segments encoding nonstructural polypeptides. *Virology* 171, 222-228.

Argos, P. (1988). A sequence motif in many polymerases. *Nucleic Acids Res.* 16, 9909-9916.

Arshava, N. V., Konareva, T. N., Riabov, E. V., and Zavriev, S. K. (1995). 42K protein of shallot X virus is expressed in infected Allium species plants. *Molekuliarnaia Biologiia* 29, 192-198 [in russian].

Ashoub, A., Rohde, W., and Prufer, D. (1998). In planta transcription of a second subgenomic RNA increases the complexity of the subgroup 2 luteovirus genome. *Nucleic Acids Res.* 26, 420-426.

Atabekov, J. G., and Morozov, S. Yu. (1979). Translation of plant virus messenger RNAs. *Adv. Virus Res.* 25, 1-91.

Atreya, P. L., Lopez-Moya, J. J., Chu, M., Atreya, C. D., and Pirone, T. P. (1995). Mutational analysis of the coat protein N-terminal amino acids involved in potyvirus transmission by aphids. *J. Gen. Virol.* 76, 265-270.

Badge, J., Robinson, D. J., Brunt, A. A., and Foster, G. D. (1997). 3'-Terminal sequences of the RNA genomes of narcissus latent and Maclura mosaic viruses suggest that they represent a new genus of the *Potyviridae. J. Gen. Virol.* 78, 253-257.

Bancroft, J. B., Rouleau, M., Johnston, R., Prins, L., and Mackie, G. A. (1991). The entire nucleotide sequence of foxtail mosaic virus RNA. *J. Gen. Virol.* 72, 2173-2181.

Bandla, M. D., Campbell, L. R., Ullman, D. E., and Sherwood, J. L. (1998). Interaction of tomato spotted wilt tospovirus (TSWV) glycoproteins with a thrips midgut protein, a potential cellular receptor for TSWV. *Phytopathology* 88, 98-104.

Blanc, S., Lopez-Moya, J. J., Wang, R., Garcia-Lampasona, S., Thornbury, D. W., and Pirone, T. P. (1997). A specific interaction between coat protein and helper component correlates with aphid transmission of a potyvirus. *Virology* 231, 141-147.

Bleykasten, C., Gilmer, D., Guilley, H., Richards, K. E., and Jonard, G. (1996). Beet necrotic yellow vein virus 42 kDa triple gene block protein binds nucleic acid in vitro. *J. Gen. Virol.* 77, 889-897.

Boccard, F., and Baulcombe, D. (1993). Mutational analysis of cis-acting sequences and gene function in RNA3 of cucumber mosaic virus. *Virology* 193, 563-578.

Boccardo, G., Lisa, V., Luisoni, E., and Milne, R. G. (1987). Cryptic plant viruses. *Adv. Virus Res.* 32, 171-214.

Bonneau, C., Brugidou, C., Chen, L., Beachy, R. N., and Fauquet, C. (1998). Expression of the rice yellow mottle virus P1 protein in vitro and in vivo and its involvement in virus spread. *Virology* 244, 79-86.

Boonham, N., Henry, C. M., and Wood, K. R. (1995). The nucleotide sequence and proposed genome organization of oat chlorotic stunt virus, a new soil-borne virus of cereals. *J. Gen. Virol.* 76, 2025-2034.

Boyko, V. P., Karasev, A. V., Agranovsky, A. A., Koonin, E. V., and Dolja, V. V. (1992). Coat protein gene duplication in a filamentous RNA virus of plants. *Proc. Natl. Acad. Sci. USA* 89, 9156-9160.

Bozarth, C. S., Weiland, J. J., and Dreher, T. W. (1992). Expression of ORF-69 of turnip yellow mosaic virus is necessary for viral spread in plants. *Virology* 187, 124-130.

Bransom, K. L., and Dreher, T. W. (1994). Identification of the essential cysteine and histidine residues of the turnip yellow mosaic virus protease. *Virology* 198, 148-154.

Bransom, K. L., Weiland, J. J., Tsai, C. H., and Dreher, T. W. (1995). Coding density of the turnip yellow mosaic virus genome: roles of the overlapping coat protein and p206-readthrough coding regions. *Virology* 206, 403-412.

Bransom, K. L., Wallace, S. E., and Dreher, T. W. (1996). Identification of the cleavage site recognized by the turnip yellow mosaic virus protease. *Virology* 217, 404-406.

Brault, V., and Miller, W. A. (1992). Translational frameshifting mediated by a viral sequence in plant cells. *Proc. Natl. Acad. Sci. USA* 89, 2262-2266.

Bruyere, A., Brault, V., Ziegler-Graff, V., Simonis, M. T., Van den Heuvel, J. F., Richards, K., Guilley, H., Jonard, G., and Herrbach, E. (1997). Effects of mutations in the beet western yellows virus readthrough protein on its expression and packaging and on virus accumulation, symptoms, and aphid transmission. *Virology* 230, 323-234.

Buck, K. W. (1996). Comparison of the replication of positive-stranded RNA viruses of plants and animals. _Adv. Virus Res._ 47, 159-251.

Burgyan, J., Rubino, L., and Russo, M. (1996). The 5'-terminal region of a tombusvirus genome determines the origin of multivesicular bodies. _J. Gen. Virol._ 77, 1967-1974.

Campbell, R. N. (1996). Fungal transmission of plant viruses. _Annu. Rev. Phytopathol._ 34, 87-108.

Canto, T., Prior, D. A. M., Hellwald, K.-H., Oparka, K. J., and Palukaitis, P. (1997) Characterization of cucumber mosaic virus. IV. Movement protein and coat protein are both essential for cell-to-cell movement of cucumber mosaic virus. _Virology_ 237, 237-248.

Carrington, J. C., Kasschau, K. D., Mahajan, S. K., and Schaad, M. C. (1996). Cell-to-cell and long-distance transport of viruses in plants. _Plant Cell_ 8, 1669-1681.

Cavileer, T. D., Halpern, B. T., Lawrence, D. M., Podleckis, E. V., Martin, R. R., and Hillman, B. I. (1994). Nucleotide sequence of the carlavirus associated with blueberry scorch and similar diseases. _J. Gen. Virol._ 75, 711-720.

Chang, Y. C., Borja, M., Scholthof, H. B., Jackson, A. O., and Morris, T. J. (1995). Host effects and sequences essential for accumulation of defective interfering RNAs of cucumber necrosis and tomato bushy stunt tombusviruses. _Virology_ 210, 41-53.

Chay, C. A., Gunasinge, U. B., Dinesh-Kumar, S. P., Miller, W. A., and Gray, S. M. (1996). Aphid transmission and systemic plant infection determinants of barley yellow dwarf luteovirus-PAV are contained in the coat protein readthrough domain and 17-kDa protein, respectively. _Virology_ 219, 57-65.

Chen, J., MacFarlane, S. A., and Wilson, T. M. (1995). An analysis of spontaneous deletion sites in soil-borne wheat mosaic virus RNA2. _Virology_ 209, 213-217.

Cheng, S. L., Domier, L. L., and D'Arcy, C. J. (1994). Detection of the readthrough protein of barley yellow dwarf virus. _Virology_ 202, 1003-1006.

Chin, L. S., Foster, J. L., and Falk, B. W. (1993). The beet western yellows virus ST9-associated RNA shares structural and nucleotide sequence homology with carmo-like viruses. _Virology_ 192, 473-482.

Chng, C. G., Wong, S. M., Mahtani, P. H., Loh, C. S., Goh, C. J., Kao, M. C., Chung, M. C., and Watanabe, Y. (1996). The complete sequence of a Singapore isolate of odontoglossum ringspot virus and comparison with other tobamoviruses. _Gene_ 171, 155-161.

Chu, F.-H., and Yeh, S.-D. (1998). Comparison of ambisense mRNA of watermelon silver mottle virus with other tospoviruses. _Phytopathology_ 88, 351-358.

Ciuffreda, P., Rubino, L., and Russo, M. (1998). Molecular cloning and complete nucleotide sequence of galinsoga mosaic virus genomic RNA. _Arch. Virol._ 143, 173-180.

Colinet, D., Kummert, J., and Lepoivre, P. (1998). The nucleotide sequence and genome organization of the whitefly transmitted sweetpotato mild mottle virus: a close relationship with members of the family Potyviridae. _Virus Res._ 53, 187-196.

Cowan, G. H., Torrance, L., and Reavy, B. (1997). Detection of potato mop-top virus capsid readthrough protein in virus particles. _J. Gen. Virol._ 78, 1779-1783.

Dalmay, T., Rubino, L., Burgyan, J., Kollar, A., and Russo, M. (1993). Functional analysis of cymbidium ringspot virus genome. _Virology_ 194, 697-704.

Davenport, G. F., and Baulcombe, D. C. (1997). Mutation of the GKS motif of the RNA-dependent RNA polymerase from potato virus X disables or eliminates virus replication. _J. Gen. Virol._ 78, 1247-1251.

Dawson, W. O., and Lehto, K. M. (1990). Regulation of tobamovirus gene expression. _Adv. Virus Res._ 38, 307-342.

Dawson, W. O. (1992). Tobamovirus-plant interactions. _Virology_ 186, 359-367.

De Graaff, M., Man in't Veld, M. R., and Jaspars, E. M. (1995). _In vitro_ evidence that the coat protein of alfalfa mosaic virus plays a direct role in the regulation of plus and minus RNA synthesis: Implications for the life cycle of alfalfa mosaic virus. _Virology_ 208, 583-589.

De Jong, W., Chu, A., and Ahlquist, P. (1995). Coding changes in the 3a cell-to-cell movement gene can extend the host range of brome mosaic virus systemic infection. _Virology_ 214, 464-474.

Deiman, B. A., Kortlever, R. M., and Pleij, C. W. (1997). The role of the pseudoknot at the 3' end of turnip yellow mosaic virus RNA in minus-strand synthesis by the viral RNA-dependent RNA polymerase. _J. Virol._ 71, 5990-5996.

Deiman, B. A., Koenen, A. K., Verlaan, P. W., and Pleij, C. W. (1998). Minimal template requirements for initiation of minus-strand synthesis in vitro by the RNA-dependent RNA polymerase of turnip yellow mosaic virus. _J. Virol._ 72, 3965-3972.

Delarue, M., Poch, O., Tordo, N., Moras, D., and Argos, P. (1990). An attempt to unify the structure of polymerases. _Protein Eng._ 3, 461-467.

Demler, S. A., Rucker, D. G., and de Zoeten, G. A. (1993). The chimeric nature of the genome of pea enation mosaic virus: the independent replication of RNA 2. *J. Gen. Virol.* 74, 1-14.

Demler, S. A., Borkhsenious, O. N., Rucker, D. G., and de Zoeten, G. A. (1994). Assessment of the autonomy of replicative and structural functions encoded by the luteo-phase of pea enation mosaic virus. *J. Gen. Virol.* 75, 997-1007.

Demler, S. A., Rucker-Feeney, D. G., Skaf, J. S., and de Zoeten, G. A. (1997). Expression and suppression of circulative aphid transmission in pea enation mosaic virus. *J. Gen. Virol.* 78, 511-523.

Deom, C. M., Quan, S., and He, X. Z. (1997). Replicase proteins as determinants of phloem-dependent long-distance movement of tobamoviruses in tobacco. *Protoplasma* 199, 1-8.

Derrick, P. M., Carter, S. A., and Nelson, R. S. (1997). Mutation of the tobacco mosaic tobamovirus 126- and 183-kDa proteins: effects on phloem-dependent virus accumulation and synthesis of viral proteins. *Mol. Plant-Microbe Interact.* 10, 589-596.

Dessens, J. T., Nguyen, M., and Meyer, M. (1995). Primary structure and sequence analysis of RNA2 of a mechanically transmitted barley mild mosaic virus isolate: an evolutionary relationship between bymo- and furoviruses. *Arch. Virol.* 140, 325-333.

Dinesh-Kumar, S. P., Brault, V., and Miller, W. A. (1992). Precise mapping and in vitro translation of a trifunctional subgenomic RNA of barley yellow dwarf virus. *Virology* 187, 711-722.

Ding, B., Li, Q., Nguyen, L., Palukaitis, P., and Lucas, W. J. (1995). Cucumber mosaic virus 3a protein potentiates cell-to-cell trafficking of CMV RNA in tobacco plants. *Virology* 207, 345-353.

Ding, S., Keese, P., and Gibbs, A. (1990). The nucleotide sequence of the genomic RNA of kennedya yellow mosaic tymovirus - Jervis Bay isolate: relationships with potex- and carlaviruses. *J. Gen. Virol.* 71, 925-931.

Dolja, V. V., Karasev, A. V., and Koonin, E. V. (1994). Molecular biology and evolution of closteroviruses: sophisticated build-up of large RNA genomes. *Annu. Rev. Phytopathol.* 32, 261-285.

Dolja, V. V., Haldeman-Cahill, R., Montgomery, A. E., Vandenbosch, K. A., and Carrington, J. C. (1995). Capsid protein determinants involved in cell-to-cell and long distance movement of tobacco etch potyvirus. *Virology* 206, 1007-1016.

Donald, R. G. K., Petty, I. T. D., Zhou, H., and Jackson, A. O. (1995). Properties of genes influencing barley stripe mosaic virus movement phenotypes. *In* "Fifth International Symposium on Biotechnology and Plant Protection: Viral Pathogenesis and Disease Resistance", pp. 135-147. World Scientific, Singapore.

Donald, R. G., and Jackson, A. O. (1994). The barley stripe mosaic virus γb gene encodes a multifunctional cysteine-rich protein that affects pathogenesis. *Plant Cell* 6, 1593-1606.

Donald, R. G., and Jackson, A. O. (1996). RNA-binding activities of barley stripe mosaic virus γb fusion proteins. *J. Gen. Virol.* 77, 879-888.

Donald, R. G., Lawrence, D. M., and Jackson, A. O. (1997). The barley stripe mosaic virus 58-kilodalton βb protein is a multifunctional RNA binding protein. *J. Virol.* 71, 1538-1546.

Dorokhov Yu. L., Ivanov, P. A., Novikov, V. K., Agranovsky, A. A., Morozov, S., Efimov, V. A., Casper, R., and Atabekov, J. G. (1994). Complete nucleotide sequence and genome organization of a tobamovirus infecting cruciferae plants. *FEBS Lett.* 350, 5-8.

Doronin, S. V., and Hemenway, C. (1996). Synthesis of potato virus X RNAs by membrane-containing extracts. *J. Virol.* 70, 4795-4799.

Dougherty, W. G., and Semler, B. L. (1993). Expression of virus-encoded proteinases: functional and structural similarities with cellular enzymes. *Microbiol. Rev.* 57, 781-822.

Dreher, T. W., and Weiland, J. J. (1994). Preferential replication of defective turnip yellow mosaic virus RNAs that express the 150-kDa protein in cis. *Arch. Virol. Suppl.* 9, 195-204.

Duggal, R., Lahser, F.C., and Hall, T.C. (1994). *cis*-Acting sequences in the replication of plant viruses with plus-sense RNA genomes. *Annu. Rev. Phytopathol.* 32, 287-309.

Edwards, M. C., Zhang, Z., and Weiland, J. J. (1997). Oat blue dwarf marafivirus resembles the tymoviruses in sequence, genome organization, and expression strategy. *Virology* 232, 217-229.

Eggen, R., and van Kammen, A. (1988). RNA replication in comoviruses. In "RNA Genetics" (P. Ahlquist, J. Holland, and E. Domingo, Eds.), Vol. I, pp.49-69. CRC Press, Boca Raton, FL.

Estabrook, E. M., Suyenaga, K., Tsai, J. H., and Falk, B. W. (1996). Maize stripe tenuivirus RNA2 transcripts in plant and insect hosts and analysis of pvc2, a protein similar to the *Phlebovirus* virion membrane glycoproteins. *Virus Genes* 12, 239-47.

Fang, R. X., Wang, Q., Xu, B. Y., Pang, Z., Zhu, H. T., Mang, K. Q., Gao, D. M., Qin, W. S., and Chua, N. H. (1994). Structure of the nucleocapsid protein gene of rice yellow stunt rhabdovirus. *Virology* 204, 367-375.

Fang, R., Luo, Z., and Zhao, H. (1998). Novel structure of the rice yellow stunt virus genome: a plant rhabdovirus encodes seven genes. *DDBJ/EMBL/GenBank databases*. Accession number

[AB011257].

Fedorkin, O. N., Denisenko, O. N., Sitikov, A. S., Zelenina, D. A., Lukasheva, L. I., Morozov, S. I., and Atabekov, I. G. (1995). A protein product of the tobamovirus open translation frame forms a stable complex with translation elongation factor eEF- 1α. *Doklady Akademii Nauk* 343, 703-704 [in Russian].

Fenczik, C. A., Padgett, H. S., Holt, C. A., Casper, S. J., and Beachy, R. N. (1995). Mutational analysis of the movement protein of odontoglossum ringspot virus to identify a host-range determinant. *Mol Plant-Microbe Interact.* 8, 666-673.

Finnen, R. L., and Rochon, D. M. (1993). Sequence and structure of defective interfering RNA associated with cucumber necrosis virus infections. *J. Gen. Virol.* 74, 1715-1720.

Forster, R. L., Beck, D. L., Guilford, P. J., Voot, D. M., Van Dolleweerd, C. J., and Andersen, M. 1 (1992). The coat protein of white clover mosaic potexvirus has a role in facilitating cell-to-cel transport in plants. *Virology* 191, 480-484.

Fowler, B. C., Sivakumaran, K., and Hacker, D. L. (1998). Subcellular localization of southern bea mosaic virus proteins involved in cell-to-cell movement. P.86, *Abstracts of 17th Annu. Meeting (American Society of Virology*, University of British Columbia, Vancouver, Canada, July 1998.

Fuentes, A. L., and Hamilton, R. I. (1993). Failure of long-distance movement of southern bean mosai virus in a resistant host is correlated with lack of normal virion formation. *J. Gen. Virol.* 74, 1903 1910.

Futterer, J., and Hohn, T. (1996). Translation in plants--rules and exceptions. *Plant Mol. Biol.* 32, 159 189.

Gallie, D. R., and Kobayashi, M. (1994). The role of the 3'-untranslated region of non-polyadenylated plant viral mRNAs in regulating translational efficiency. *Gene* 142, 159-165.

Gallie, D. R. (1996). Translational control of cellular and viral mRNAs. *Plant Mol. Biol.* 32, 145-158.

Gargouri, R., Joshi, R. L., Bol, J. F., Astier-Manifacier, S., and Haenni, A. L. (1989). Mechanism of synthesis of turnip yellow mosaic virus coat protein subgenomic RNA *in vivo*. *Virology* 171, 386-393.

Ge, X., Scott, S. W., and Zimmerman, M. T. (1997). The complete sequence of the genomic RNAs of spinach latent virus. *Arch. Virol.* 142, 1213-1226.

German-Retana, S., Bergey, B., Delbos, R. P., Candresse, T., and Dunez, J. (1997). Complete nucleotide sequence of the genome of a severe cherry isolate of apple chlorotic leaf spot trichovirus (ACLSV). *Arch. Virol.* 142, 833-841.

Gibbs, A. (1987). Molecular evolution of viruses; 'trees', 'clocks' and 'modules'. *J. Cell. Sci. Suppl.* 7, 319-337.

Gibbs, A. J. (1994). Tymoviruses. In *Encyclopedia of Virology*, vol.3, pp. 1500-1502. Edited by A. Granoff and R. G. Webster. London: Academic Press.

Gibbs, M. J., Cooper, J. I., and Waterhouse, P. M. (1996). The genome organization and affinities of an Australian isolate of carrot mottle umbravirus. *Virology* 224, 310-313.

Giege, R., Florentz, C., and Dreher, T. W. (1993). The TYMV tRNA-like structure. *Biochimie* 75, 569-582.

Giesman-Cookmeyer, D., and Lommel, S. A. (1993). Alanine scanning mutagenesis of a plant virus movement protein identifies three functional domains. *Plant Cell* 5, 973-982.

Gilmer, D., Bouzoubaa, S., Hehn, A., Guilley, H., Richards, K., and Jonard, G. (1992a). Efficient cell-to-cell movement of beet necrotic yellow vein virus requires 3' proximal genes located on RNA 2. *Virology* 189, 40-47.

Gilmer, D., Richards, K., Jonard, G., and Guilley, H. (1992b). cis-Active sequences near the 5'-termini of beet necrotic yellow vein virus RNAs 3 and 4. *Virology* 190, 55-67.

Gilmer, D., Allmang, C., Ehresmann, C., Guilley, H., Richards, K., Jonard, G., and Ehresmann, B. (1993). The secondary structure of the 5'-noncoding region of beet necrotic yellow vein virus RNA 3: evidence for a role in viral RNA replication. *Nucleic Acids Res.* 21, 1389-1395.

Goldbach, R. (1987). Genome similarities between plant and animal RNA viruses. *Microbiol. Sci.* 4, 197-202.

Goodwin, J. B., Skuzeski, J. M., and Dreher, T. W. (1997). Characterization of chimeric turnip yellow mosaic virus genomes that are infectious in the absence of aminoacylation. *Virology* 230, 113-124.

Goodwin, J. B., and Dreher, T. W. (1998). Transfer RNA mimicry in a new group of positive-strand RNA plant viruses, the furoviruses: differential aminoacylation between the RNA components of one genome. *Virology* 246, 170-178.

Gorbalenya, A. E., Koonin, E. V., Blinov, V. M., and Donchenko, A. P. (1988). Sobemovirus genome appears to encode a serine protease related to cysteine proteases of picornaviruses. *FEBS Lett.* 236,

287-290.

Gorbalenya, A. E., Blinov, V. M., Donchenko, A. P., and Koonin, E. V. (1989a). An NTP-binding motif is the most conserved sequence in a highly diverged monophyletic group of proteins involved in positive strand RNA viral replication. *J. Mol. Evol.* 28, 256-268.

Gorbalenya, A. E., Koonin, E. V., Donchenko, A. P., and Blinov, V. M. (1989b). Two related superfamilies of putative helicases involved in replication, recombination, repair and expression of DNA and RNA genomes. *Nucleic Acids Res.* 17, 4713-4730.

Gorbalenya, A. E., and Koonin, E.V. (1993). Helicases. Amino acid sequence comparisons and beyond. *Curr. Opin. Struct. Biol.* 3, 419-429.

Gotz, R., and Maiss, E. (1995). The complete nucleotide sequence and genome organization of the mite-transmitted brome streak mosaic rymovirus in comparison with those of potyviruses. *J. Gen. Virol.* 76, 2035-2042.

Grieco, F., Dell'Orco, M., and Martelli, G. P. (1996). The nucleotide sequence of RNA1 and RNA2 of olive latent virus 2 and its relationships in the family *Bromoviridae. J. Gen. Virol.* 77, 2637-2644.

Guan, H., Song, C., and Simon, A. E. (1997). RNA promoters located on (-)-strands of a subviral RNA associated with turnip crinkle virus. *RNA* 3, 1401-1412.

Gustafson, G., Armour, S. L., Gamboa, G. C., Burgett, S. G., and Shepherd, J. W. (1989). Nucleotide sequence of barley stripe mosaic virus RNAα: RNAα encodes a single polypeptide with homology to corresponding proteins from other viruses. *Virology* 170, 370-377.

Hacker, D. L., Petty, I. T., Wei, N., and Morris, T. J. (1992). Turnip crinkle virus genes required for RNA replication and virus movement. *Virology* 186, 1-8.

Hacker, D. L., and Sivakumaran, K. (1997). Mapping and expression of southern bean mosaic virus genomic and subgenomic RNAs. *Virology* 234, 317-327.

Haldemann-Cahill, R., Daros, A.-J., and Carrington, J. C. (1998). Secondary structures in the capsid protein coding sequence and 3' nontranslated region involved in amplification of the tobacco etch virus genome. *J. Virol.* 72, 4072-4079.

Hall, T. C., Miller, W. A., and Bujarski, J.J. (1982). Enzymes involved in the replication of plant viral RNAs. *Adv. Plant Pathol.* 1, 179-219.

Hamilton, W. D., Boccara, M., Robinson, D. J., and Baulcombe, D. C. (1987). The complete nucleotide sequence of tobacco rattle virus RNA-1. *J. Gen. Virol.* 68, 2563-2575.

Havelda, Z., and Burgyan, J. (1995). 3' Terminal putative stem-loop structure required for the accumulation of cymbidium ringspot viral RNA. *Virology* 214, 269-272.

Hayden, C. M., Mackenzie, A. M., Skotnicki, M. L., and Gibbs, A. (1998). Turnip yellow mosaic virus isolates with experimentally produced recombinant virion proteins. *J. Gen. Virol.* 79, 395-403.

Hefferon, K. L., Doyle, S., and AbouHaidar, M. G. (1997). Immunological detection of the 8K protein of potato virus X (PVX) in cell walls of PVX-infected tobacco and transgenic potato. *Arch. Virol.* 142, 425-433.

Hehn, A., Bouzoubaa, S., Bate, N., Twell, D., Marbach, J., Richards, K., Guilley, H., and Jonard, G. (1995). The small cysteine-rich protein P14 of beet necrotic yellow vein virus regulates accumulation of RNA 2 *in cis* and coat protein *in trans. Virology* 210, 73-81.

Hehn, A., Fritsch, C., Richards, K. E., Guilley, H., and Jonard, G. (1997). Evidence for *in vitro* and *in vivo* autocatalytic processing of the primary translation product of beet necrotic yellow vein virus RNA 1 by a papain-like proteinase. *Arch. Virol.* 142, 1051-1058.

Heinlein, M., Padgett, H. S., Gens, J. S., Pickard, B. G., Casper, S. J., Epel, B. L., and Beachy, R. N. (1998). Changing patterns of localization of the tobacco mosaic virus movement protein and replicase to the endoplasmic reticulum and microtubules during Infection. *Plant Cell* 10, 1107-1120.

Hellendoorn, K., Verlaan, P. W., and Pleij, C. W. (1997). A functional role for the conserved protonatable hairpins in the 5' untranslated region of turnip yellow mosaic virus RNA. *J. Virol.* 71, 8774-8779.

Hernandez, C., Visser, P. B., Brown, D. J., and Bol, J. F. (1997). Transmission of tobacco rattle virus isolate PpK20 by its nematode vector requires one of the two non-structural genes in the viral RNA 2. *J. Gen. Virol.* 78, 465-467.

Herzog, E., Guilley, H., Manohar, S. K., Dollet, M., Richards, K., Fritsch, C., and Jonard, G. (1994). Complete nucleotide sequence of peanut clump virus RNA 1 and relationships with other fungus-transmitted rod-shaped viruses. *J. Gen. Virol.* 75, 3147-3155.

Hilf, M. E., and Dawson, W. O. (1993). The tobamovirus capsid protein functions as a host-specific determinant of long-distance movement. *Virology* 193, 106-114.

Hong, Y., and Hunt, A. G. (1996). RNA polymerase activity catalyzed by a potyvirus-encoded RNA-dependent RNA polymerase. *Virology* 226, 146-151.

Hull, R. (1996). Molecular biology of rice tungro viruses. *Annu. Rev. Phytopathol.* 34, 275-297.

Ishikawa, M., Janda, M., Krol, M. A., and Ahlquist, P. (1997). In vivo DNA expression of functional brome mosaic virus RNA replicons in *Saccharomyces cerevisiae. J. Virol.* 71, 7781-7790.

Isogai, M., Uyeda, I., and Lindsten, K. (1998). Taxonomic characteristics of fijiviruses based on nucleotide sequences of the oat sterile dwarf virus genome. *J. Gen. Virol.* 79, 1479-1485.

Itaya, A., Hickman, H., Bao, Y., Nelson, R., and Ding, B. (1997). Cell-to-cell trafficking of cucumber mosaic virus movement protein: green fluorescent protein produced by biolistic gene bombardment in tobacco. *Plant J.* 12, 1223-1230.

Ivanov, P. A., Karpova, O. V., Skulachev, M. V., Tomashevskaya, O. L., Rodionova, N. P., Dorokhov Yu, L., and Atabekov, J. G. (1997). A tobamovirus genome that contains an internal ribosome entry site functional *in vitro. Virology* 232, 32-43.

Jackson, A.O., Petty, I.T.D., Jones, R.W., Edwards, M.C., and French, R. (1991). Analysis of barley stripe mosaic virus pathogenicity. *Semin. Virol.* 2: 107-119

Jackson, A.O., Goodin, M., Moreno, I., Johnson, J., and Lawrence, D.M. (1998). Plant rhabdoviruses. In *Encyclopedia of Virology,* in press.

Jelkmann, W. (1994). Nucleotide sequences of apple stem pitting virus and of the coat protein gene of a similar virus from pear associated with vein yellows disease and their relationship with potex- and carlaviruses. *J. Gen. Virol.* 75, 1535-1542.

Jelkmann, W. (1995). Cherry virus A: cDNA cloning of dsRNA, nucleotide sequence analysis and serology reveal a new plant capillovirus in sweet cherry. *J. Gen. Virol.* 76, 2015-2024.

Jelkmann, W., Fechtner, B., and Agranovsky, A. A. (1997). Complete genome structure and phylogenetic analysis of little cherry virus, a mealybug-transmissible closterovirus. *J. Gen. Virol.* 78, 2067-2071.

Johnston, J. C., and Rochon, D. M. (1995). Deletion analysis of the promoter for the cucumber necrosis virus 0.9-kb subgenomic RNA. *Virology* 214, 100-109.

Johnston, J. C., and Rochon, D. M. (1996). Both codon context and leader length contribute to efficient expression of two overlapping open reading frames of a cucumber necrosis virus bifunctional subgenomic mRNA. *Virology* 221, 232-239.

Jupin, I., Guilley, H., Richards, K. E., and Jonard, G. (1992). Two proteins encoded by beet necrotic yellow vein virus RNA 3 influence symptom phenotype on leaves. *EMBO J.* 11, 479-488.

Kadare, G., Rozanov, M., and Haenni, A. L. (1995). Expression of the turnip yellow mosaic virus proteinase in *Escherichia coli* and determination of the cleavage site within the 206 kDa protein. *J. Gen. Virol.* 76, 2853-2857.

Kadare, G., and Haenni, A. L. (1997). Virus-encoded RNA helicases. *J. Virol.* 71, 2583-2590.

Kalinina, N. O., Fedorkin, O. N., Samuilova, O. V., Maiss, E., Korpela, T., Morozov, S., and Atabekov, J. G. (1996). Expression and biochemical analyses of the recombinant potato virus X 25K movement protein. *FEBS Lett.* 397, 75-78.

Kamer, G., and Argos, P. (1984). Primary structural comparison of RNA-dependent polymerases from plant, animal and bacterial viruses. *Nucleic Acids Res.* 12, 7269-7282.

Kanyuka, K. V., Vishnichenko, V. K., Levay, K. E., Kondrikov, D. Y., Ryabov, E. V., and Zavriev, S. K. (1992). Nucleotide sequence of shallot virus X RNA reveals a 5'-proximal cistron closely related to those of potexviruses and a unique arrangement of the 3'-proximal cistrons. *J. Gen. Virol.* 73, 2553-2560.

Kaplan, I. B., Gal-On, A., and Palukaitis, P. (1997). Characterization of cucumber mosaic virus. III. Localization of sequences in the movement protein controlling systemic infection in cucurbits. *Virology* 230, 343-349.

Kaplan, I. B., Shintaku, M. H., Li, Q., Zhang, L., Marsh, L. E., and Palukaitis, P. (1995). Complementation of virus movement in transgenic tobacco expressing the cucumber mosaic virus 3a gene. *Virology* 209, 188-199.

Karasev, A. V., Kashina, A. S., Gelfand, V. I., and Dolja, V. V. (1992). HSP70-related 65 kDa protein of beet yellows closterovirus is a microtubule-binding protein. *FEBS Lett.* 304, 12-14.

Karasev, A. V., Boyko, V. P., Gowda, S., Nikolaeva, O. V., Hilf, M. E., Koonin, E. V., Niblett, C. L., Cline, K., Gumpf, D. J., Lee, R. F., and al, e. (1995). Complete sequence of the citrus tristeza virus RNA genome. *Virology* 208, 511-520.

Karasev, A. V., Nikolaeva, O. V., Mushegian, A. R., Lee, R. F., and Dawson, W. O. (1996). Organization of the 3'-terminal half of beet yellow stunt virus genome and implications for the evolution of closteroviruses. *Virology* 221, 199-207.

Karasev, A. V., Hilf, M. E., Garnsey, S. M., and Dawson, W. O. (1997). Transcriptional strategy of closteroviruses: mapping the 5' termini of the citrus tristeza virus subgenomic RNAs. *J. Virol.* 71, 6233-6236.

Karpova, O. V., Ivanov, K. I., Rodionova, N. P., Dorokhov Yu, L., and Atabekov, J. G. (1997). Nontranslatability and dissimilar behavior in plants and protoplasts of viral RNA and movement protein complexes formed *in vitro*. *Virology* 230, 11-21.

Kashiwazaki, S., Minobe, Y., Omura, T., and Hibino, H. (1990). Nucleotide sequence of barley yellow mosaic virus RNA 1: a close evolutionary relationship with potyviruses. *J. Gen. Virol.* 71, 2781-2790.

Kashiwazaki, S., Minobe, Y., and Hibino, H. (1991). Nucleotide sequence of barley yellow mosaic virus RNA 2. *J. Gen. Virol.* 72, 995-999.

Kashiwazaki, S., Scott, K. P., Reavy, B., and Harrison, B. D. (1995). Sequence analysis and gene content of potato mop-top virus RNA 3: further evidence of heterogeneity in the genome organization of furoviruses. *Virology* 206, 701-706.

Kasschau, K. D., Cronin, S., and Carrington, J. C. (1997). Genome amplification and long-distance movement functions associated with the central domain of tobacco etch potyvirus helper component-proteinase. *Virology* 228, 251-62.

Kasteel, D. T., Perbal, M. C., Boyer, J. C., Wellink, J., Goldbach, R. W., Maule, A. J., and van Lent, J. W. (1996). The movement proteins of cowpea mosaic virus and cauliflower mosaic virus induce tubular structures in plant and insect cells. *J. Gen. Virol.* 77, 2857-2864.

Kasteel, D. T., van der Wel, N. N., Jansen, K. A., Goldbach, R. W., and van Lent, J. W. (1997). Tubule-forming capacity of the movement proteins of alfalfa mosaic virus and brome mosaic virus. *J. Gen. Virol.* 78, 2089-2093.

Kim, K. H., and Lommel, S. A. (1994). Identification and analysis of the site of -1 ribosomal frameshifting in red clover necrotic mosaic virus. *Virology* 200, 574-582.

Kim, K. H., and Hemenway, C. (1996). The 5' nontranslated region of potato virus X RNA affects both genomic and subgenomic RNA synthesis. *J. Virol.* 70, 5533-5540.

Kim, K. H., and Hemenway, C. (1997). Mutations that alter a conserved element upstream of the potato virus X triple block and coat protein genes affect subgenomic RNA accumulation. *Virology* 232, 187-197.

Klaassen, V. A., Boeshore, M. L., Koonin, E. V., Tian, T., and Falk, B. W. (1995). Genome structure and phylogenetic analysis of lettuce infectious yellows virus, a whitefly-transmitted, bipartite closterovirus. *Virology* 208, 99-110.

Koenig, R., and Loss, S. (1997). Beet soil-borne virus RNA 1: genetic analysis enabled by a starting sequence generated with primers to highly conserved helicase-encoding domains. *J. Gen. Virol.* 78, 3161-3165.

Koenig, R., Pleij, C. W., Beier, C., and Commandeur, U. (1998). Genome properties of beet virus Q, a new furo-like virus from sugarbeet, determined from unpurified virus. *J. Gen. Virol.* 79, 2027-2036.

Koev, G., Mohan, B. R., and Miller, W. A. (1998). Characterization of the subgenomic RNA1 promoter of barley yellow dwarf virus-PAV. P.132, *Abstracts of 17th Annu.Meeting of American Society of Virology*, University of British Columbia, Vancouver, Canada, July 1998.

Kolk, M. H., van der Graaf, M., Wijmenga, S. S., Pleij, C. W., Heus, H. A., and Hilbers, C. W. (1998). NMR structure of a classical pseudoknot: interplay of single- and double-stranded RNA. *Science* 280, 434-438.

Kollar, A., and Burgyan, J. (1994). Evidence that ORF 1 and 2 are the only virus-encoded replicase genes of cymbidium ringspot tombusvirus. *Virology* 201, 169-172.

Koonin, E. V., Gorbalenya, A. E., and Chumakov, K. M. (1989). Tentative identification of RNA-dependent RNA polymerases of dsRNA viruses and their relationship to positive strand RNA viral polymerases. *FEBS Lett.* 252, 42-46.

Koonin, E. V. (1991). The phylogeny of RNA-dependent RNA polymerases of positive-strand RNA viruses. *J. Gen. Virol.* 72, 2197-2206.

Koonin, E.V. (1992). Evolution of double-stranded RNA viruses: a case for polyphyletic origin from different groups of positive-stranded RNA viruses. *Semin. Virol.* 3, 327-339.

Koonin, E. V., and Dolja, V. V. (1993). Evolution and taxonomy of positive-strand RNA viruses: implications of comparative analysis of amino acid sequences. *Crit. Rev. Biochem. Mol. Biol.* 28, 375-430.

Lartey, R. T., Voss, T. C., and Melcher, U. (1996). Tobamovirus evolution: gene overlaps, recombination, and taxonomic implications. *Mol. Biol. Evol.* 13, 1327-1338.

Lauber, E., Guilley, H., Richards, K., Jonard, G., and Gilmer, D. (1997). Conformation of the 3'-end of beet necrotic yellow vein benevirus RNA 3 analysed by chemical and enzymatic probing and mutagenesis. *Nucleic Acids Res.* 25, 4723-4729.

Lauber, E., Bleykasten-Grosshans, C., Erhardt, M., Bouzoubaa, S., Jonard, G., Richards, K. E., and

Guilley, H. (1998). Cell-to-cell movement of beet necrotic yellow vein virus: I. Heterologous complementatiuon experiments provide evidence for specific interactions among the triple gene block proteins. *Mol. Plant-Microbe Interact.* 11, 618-625.

Lawrence, D. M., Rozanov, M. N., and Hillman, B. I. (1995). Autocatalytic processing of the 223-kDa protein of blueberry scorch carlavirus by a papain-like proteinase. *Virology* 207, 127-135.

Lehto, K., Grantham, G. L., and Dawson, W. O. (1990). Insertion of sequences containing the coat protein subgenomic RNA promoter and leader in front of the tobacco mosaic virus 30K ORF delays its expression and causes defective cell-to-cell movement. *Virology* 174, 145-157.

Li, Q., and Palukaitis, P. (1996). Comparison of the nucleic acid- and NTP-binding properties of the movement protein of cucumber mosaic cucumovirus and tobacco mosaic tobamovirus. *Virology* 216, 71-79.

Li, X. H., Valdez, P., Olvera, R. E., and Carrington, J. C. (1997). Functions of the tobacco etch virus RNA polymerase (NIb): subcellular transport and protein-protein interaction with VPg/proteinase (NIa). *J. Virol.* 71, 1598-1607.

Li, Z., Upadhyaya, N. M., Kositratana, W., Gibbs, A. J., and Waterhouse, P. M. (1996). Genome segment 5 of rice ragged stunt virus encodes a virion protein. *J. Gen. Virol.* 77, 3155-3160.

Lot, H., Rubino, L., Delecolle, B., Jacquemond, M., Turturo, C., and Russo, M. (1996). Characterization, nucleotide sequence and genome organization of leek white stripe virus, a putative new species of the genus Necrovirus. *Arch. Virol.* 141, 2375-2386.

Lough, T. J., Shash, K., Xoconostle-Cazares, B., Hofstra, K. R., Beck, D. L., Balmori, E., Forster, R. L., and Lucas, W. J. (1998). Molecular dissection of the mechanism by which potexvirus triple gene block proteins mediate cell-to-cell transport of infectious RNA. *Mol. Plant-Microbe Interact.* 11, 801-814.

Lu, X., Yamamoto, S., Tanaka, M., Hibi, T., and Namba, S. (1998). The genome organization of the broad bean necrosis virus (BBNV). *Arch. Virol.* 143, 1335-1348.

MacFarlane, S. A., Taylor, S. C., King, D. I., Hughes, G., and Davies, J. W. (1989). Pea early browning virus RNA1 encodes four polypeptides including a putative zinc-finger protein. *Nucleic Acids Res.* 17, 2245-2260.

Maia, I. G., Haenni, A., and Bernardi, F. (1996). Potyviral HC-Pro: a multifunctional protein. *J. Gen. Virol.* 77, 1335-1341.

Makinen, K., Naess, V., Tamm, T., Truve, E., Aaspollu, A., and Saarma, M. (1995a). The putative replicase of the cocksfoot mottle sobemovirus is translated as a part of the polyprotein by -1 ribosomal frameshift. *Virology* 207, 566-571.

Makinen, K., Tamm, T., Naess, V., Truve, E., Puurand, U., Munthe, T., and Saarma, M. (1995b). Characterization of cocksfoot mottle sobemovirus genomic RNA and sequence comparison with related viruses. *J. Gen. Virol.* 76, 2817-2825.

Manohar, S. K., Guilley, H., Dollet, M., Richards, K., and Jonard, G. (1993). Nucleotide sequence and genetic organization of peanut clump virus RNA 2 and partial characterization of deleted forms. *Virology* 195, 33-41.

Martelli, G. P., Minafra, A., and Saldarelli, P. (1997). Vitivirus, a new genus of plant viruses. *Arch. Virol.* 142, 1929-1932.

Martelli, G. P., and Jelkmann, W. (1998). Foveavirus, a new plant virus genus. *Arch. Virol.* 143, 1245-1249.

Maruyama, W., Ichimi, K., Fukui, Y., Yan, J., Zhu, Y., Kamiunten, H., and Omura, T. (1997). The minor outer capsid protein P2 of rice gall dwarf virus has a primary structure conserved with, yet is chemically dissimilar to, rice dwarf virus P2, a protein associated with virus infectivity. *Arch. Virol.* 142, 2011-2019.

Mayo, M. A., and Ziegler-Graff, V. (1996). Molecular biology of luteoviruses. *Adv. Virus. Res.* 46, 413-460.

Mayo, M. A., and Pringle, C. R. (1998). Virus taxonomy - 1997. *J. Gen. Virol.* 79, 649-657.

Meng, B., Pang, S., Forsline, P. L., McFerson, J. R., and Gonsalves, D. (1998). Nucleotide sequence and genome structure of grapevine rupestris stem pitting associated virus-1 reveal similarities to apple stem pitting virus. *J. Gen. Virol.* 79, 2059-2069.

Miller, J. S., Wesley, S. V., Naidu, R. A., Reddy, D. V., and Mayo, M. A. (1996). The nucleotide sequence of RNA-1 of Indian peanut clump furovirus. *Arch. Virol.* 141, 2301-2312.

Miller, J. S., Damude, H., Robbins, M. A., Reade, R. D., and Rochon, D. M. (1997). Genome structure of cucumber leaf spot virus: sequence analysis suggests it belongs to a distinct species within the Tombusviridae. *Virus Res.* 52, 51-60.

Miller, W. A., and Rasochova, L. (1997). Barley yellow dwarf viruses. *Annu. Rev. Phytopathol.* 35, 167-

190.

Minafra, A., Saldarelli, P., Grieco, F., and Martelli, G. P. (1994). Nucleotide sequence of the 3' terminal region of the RNA of two filamentous grapevine viruses. *Arch. Virol.* 137, 249-261.

Minafra, A., Saldarelli, P., and Martelli, G. P. (1997). Grapevine virus A: Nucleotide sequence, genome organization, and relationship in the Trichovirus genus. *Arch. Virol.* 142, 417-423.

Mise, K., and Ahlquist, P. (1995). Host-specificity restriction by bromovirus cell-to-cell movement protein occurs after initial cell-to-cell spread of infection in nonhost plants. *Virology* 206, 276-286.

Mohan, B. R., Dinesh-Kumar, S. P., and Miller, W. A. (1995). Genes and *cis*-acting sequences involved in replication of barley yellow dwarf virus-PAV RNA. *Virology* 212, 186-195.

Molinari, P., Marusic, C., Lucioli, A., Tavazza, R., and Tavazza, M. (1998). Identification of artichoke mottled crinkle virus (AMCV) proteins required for virus replication: complementation of AMCV p33 and p92 replication-defective mutants. *J. Gen. Virol.* 79, 639-647.

Molnar, A., Havelda, Z., Dalmay, T., Szutorisz, H., and Burgyan, J. (1997). Complete nucleotide sequence of tobacco necrosis virus strain DH and genes required for RNA replication and virus movement. *J. Gen. Virol.* 78, 1235-1239.

Morozov, S. Yu., and Rupasov, V. V. (1985). Possible commonality of origin of the RNA polymerase genes of plus-RNA-containing viruses of bacteria, plants and animals. *Biologicheskie Nauki*, 19-23 [in Russian].

Morozov, S. Y. (1989). A possible relationship of reovirus putative RNA polymerase to polymerases of positive-strand RNA viruses. *Nucleic Acids Res.* 17, 5394.

Morozov, S. Yu., Dolja, V. V., and Atabekov, J. G. (1989). Probable reassortment of genomic elements among elongated RNA-containing plant viruses. *J. Mol. Evol.* 29, 52-62.

Morozov, S. Y., Kanyuka, K. V., Levay, K. E., and Zavriev, S. K. (1990). The putative RNA replicase of potato virus M: obvious sequence similarity with potex- and tymoviruses. *Virology* 179, 911-914.

Morozov, S. Y., Miroshnichenko, N. A., Solovyev, A. G., Fedorkin, O. N., Zelenina, D. A., Lukasheva, L. I., Karasev, A. V., Dolja, V. V., and Atabekov, J. G. (1991). Expression strategy of the potato virus X triple gene block. *J. Gen. Virol.* 72, 2039-2042.

Morozov, S. Y., Denisenko, O. N., Zelenina, D. A., Fedorkin, O. N., Solovyev, A. G., Maiss, E., Casper, R., and Atabekov, J. G. (1993). A novel open reading frame in tobacco mosaic virus genome coding for a putative small, positively charged protein. *Biochimie* 75, 659-665.

Morozov, S., Fedorkin, O. N., Juttner, G., Schiemann, J., Baulcombe, D. C., and Atabekov, J. G. (1997). Complementation of a potato virus X mutant mediated by bombardment of plant tissues with cloned viral movement protein genes. *J. Gen. Virol.* 78, 2077-2083.

Murphy, J. F., Klein, P. G., Hunt, A. G., and Shaw, J. G. (1996). Replacement of the tyrosine residue that links a potyviral VPg to the viral RNA is lethal. *Virology* 220, 535-538.

Mushegian, A. R., and Koonin, E. V. (1993). Cell-to-cell movement of plant viruses. Insights from amino acid sequence comparisons of movement proteins and from analogies with cellular transport systems. *Arch. Virol.* 133, 239-257.

Namba, S., Kashiwazaki, S., Lu, X., Tamura, M., and Tsuchizaki, T. (1998). Complete nucleotide sequence of wheat yellow mosaic bymovirus genomic RNAs. *Arch. Virol.* 143, 631-643.

Ngon A Yassi, M., Ritzenthaler, C., Brugidou, C., Fauquet, C., and Beachy, R. N. (1994). Nucleotide sequence and genome characterization of rice yellow mottle virus RNA. *J. Gen. Virol.* 75, 249-257.

Offei, S. K., Coffin, R. S., and Coutts, R. H. (1995). The tobacco necrosis virus p7a protein is a nucleic acid-binding protein. *J. Gen. Virol.* 76, 1493-1496.

Ogawa, T., Watanabe, Y., and Okada, Y. (1992). *cis*-Acting elements for *in trans* complementation of replication-defective mutant of tobacco mosaic virus. *Virology* 191, 454-458.

Ohira, K., Namba, S., Rozanov, M., Kusumi, T., and Tsuchizaki, T. (1995). Complete sequence of an infectious full-length cDNA clone of citrus tatter leaf capillovirus: comparative sequence analysis of capillovirus genomes. *J. Gen. Virol.* 76, 2305-2309.

Okada, Y. (1986). Molecular assembly of tobacco mosaic virus in vitro. *Adv.Biophys.* 22, 95-149.

Oparka, K. J., Prior, D. A., Santa Cruz, S., Padgett, H. S., and Beachy, R. N. (1997). Gating of epidermal plasmodesmata is restricted to the leading edge of expanding infection sites of tobacco mosaic virus (TMV). *Plant J.* 12, 781-789.

O'Reilly, E. K., Tang, N., Ahlquist, P., and Kao, C. C. (1995). Biochemical and genetic analyses of the interaction between the helicase-like and polymerase-like proteins of the brome mosaic virus. *Virology* 214, 59-71.

Oster, S. K., Wu, B., and White, K. A. (1998). Uncoupled expression of p33 and p92 permits amplification of tomato bushy stunt virus RNAs. *J. Virol.* 72, 5845-5851.

Pacha, R. F., Allison, R. F., and Ahlquist, P. (1990). cis-acting sequences required for in vivo amplification of genomic RNA3 are organized differently in related bromoviruses. *Virology* 174, 436-443.

Padgett, H. S., Epel, B. L., Kahn, T. W., Heinlein, M., Watanabe, Y., and Beachy, R. N. (1996). Distribution of tobamovirus movement protein in infected cells and implications for cell-to-cell spread of infection. *Plant J.* 10, 1079-1088.

Palukaitis, P., Roossinck, M. J., Dietzgen, R. G., and Francki, R. I. (1992). Cucumber mosaic virus. *Adv. Virus Res.* 41, 281-348.

Peremyslov, V. V., Hagiwara, Y., and Dolja, V. V. (1998). Genes required for replication of the 15.5-kilobase RNA genome of a plant closterovirus. *J. Virol.* 72, 5870-5876.

Petty, I. T., and Jackson, A. O. (1990). Mutational analysis of barley stripe mosaic virus RNAβ. *Virology* 179, 712-718.

Petty, I. T., French, R., Jones, R. W., and Jackson, A. O. (1990). Identification of barley stripe mosaic virus genes involved in viral RNA replication and systemic movement. *EMBO J.* 9, 3453-3457.

Poch, O., Sauvaget, I., Delarue, M., and Tordo, N. (1989). Identification of four conserved motifs among the RNA-dependent polymerase encoding elements. *EMBO J.* 8, 3867-3874.

Pogue, G. P., and Hall, T. C. (1992). The requirement for a 5' stem-loop structure in brome mosaic virus replication supports a new model for viral positive-strand RNA initiation. *J. Virol.* 66, 674-684.

Pogue, G. P., Marsh, L. E., Connell, J. P., and Hall, T. C. (1992). Requirement for ICR-like sequences in the replication of brome mosaic virus genomic RNA. *Virology* 188, 742-753.

Pringle, C. R. (1998). The universal system of virus taxonomy of the International Committee on Virus Taxonomy (ICTV), including new proposals ratified since publication of the Sixth ICTV Report in 1995. *Arch. Virol.* 143, 203-210.

Prufer, D., Tacke, E., Schmitz, J., Kull, B., Kaufmann, A., and Rohde, W. (1992). Ribosomal frameshifting in plants: a novel signal directs the -1 frameshift in the synthesis of the putative viral replicase of potato leafroll luteovirus. *EMBO J.* 11, 1111-1117.

Qu, F., and Morris, T. J. (1997). Encapsidation of turnip crinkle virus is defined by a specific packaging signal and RNA size. *J. Virol.* 71, 1428-1435.

Qui, W. P., Geske, S. M., Hickey, C. M., and Moyer, J. W. (1998). Tomato spotted wilt *Tospovirus* genome reassortment and genome segment-specific adaptation. *Virology* 244, 186-194.

Ramirez, B. C., and Haenni, A. L. (1994). Molecular biology of tenuiviruses, a remarkable group of plant viruses. *J. Gen. Virol.* 75, 467-475.

Ranjith-Kumar, C. T., Haenni, A. L., and Savithri, H. S. (1998). Interference with Physalis mottle tymovirus replication and coat protein synthesis by transcripts corresponding to the 3'-terminal region of the genomic RNA - role of the pseudoknot structure. *J. Gen. Virol.* 79, 185-189.

Rao, A. L. (1997). Molecular studies on bromovirus capsid protein. III. Analysis of cell- to-cell movement competence of coat protein defective variants of cowpea chlorotic mottle virus. *Virology* 232, 385-395.

Rathjen, J. P., Karageorgos, L. E., Habili, N., Waterhouse, P. M., and Symons, R. H. (1994). Soybean dwarf luteovirus contains the third variant genome type in the luteovirus group. *Virology* 198, 671-679.

Reavy, B., Arif, M., Cowan G. H., and Torrance L. (1998). Association of sequences in the coat protein/readthrough domain of potato mop-top virus with transmission by *Spongospora subterranea*. *J. Gen. Virol.* 78, 2343 - 2347.

Reddick, B. B., Habera, L. F., and Law, M. D. (1997). Nucleotide sequence and taxonomy of maize chlorotic dwarf virus within the family Sequiviridae. *J. Gen. Virol.* 78, 1165-1174.

Reutenauer, A., Ziegler-Graff, V., Lot, H., Scheidecker, D., Guilley, H., Richards, K., and Jonard, G. (1993). Identification of beet western yellows luteovirus genes implicated in viral replication and particle morphogenesis. *Virology* 195, 692-699.

Richards, K.E., and Tamada, T. (1992). Mapping functions on the multipartite genome of beet necrotic yellow vein virus. *Annu. Rev. Phytopathol.* 30, 291-313.

Riechmann, J. L., Lain, S., and Garcia, J. A. (1992). Highlights and prospects of potyvirus molecular biology. *J. Gen. Virol.* 73, 1-16.

Ritzenthaler, C., Pinck, M., and Pinck, L. (1995). Grapevine fanleaf nepovirus P38 putative movement protein is not transiently expressed and is a stable final maturation product in vivo. *J. Gen. Virol.* 76, 907-915.

Robbins, M. A., Reade, R. D., and Rochon, D. M. (1997). A cucumber necrosis virus variant deficient in fungal transmissibility contains an altered coat protein shell domain. *Virology* 234, 138-146.

Roberts, I. M., Wang, D., Findlay, K., and Maule, A. J. (1998). Ultrastructural and temporal observations of the potyvirus cylindrical inclusions (CIs) show that the CI protein acts transiently in aiding virus movement. *Virology* 245, 173-181.

Rodriguez-Cerezo, E., Findlay, K., Shaw, J. G., Lomonossoff, G. P., Qiu, S. G., Linstead, P., Shanks, M., and Risco, C. (1997). The coat and cylindrical inclusion proteins of a potyvirus are associated with connections between plant cells. *Virology* 236, 296-306.

Rohde, W., Gramstat, A., Schmitz, J., Tacke, E., and Prufer, D. (1994). Plant viruses as model systems for the study of non-canonical translation mechanisms in higher plants. *J. Gen. Virol.* 75, 2141-2149.

Rojas, M. R., Zerbini, F. M., Allison, R. F., Gilbertson, R. L., and Lucas, W. J. (1997). Capsid protein and helper component-proteinase function as potyvirus cell-to-cell movement proteins. *Virology* 237, 283-295.

Roossinck, M. J. (1997). Mechanisms of plant virus evolution. *Annu. Rev. Phytopathol.* 35, 191-209.

Rott, M. E., Gilchrist, A., Lee, L., and Rochon, D. (1995). Nucleotide sequence of tomato ringspot virus RNA1. *J. Gen. Virol.* 76, 465-473.

Rozanov, M. N., Morozov, S. Yu., and Skryabin, K. G. (1990). Unexpected close relationship between the large nonvirion proteins of filamentous potexviruses and spherical tymoviruses. *Virus Genes* 3, 373-379.

Rozanov, M. N., Koonin, E. V., and Gorbalenya, A. E. (1992). Conservation of the putative methyltransferase domain: a hallmark of the 'Sindbis-like' supergroup of positive-strand RNA viruses. *J. Gen. Virol.* 73, 2129-2134.

Rozanov, M. N., Drugeon, G., and Haenni, A. L. (1995). Papain-like proteinase of turnip yellow mosaic virus: a prototype of a new viral proteinase group. *Arch. Virol.* 140, 273-288.

Rubino, L., and Russo, M. (1997). Molecular analysis of the pothos latent virus genome. *J. Gen. Virol.* 78, 1219-1226.

Rupasov, V. V., Morozov, S. Y., Kanyuka, K. V., and Zavriev, S. K. (1989). Partial nucleotide sequence of potato virus M RNA shows similarities to potexviruses in gene arrangement and the encoded amino acid sequences. *J. Gen. Virol.* 70, 1861-1869.

Russo, M., Burgyan, J., and Martelli, G. P. (1994). Molecular biology of tombusviridae. *Adv. Virus Res.* 44, 381-428.

Ryabov, E. V., Krutov, A. A., Novikov, V. K., Zheleznikova, O. V., Morozov, S. Yu., and Zavriev, S. K. (1996). Nucleotide sequence of RNA from the sobemovirus found in infected cocksfoot shows a luteovirus-like arrangement of the putative replicase and protease genes. *Phytopathology* 86, 391-397.

Ryabov, E. V., Oparka, K. J., Santa Cruz, S., Robinson, D. J., and Taliansky, M. E. (1998). Intracellular location of two groundnut rosette umbravirus proteins delivered by PVX and TMV vectors. *Virology* 242, 303-313.

Ryan, M. D., and Flint, M. (1997). Virus-encoded proteinases of the picornavirus super-group. *J. Gen. Virol.* 78, 699-723.

Salm, S. N., Rey, M. E., and Rybicki, E. P. (1996). Phylogenetic justification for splitting the Rymovirus genus of the taxonomic family Potyviridae. *Arch. Virol.* 141, 2237-2242.

Samuilova, O. V., Kalinina, N. O., Solovyev, A. G., Savenkov, E. I., and Morozov, S. Yu. (1998). Mapping of NTPase domain in TGBp1 protein of hordei- and potexviruses. P.167, *Abstracts of 17th Annu.Meeting of American Society of Virology*, University of British Columbia, Vancouver, Canada, July 1998.

Sanger, M., Passmore, B., Falk, B. W., Bruening, G., Ding, B., and Lucas, W. J. (1994). Symptom severity of beet western yellows virus strain ST9 is conferred by the ST9-associated RNA and is not associated with virus release from the phloem. *Virology* 200, 48-55.

Santa Cruz, S., Roberts, A. G., Prior, D. A. M., Chapman, S., and Oparka, K. J. (1998). Cell-to-cell and phloem-mediated transport of potato virus X: The role of virions. *Plant Cell* 10, 495-510.

Sato, K., Yoshikawa, N., Takahashi, T., and Taira, H. (1995). Expression, subcellular location and modification of the 50 kDa protein encoded by ORF2 of the apple chlorotic leaf spot trichovirus genome. *J. Gen. Virol.* 76, 1503-1507.

Satyanarayana, T., Gowda, S., Reddy, K. L., Mitchell, S. E., Dawson, W.O., and Reddy, D. V. R. (1998). Peanut yellow spot virus is a member of a new serogroup of *Tospovirus* genus based on small (S) RNA sequence and organization. *Arch. Virol.* 143, 353-364.

Savenkov, E. I., Solovyev, A. G., and Morozov, S. Y. (1998). Genome sequences of poa semilatent and lychnis ringspot hordeiviruses. *Arch. Virol.* 143, 1379-1393.

Schaad, M. C., Haldeman-Cahill, R., Cronin, S., and Carrington, J. C. (1996). Analysis of the VPg-proteinase (NIa) encoded by tobacco etch potyvirus: effects of mutations on subcellular transport, proteolytic processing, and genome amplification. *J. Virol.* 70, 7039-7048.

Schaad, M. C., Jensen, P. E., and Carrington, J. C. (1997). Formation of plant RNA virus replication complexes on membranes: role of an endoplasmic reticulum-targeted viral protein. *EMBO J.* 16, 4049-4059.

Schmitt, C., Balmori, E., Jonard, G., Richards, K. E., and Guilley, H. (1992). In vitro mutagenesis of biologically active transcripts of beet necrotic yellow vein virus RNA 2: evidence that a domain of the 75-kDa readthrough protein is important for efficient virus assembly. *Proc. Natl. Acad. Sci. USA* 89, 5715-5719.

Schmitt, C., Mueller, A. M., Mooney, A., Brown, D., and MacFarlane, S. (1998). Immunological detection and mutational analysis of the RNA2-encoded nematode transmission proteins of pea early browning virus. *J. Gen. Virol.* 79, 1281-1288.

Schmitz, I., and Rao, A. L. (1996). Molecular studies on bromovirus capsid protein. I. Characterization of cell-to-cell movement-defective RNA3 variants of brome mosaic virus. *Virology* 226, 281-293.

Schmitz, J., Stussi-Garaud, C., Tacke, E., Prufer, D., Rohde, W., and Rohfritsch, O. (1997). In situ localization of the putative movement protein (pr17) from potato leafroll luteovirus (PLRV) in infected and transgenic potato plants. *Virology* 235, 311-322.

Schneider, W. L., Greene, A. E., and Allison, R. F. (1997). The carboxy-terminal two-thirds of the cowpea chlorotic mottle bromovirus capsid protein is incapable of virion formation yet supports systemic movement. *J. Virol.* 71, 4862-4865.

Scholthof, H. B., Morris, T. J., and Jackson, A. O. (1993). The capsid protein gene of tomato bushy stunt virus is dispensable for systemic movement and can be replaced for localized expression of foreign genes. *Mol. Plant-Microbe Interact.* 6, 309-322.

Scholthof, H. B., Scholthof, K. B., Kikkert, M., and Jackson, A. O. (1995a). Tomato bushy stunt virus spread is regulated by two nested genes that function in cell-to-cell movement and host-dependent systemic invasion. *Virology* 213, 425-438.

Scholthof, H. B., Scholthof, K. B., Kikkert, M., and Jackson, A. O. (1995b). Tomato bushy stunt virus spread is regulated by two nested genes that function in cell-to-cell movement and host-dependent systemic invasion. *Virology* 213, 425-438.

Scholthof, H. B., and Jackson, A. O. (1997). The enigma of pX: A host-dependent cis-acting element with variable effects on tombusvirus RNA accumulation. *Virology* 237, 56-65.

Schubert, J., Merits, A., Jaervekuelg, L., Paulin, L., and Rubenstein, F. (1996) The complete nucleotide sequence of the Ryegrass mosaic virus genomic RNA. *DDBJ/EMBL/GenBank databases.* Accession number [Y09854].

Scott, K. P., Kashiwazaki, S., Reavy, B., and Harrison, B. D. (1994). The nucleotide sequence of potato mop-top virus RNA 2: a novel type of genome organization for a furovirus. *J. Gen. Virol.* 75, 3561-3568.

Shams-bakhsh, M., and Symons, R. H. (1997). Barley yellow dwarf virus-PAV RNA does not have a VPg. *Arch. Virol.* 142, 2529-2535.

Shanks, M., Dessens, J. T., and Lomonossoff, G. P. (1996). The 24 kDa proteinases of comoviruses are virus-specific *in cis* as well as *in trans. J. Gen. Virol.* 77, 2365-2369.

Shen, P., Kaniewska, M., Smith, C., and Beachy, R. N. (1993). Nucleotide sequence and genomic organization of rice tungro spherical virus. *Virology* 193, 621-630.

Shi, B. J., Ding, S. W., and Symons, R. H. (1997). In vivo expression of an overlapping gene encoded by the cucumoviruses. *J. Gen. Virol.* 78, 237-241.

Shirako, Y., and Wilson, T. M. (1993). Complete nucleotide sequence and organization of the bipartite RNA genome of soil-borne wheat mosaic virus. *Virology* 195, 16-32.

Shirako, Y. (1998). Non-AUG translation initiation in a plant RNA virus: a forty-amino-acid extension is added to the N terminus of the soil-borne wheat mosaic virus capsid protein. *J. Virol.* 72, 1677-1682.

Shikata, E. (1989). Plant reoviruses. *In* "Plant Viruses. Vol. II. Structure and Replication" (C. L. Mandahar, Ed.) pp.207-234. CRC Press, Boca Raton.

Siegel, R. W., Adkins, S., and Kao, C. C. (1997). Sequence-specific recognition of a subgenomic RNA promoter by a viral RNA polymerase. *Proc. Natl. Acad. Sci. USA* 94, 11238-11243.

Simon, A. E., and Bujarski, J. J. (1994). RNA-RNA recombination and evolution in virus-infected plants. *Annu. Rev. Phytopathol.* 32, 337-362.

Singh, R. N., and Dreher, T. W. (1997). Turnip yellow mosaic virus RNA-dependent RNA polymerase: initiation of minus strand synthesis *in vitro. Virology* 233, 430-439.

Sit, T. L., Leclerc, D., and Abouhaidar, M. G. (1994). The minimal 5' sequence for in vitro initiation of papaya mosaic potexvirus assembly. *Virology* 199, 238-242.

Sit, T. L., Vaewhongs, A. A., and Lommel, S. A. (1998). RNA-mediated trans-activation of transcription from a viral RNA. *Science* 281, 829-832.

Sivakumaran, K., and Hacker, D. L. (1998). The 105-kDa polyprotein of southern bean mosaic virus is translated by scanning ribosomes. *Virology* 246, 34-44.

Skaf, J. S., Rucker, D. G., Demler, S. A., Wobus, C. E., and de Zoeten, G. A. (1997). The coat protein is dispensable for the establishment of systemic infections by pea enation mosaic virus. *Mol. Plant-Microbe Interact.* 10, 929-932.

Smirnyagina, E., Hsu, Y. H., Chua, N., and Ahlquist, P. (1994). Second-site mutations in the brome mosaic virus RNA3 intercistronic region partially suppress a defect in coat protein mRNA transcription. *Virology* 198, 427-436.

Sokolova, M., Prufer, D., Tacke, E., and Rohde, W. (1997). The potato leafroll virus 17K movement protein is phosphorylated by a membrane-associated protein kinase from potato with biochemical features of protein kinase C. *FEBS Lett.* 400, 201-205.

Solovyev, A. G., Novikov, V. K., Merits, A., Savenkov, E. I., Zelenina, D. A., Tyulkina, L. G., and Morozov, S. Yu. (1994). Genome characterization and taxonomy of *Plantago asiatica* mosaic potexvirus. *J. Gen. Virol.* 75, 259-267.

Solovyev, A. G., Savenkov, E. I., Agranovsky, A. A., and Morozov, S. Y. (1996). Comparisons of the genomic cis-elements and coding regions in RNAβ components of the hordeiviruses barley stripe mosaic virus, lychnis ringspot virus, and poa semilatent virus. *Virology* 219, 9-18.

Solovyev, A. G., Savenkov, E. I., Grdzelishvili, V. Z., Kalinina, N. O., Morozov, S. Yu., Schiemann, J., and Atabekov, J. G. (1998). Movement of hordeivirus hybrids with exchanges in the triple gene block. *Virology*, in press.

Song, C., and Simon, A. E. (1995). Requirement of a 3'-terminal stem-loop in *in vitro* transcription by an RNA-dependent RNA polymerase. *J. Mol. Biol.* 254, 6-14.

Song, S. I., Song, J. T., Kim, C. H., Lee, J. S., and Choi, Y. D. (1998). Molecular characterization of the garlic virus X genome. *J. Gen. Virol.* 79, 155-159.

Sriskanda, V. S., Pruss, G., Ge, X., and Vance, V. B. (1996). An eight-nucleotide sequence in the potato virus X 3' untranslated region is required for both host protein binding and viral multiplication. *J. Virol.* 70, 5266-5271.

Stenger, D. C., Hall, J. S., Choi, I.-R., and French, R. (1998). Phylogenetic relationships within the family *Potyviridae*: wheat streak mosaic virus and brome mosaic virus are not members of the genus *Rymovirus*. *Phytopathology*, in press.

Storms, M. M., Kormelink, R., Peters, D., van Lent, J. W., and Goldbach, R. W. (1995). The nonstructural NSm protein of tomato spotted wilt virus induces tubular structures in plant and insect cells. *Virology* 214, 485-493.

Sudarshana, M. R., and Berger, P. H. (1998). Nucleotide sequence of both genomic RNAs of a North American tobacco rattle virus isolate. *Arch. Virol.* 143, 1535-1544.

Suzuki, N., Tanimura, M., Watanabe, Y., Kusano, T., Kitagawa, Y., Suda, N., Kudo, H., Uyeda, I., and Shikata, E. (1992). Molecular analysis of rice dwarf phytoreovirus segment S1: interviral homology of the putative RNA-dependent RNA polymerase between plant- and animal-infecting reoviruses. *Virology* 190, 240-247.

Suzuki, N., Kusano, T., Matsuura, Y., and Omura, T. (1996). Novel NTP binding property of rice dwarf phytoreovirus minor core protein P5. *Virology* 219, 471-474.

Takamatsu, N., Watanabe, Y., Meshi, T., and Okada, Y. (1990). Mutational analysis of the pseudoknot region in the 3' noncoding region of tobacco mosaic virus RNA. *J. Virol.* 64, 3686-3693.

Takamatsu, N., Watanabe, Y., Iwasaki, T., Shiba, T., Meshi, T., and Okada, Y. (1991). Deletion analysis of the 5' untranslated leader sequence of tobacco mosaic virus RNA. *J. Virol.* 65, 1619-1622.

Taliansky, M. E., and Garcia-Arenal, F. (1995). Role of cucumovirus capsid protein in long-distance movement within the infected plant. *J. Virol.* 69, 916-922.

Taliansky, M. E., Robinson, D. J., and Murant, A. F. (1996). Complete nucleotide sequence and organization of the RNA genome of groundnut rosette umbravirus. *J. Gen. Virol.* 77, 2335-2345.

Tamada, T., Schmitt, C., Saito, M., Guilley, H., Richards, K., and Jonard, G. (1996). High resolution analysis of the readthrough domain of beet necrotic yellow vein virus readthrough protein: a KTER motif is important for efficient transmission of the virus by *Polymyxa betae*. *J. Gen. Virol.* 77, 1359-1367.

Tavazza, M., Lucioli, A., Calogero, A., Pay, A., and Tavazza, R. (1994). Nucleotide sequence, genomic organization and synthesis of infectious transcripts from a full-length clone of artichoke mottle crinkle virus. *J. Gen. Virol.* 75, 1515-1524.

Tomashevskaya, O. L., Solovyev, A. G., Karpova, O. V., Fedorkin, O. N., Rodionova, N. P., Morozov, S. Y., and Atabekov, J. G. (1993). Effects of sequence elements in the potato virus X RNA 5' non-translated αβ-leader on its translation enhancing activity. *J. Gen. Virol.* 74, 2717-2724.

Tordo, N., de Haan, P., Goldbach, R., and Poch, O. (1992). Evolution of negative-stranded RNA viruses.

Semin. Virol. 3, 341-357.

Toriyama, S., Takahashi, M., Sano, Y., Shimizu, T., and Ishihama, A. (1994). Nucleotide sequence of RNA 1, the largest genomic segment of rice stripe virus, the prototype of the tenuiviruses. *J. Gen. Virol.* 75, 3569-3579.

Toriyama, S., Kimishima, T., and Takahashi, M. (1997). The proteins encoded by rice grassy stunt virus RNA5 and RNA6 are only distantly related to the corresponding proteins of other members of the genus Tenuivirus. *J. Gen. Virol.* 78, 2355-2363.

Toriyama, S., Kimishima, T., Takahashi, M., Shimizu, T., Minaka, N., and Akutsu, K. (1998). The complete nucleotide sequence of the rice grassy stunt virus genome and genomic comparisons with viruses of the genus *Tenuivirus*. *J. Gen. Virol.* 79, 2051–2058.

Torrance, L., and Mayo, M. A. (1997). Proposed re-classification of furoviruses. *Arch. Virol.* 142, 435-439.

Turina, M., Maruoka, M., Monis, J., Jackson, A. O., and Scholthof, K. B. (1998). Nucleotide sequence and infectivity of a full-length cDNA clone of panicum mosaic virus. *Virology* 241, 141-155.

Turnbull-Ross, A. D., Mayo, M. A., Reavy, B., and Murant, A. F. (1993). Sequence analysis of the parsnip yellow fleck virus polyprotein: evidence of affinities with picornaviruses. *J. Gen. Virol.* 74, 555-561.

Turnbull-Ross, A. D., Reavy, B., Mayo, M. A., and Murant, A. F. (1992). The nucleotide sequence of parsnip yellow fleck virus: a plant picorna-like virus. *J. Gen. Virol.* 73, 3203-3211.

Uhde, K., Koenig, R., and Lesemann, D. E. (1998). An onion isolate of tobacco rattle virus: reactivity with an antiserum to Hypochoeris mosaic virus, a putative furovirus, and molecular analysis of its RNA 2. *Arch. Virol.* 143, 1041-1053.

Urban, C., and Beier, H. (1995). Cysteine tRNAs of plant origin as novel UGA suppressors. *Nucleic Acids Res.* 23, 4591-4597.

Uyeda, I., Kimura, I., and Shikata, E. (1995). Characterization of genome structure and establishment of vector cell lines for plant reoviruses. *Adv. Virus Res.* 45, 249-279.

van den Heuvel, J. F., Bruyere, A., Hogenhout, S. A., Ziegler-Graff, V., Brault, V., Verbeek, M., van der Wilk, F., and Richards, K. (1997). The N-terminal region of the luteovirus readthrough domain determines virus binding to Buchnera GroEL and is essential for virus persistence in the aphid. *J. Virol.* 71, 7258-7265.

van der Vossen, E. A., Neeleman, L., and Bol, J. F. (1994). Early and late functions of alfalfa mosaic virus coat protein can be mutated separately. *Virology* 202, 891-903.

van der Wel, N. N., Goldbach, R. W., and van Lent, J. W. (1998). The movement protein and coat protein of alfalfa mosaic virus accumulate in structurally modified plasmodesmata. *Virology* 244, 322-329.

van der Wilk, F., Houterman, P., Molthoff, J., Hans, F., Dekker, B., van den Heuvel, J., Huttinga, H., and Goldbach, R. (1997a). Expression of the potato leafroll virus ORF0 induces viral-disease-like symptoms in transgenic potato plants. *Mol. Plant-Microbe Interact.* 10, 153-159.

van der Wilk, F., Verbeek, M., Dullemans, A. M., and van den Heuvel, J. F. (1997b). The genome-linked protein of potato leafroll virus is located downstream of the putative protease domain of the ORF1 product. *Virology* 234, 300-303.

van Poelwijk, F., Prins, M., and Goldbach, R. (1997). Completion of the impatiens necrotic spot virus genome sequence and genetic comparison of the L proteins within the family *Bunyaviridae*. *J. Gen. Virol.* 78, 543-546.

van Rossum, C. M., Brederode, F. T., Neeleman, L., and Bol, J. F. (1997a). Functional equivalence of common and unique sequences in the 3' untranslated regions of alfalfa mosaic virus RNAs 1, 2, and 3. *J. Virol.* 71, 3811-3816.

van Rossum, C. M., Neeleman, L., and Bol, J. F. (1997b). Comparison of the role of 5' terminal sequences of alfalfa mosaic virus RNAs 1, 2, and 3 in viral RNA replication. *Virology* 235, 333-341.

Vaquero, C., Turner, A. P., Demangeat, G., Sanz, A., Serra, M. T., Roberts, K., and Garcia-Luque, I. (1994). The 3a protein from cucumber mosaic virus increases the gating capacity of plasmodesmata in transgenic tobacco plants. *J. Gen. Virol.* 75, 3193-3197.

Vaquero, C., Liao, Y. C., Nahring, J., and Fischer, R. (1997). Mapping of the RNA-binding domain of the cucumber mosaic virus movement protein. *J. Gen. Virol.* 78, 2095-2099.

Verchot, J., and Carrington, J. C. (1995). Evidence that the potyvirus P1 proteinase functions in trans as an accessory factor for genome amplification. *J. Virol.* 69, 3668-3674.

Verver, J., Le Gall, O., van Kammen, A., and Wellink, J. (1991). The sequence between nucleotides 161 and 512 of cowpea mosaic virus M RNA is able to support internal initiation of translation in vitro. *J. Gen. Virol.* 72, 2339-2345.

Verver, J., Wellink, J., Van Lent, J., Gopinath, K., and Van Kammen, A. (1998). Studies on the movement of cowpea mosaic virus using the jellyfish green fluorescent protein. *Virology* 242, 22-27.

Wagner, J. D., and Jackson, A. O. (1997). Characterization of the components and activity of sonchus yellow net rhabdovirus polymerase. *J. Virol.* 71, 2371-2382.

Wang, H., Culver, J. N., and Stubbs, G. (1997). Structure of ribgrass mosaic virus at 2.9 A resolution: Evolution and taxonomy of tobamoviruses. *J. Mol. Biol.* 269, 769-79.

Wang, H. L., Wang, Y., Giesman-Cookmeyer, D., Lommel, S. A., and Lucas, W. J. (1998). Mutations in viral movement protein alter systemic infection and identify an intercellular barrier to entry into the phloem long- distance transport system. *Virology* 245, 75-89.

Wang, J., and Simon, A. E. (1997). Analysis of the two subgenomic RNA promoters for turnip crinkle virus in vivo and in vitro. *Virology* 232, 174-86.

Watson, M. T., Tian, T., Estabrook, E., and Falk, B. W. (1997). A small RNA identified as a component of California carrot motley dwarf resembles the beet western yellows ST9-associated RNA. *DDBJ/EMBL/GenBank databases.* Accession number [AF020616]

Weiland, J. J., and Edwards, M. C. (1994). Evidence that the αa gene of barley stripe mosaic virus encodes determinants of pathogenicity to oat (*Avena sativa*). *Virology* 201, 116-126.

Weng, Z., and Xiong, Z. (1997). Genome organization and gene expression of saguaro cactus carmovirus. *J. Gen. Virol.* 78, 525-534.

Wetzel, T., Dietzgen, R. G., and Dale, J. L. (1994). Genomic organization of lettuce necrotic yellows rhabdovirus. *Virology* 200, 401-412.

Wieczorek, A., and Sanfacon, H. (1993). Characterization and subcellular localization of tomato ringspot nepovirus putative movement protein. *Virology* 194, 734-742.

Wintermantel, W. M., Banerjee, N., Oliver, J. C., Paolillo, D. J., and Zaitlin, M. (1997). Cucumber mosaic virus is restricted from entering minor veins in transgenic tobacco exhibiting replicase-mediated resistance. *Virology* 231, 248-257.

Wobbe, K. K., Akgoz, M., Dempsey, D. A., and Klessig, D. F. (1998). A single amino acid change in turnip crinkle virus movement protein p8 affects RNA binding and virulence on *Arabidopsis thaliana*. *J. Virol.* 72, 6247-6250.

Wobus, C. E., Skaf, J. S., Schultz, M. H., and de Zoeten1 G. A. (1998). Sequencing, genomic localization and initial characterization of the VPg of pea enation mosaic enamovirus. *J. Gen. Virol.* 79, 2023-2025.

Wong, S. M., Mahtani, P. H., Lee, K. C., Yu, H. H., Tan, Y., Neo, K. K., Chan, Y., Wu, M., and Chng, C. G. (1997). Cymbidium mosaic potexvirus RNA: complete nucleotide sequence and phylogenetic analysis. *Arch. Virol.* 142, 383-391.

Wu, X., and Shaw, J. G. (1998). Evidence that assembly of a potyvirus begins near the 5' terminus of the viral RNA. *J. Gen. Virol.* 79, 1525-1529.

Xie, W. S., Antoniw, J. F., and White, R. F. (1993). Nucleotide sequence of beet cryptic virus 3 dsRNA2 which encodes a putative RNA-dependent RNA polymerase. *J. Gen. Virol.* 74, 2303.

Xin, H. W., Ji, L. H., Scott, S. W., Symons, R. H., and Ding, S. W. (1998). Ilarviruses encode a cucumovirus-like 2b gene that is absent in other genera within the *Bromoviridae*. *J. Virol.* 72, 6956-6959.

Xiong, Z., and Lommel, S. A. (1989). The complete nucleotide sequence and genome organization of red clover necrotic mosaic virus RNA-1. *Virology* 171, 543-554.

Xiong, Z., Kim, K. H., Giesman-Cookmeyer, D., and Lommel, S. A. (1993). The roles of the red clover necrotic mosaic virus capsid and cell-to-cell movement proteins in systemic infection. *Virology* 192, 27-32.

Xu, H., Li, Y., Mao, Z., Li, Y., Wu, Z., Qu, L., An, C., Ming, X., Schiemann, J., Casper, R., and Chen, Z. (1998). Rice dwarf phytoreovirus segment S11 encodes a nucleic acid binding protein. *Virology* 240, 267-272.

Yoshikawa, N., Imaizumi, M., Takahashi, T., and Inouye, N. (1993). Striking similarities between the nucleotide sequence and genome organization of citrus tatter leaf and apple stem grooving capilloviruses. *J. Gen. Virol.* 74, 2743-2747.

Yoshikawa, N., Iida, H., Goto, S., Magome, H., Takahashi, T., and Terai, Y. (1997). Grapevine berry inner necrosis, a new trichovirus: comparative studies with several known trichoviruses. *Arch. Virol.* 142, 1351-1363.

Zaccomer, B., Haenni, A. L., and Macaya, G. (1995). The remarkable variety of plant RNA virus genomes. *J. Gen. Virol.* 76, 231-47.

Zaitlin, M. (1998). Elucidation of the genome organization of tobacco mosaic virus. *Philosophical Transactions: Biological Sciences (London),* in press.

Zalloua, P. A., Buzayan, J. M., and Bruening, G. (1996). Chemical cleavage of 5'-linked protein from tobacco ringspot virus genomic RNAs and characterization of the protein-RNA linkage. *Virology* 219, 1-8.

Zanotto, P. M., Gibbs, M. J., Gould, E. A., and Holmes, E. C. (1996). A reevaluation of the higher taxonomy of viruses based on RNA polymerases. *J. Virol.* 70, 6083-6096.

Zavriev, S. K., Hickey, C. M., and Lommel, S. A. (1996). Mapping of the red clover necrotic mosaic virus subgenomic RNA. *Virology* 216, 407-410.

Zhang, Y.-P., Kirkpatrick, B. C., Smart, C. D., and Uyemoto, J. K. (1998). cDNA cloning and molecular characterization of cherry green ring mottle virus. *J. Gen. Virol.* 79, 2275–2281.

Zhou, H., and Jackson, A. O. (1996a). Expression of the barley stripe mosaic virus RNAβ "triple gene block". *Virology* 216, 367-379.

Zhou, H., and Jackson, A.O. (1996b). Analysis of *cis*-acting elements required for replication of barley stripe mosaic virus RNAs. *Virology* 219, 150-160

Zhu, H. Y., Ling, K. S., Goszczynski, D. E., McFerson, J. R., and Gonsalves, D. (1998). Nucleotide sequence and genome organization of grapevine leafroll- associated virus-2 are similar to beet yellows virus, the closterovirus type member. *J. Gen. Virol.* 79, 1289-1298.

Ziegler, A., Mayo, M. A., and Murant, A. F. (1993). Proposed classification of the bipartite-genomed raspberry bushy dwarf idaeovirus, with tripartite-genomed viruses in the family *Bromoviridae*. *Arch. Virol.* 131, 483-488.

Ziegler-Graff, V., Guilford, P. J., and Baulcombe, D. C. (1991). Tobacco rattle virus RNA-1 29K gene product potentiates viral movement and also affects symptom induction in tobacco. *Virology* 182, 145-155.

4 GENE EXPRESSION IN POSITIVE STRAND RNA VIRUSES: CONVENTIONAL AND ABERRANT STRATEGIES

Alexey Agranovsky and Sergey Morozov

INTRODUCTION

The rapid evolution of RNA virus genomes, which is driven by high mutation rates in replication (Steinhauer and Holland, 1987) and recombinational reassortment (Gibbs, 1987; Zimmern, 1988; Morozov et al., 1989), must conciliate two opposite factors: the necessity of acquiring new genes for adaptation, and the limitations on the genome size imposed by packaging and replication constraints (Dolja et al., 1994; Agranovsky, 1996). As a result, (+)RNA genomes of plant viruses rarely exceed the 10-kb size limit, some of them containing only the three genes that suffice for the basic functions of replication, cell-to-cell movement, and encapsidation. It is not uncommon that the plant virus genomes are 'compressed', i.e., contain extensively overlapping open reading frames (ORFs). To efficiently express their genomes, RNA viruses have developed mosaic strategies, of which some conform to the rules of eukaryotic mRNA translation, and some break these rules.

The most evolutionarily stable element in (+)RNA viruses is the replicase gene coding for the conserved domain of RNA-dependent RNA polymerase (POL) (Kamer and Argos, 1984; Morozov and Rupasov, 1985; Koonin, 1991a). Paradoxically, no uniformity is seen in the POL expression patterns even in closely related viruses. While in some viral groups the conventional translation of a single uninterrupted gene results in mature replicase, in others this requires non-orthodox translational and posttranslational strategies. The only rule without exception is that the replicase gene is translated directly from the viral genomic RNA (which in most cases dictates the 5'-proximal position of the replicase gene in the genome). The internal genes in plant virus RNAs, coding for the proteins involved in encapsidation, cell-to-cell movement, and other accessory functions, may also be expressed by a variety of strategies. The most common are the use of subgenomic RNAs (sgRNAs), and proteolytic processing of a polyprotein precursor. In addition, non-orthodox translational mechanisms (leaky scanning and readthrough of leaky stop codons) may be employed.

This chapter reviews the molecular mechanisms in plant virus gene expression. A number of comprehensive reviews dealing with this subject has been published in the recent years, including those focusing on specific viral groups and/or expression mechanisms (Ahlquist, 1992; Dougherty and Semler, 1993; Farabaugh, 1993; Rohde *et al.*, 1994; Zaccomer *et al.*, 1995; Agranovsky, 1996; Buck, 1996; Fütterer and Hohn, 1996; Gallie, 1996; Snijder and Meulenberg, 1998). Here we consider the problem in the evolutionary aspect.

VIRAL GENOMIC RNA AS A SINGLE TRANSLATION UNIT

The paradigm of eukaryotic mRNA translation is that ribosomes initiate protein synthesis at the 5'-proximal AUG codon in the RNA template, and decode a fixed reading frame until termination at a stop codon. Conceivably, the majority of eukaryotic cell mRNAs are monocistronic. Viral genomic (+)RNAs, despite their polycistronic nature, conform to this basic rule in being functionally monocistronic; as a rule, only the 5'-proximal gene in a viral RNA is available for the ribosomes, whereas the other genes are translationally silent. Moreover, some viral RNAs contain the structural elements of cell mRNAs that influence their translational efficiency, i.e., the 5'-terminal cap and the 3'-terminal poly(A) tail. In cases when virus RNAs lack either the cap, or the poly(A), or both, these are substituted by other elements still capable of recruiting the cell translational machinery. Upon translation of monocistronic messengers, the rate-limiting step is the initiation.

Initiation Codon Choice and Translation in Plants

Conventional Scanning and Initiation

According to the widely accepted model of eukaryotic translation initiation, the 40S ribosomal subunit, assisted by the initiation factors, binds at the 5'-capped end of a mRNA and scans the 5'-untranslated region (UTR) (Kozak, 1986, 1991). When the initiating codon is reached, the 60S subunit binds to the 40S subunit to form the translation-competent 80S ribosome. The nucleotide sequence context of an AUG codon greatly influences the efficiency of initiation. The strongest initiating codons in plants are found in the context ACA**AUG**GC, where the most important positions are a purine and guanine at -3 and +4, respectively (Kozak, 1991; Lütcke *et al.*, 1987). When the scanning ribosome encounters an AUG in a non-optimal context, there is a strong chance of bypassing this codon, which would lead to underexpression of the gene.

Plant viruses may use the initiation codon choice as a means to quantitatively regulate the expression of their genes. Circumstantially, the fact that individual genes in a plant virus genome have variable contexts (Gustafson and Armour, 1986; Gustafson *et al.,* 1989; Lehto and Dawson, 1990; Agranovsky *et al.*, 1991; 1994a) supports their role in translation. However, the expression level of a given gene results from a fine interplay of different mechanisms, and thus a non-optimal context of the initiating codon does not necessarily mean that the gene will be poorly expressed (Lehto and Dawson, 1990).

Non-AUG Initiation Codons in Plant Virus RNAs

Translation of plant virus genes may start at a codon distinct from the conventional AUG. In some instances, the non-AUG codon(s) may compete with the downstream in-frame AUG codon, thus giving rise to proteins that differ in their N-terminal sequences. In the genome of soil-borne wheat mosaic furovirus (SBWMV), the AUG codon of the 19-kDa coat protein (CP) gene is preceded by an in-frame CUG. Using site-directed mutagenesis and *in vitro* transcription and translation, Shirako (1998) demonstrated that translation of the CP ORF may be initiated at the CUG (although less efficiently than at the AUG), thus yielding a 25-kDa CP with the N-terminal extension of 40 amino acids. Likewise, while scanning the tobamovirus 5'-UTR, the ribosomes may initiate translation at several AUU codons upstream of and in frame with the AUG of the replicase gene (Schmitz *et al.*, 1996). The efficiency of the non-canonical initiation, at least for the AUU codon at positions 63-65, was quite high both *in vitro* and *in vivo* (Schmitz *et al.*, 1996). These translation patterns indicate that leaky scanning (see later) is used to discriminate between non-AUG and AUG codons in the tobamovirus and furovirus RNAs.

In the genome of strawberry mild yellow edge-associated potexvirus (SMYEAV), the ORF2 lacks AUG codons (Jelkmann *et al.*, 1992). In the potexvirus genomes, ORF 2 is the 5'-proximal gene in the conserved 'Triple Gene Block' (TGB; Morozov *et al.*, 1989) coding for the movement proteins (MPs). It has been suggested that translation of this gene in SMYEAV might be initiated at either a CUG or an AUU codon to yield a protein of the size and sequence consistent with those of the other potexvirus MPs (25-26 kDa) (Jelkmann *et al.*, 1992). Likewise, the 3'-proximal TGB ORFs in the related genomes of lily virus X (Memelink *et al.*, 1990) and shallot virus X (Kanyuka *et al.*, 1992) do not contain AUG codons. Although the expression mode of these aberrant ORFs awaits experimental testing, these appear to be peculiar examples of plant virus genes whose expression depends on non-AUG initiation.

Role of 5'- and 3'-Noncoding Regions in Initiation of Translation

All mRNAs of higher plants contain the capped 5'-UTR and the polyadenylated 3'-UTR, which are the elements that influence the messenger activity and stability. In translation, the cap and poly(A) act synergistically and are likely to physically interact with each other via the bound initiation factors eIF-4F, eIF-4B, and poly(A)-binding proteins (Gallie, 1991; Sachs *et al.,* 1997). According to the codependent model of translation (Gallie, 1996), the closed status of mRNA molecule, maintained by RNA-protein and protein-protein interaction in the translation complex, allows recycling the 40S subunits to increase the efficiency of translation.

To be competent in translation, viral RNAs must mimic the properties of cell mRNAs. However, the plant virus RNAs that have both the cap and the poly(A) tail (e.g., those of potex-, carla-, beny-, and trichoviruses) are a minority; in most cases, viral templates lack the cap, or poly(A), or both. Nevertheless, the available data suggest that the 5'- and 3'-terminal structures in such viral RNAs may also function synergistically in translation. In the RNA genomes of tobacco mosaic tobamovirus (TMV) and brome mosaic bromovirus (BMV), the 5' end is capped and the 3'-UTR contains a tRNA-like structure (Buck, 1996). The terminal sequences in these viral RNAs stimulate translation cooperatively, similarly to the cap and poly(A) in a cell mRNA (Gallie, 1996). The specific elements in TMV RNA, the CAA repeats in the 5'-leader and the 3'-distal pseudoknot in the 3'-UTR, were recognized by a cell heat shock-inducible protein, p102, which thus serves to bring the 5' and 3' ends into proximity (Gallie, 1996). In addition, the synergistic effects of the 5'- and 3'-terminal sequences in translation were demonstrated for the potyvirus RNA having the 3'-poly(A) but no 5'-cap (Gallie *et al.*, 1995), and the RNAs of a luteovirus (Wang and Miller, 1995) and satellite tobacco necrosis virus (Timmer *et al.*, 1993) having neither cap nor poly(A). Apparently, whatever is the evolutionarily gained combination of the terminal structures in a plant virus RNA, its elements may still be capable of the interactions having impact on the initiation.

The 5'-UTRs in plant mRNAs may influence translation both directly, depending on their length and degree of secondary structure, and indirectly, owing to the presence of specific elements that constitute the binding sites for protein factors (Gallie, 1996). Translational enhancers are found in a number of plant viral RNAs, including those whose translation is cap-dependent (tobamoviruses, potexviruses, and alfalfa mosaic virus) and cap-independent (potyviruses, comoviruses, and luteoviruses) (Gallie *et al.*, 1987; Jobling and Gehrke, 1987; Carrington and Freed, 1990; Zelenina *et al.*, 1992; Tomashevskaya *et al.*, 1993; Gallie, 1996). In TMV, the enhancer (Ω leader) consists of 68 nt, with its key part comprising eight direct CAA repeats (Gallie *et al.*, 1987). In this case, the translation enhancement may be explained by the action of the cell p102 protein that locks the 5' and 3' termini of the genomic RNA. The 5'-leader of alfalfa mosaic virus RNA 4 possibly enhances translation by a different mechanism based on its reduced requirement for the translation initiation factors (Browning *et al.*, 1988).

Small upstream ORFs (uORFs) constitute yet another type of the regulatory elements in the 5'-UTR of plant and viral mRNAs. As the scanning ribosome predominantly initiates translation at the 5'-most AUG in the template and has low re-initiation capacity, the presence of uORFs must repress the initiation at the 'proper' AUG in the downstream gene. The available examples show that downregulation of the gene translation by uORFs is not straightforward, implying that these may be instruments of fine translational control (Geballe and Morris, 1994; Lovett and Rogers, 1996). Indeed, of the three uORFs in the 5'-UTR of an animal retrovirus, uORF1 serves as a unique translation enhancer, uORF3 inhibits translation of the downstream *gag* gene, and uORF2 has no effect on translation (Donze and Spahr, 1992). A single uORF artificially created in the 5'-UTR of bromovirus RNA 3 reduces translation of the MP gene *in vivo* up to 18 times (De Jong *et al.*, 1997), whereas a naturally occurring uORF in the RNAγ of the barley stripe mosaic hordeivirus (BSMV) strain CV17 caused a considerable reduction of the γa protein (POL) synthesis and the host-specific inhibition of the virus long-distance movement(Jackson *et al.*, 1991). The uORFs have also been found in the BSMV RNAα, as well as in the genomic RNAs of some members of the *Bromoviridae* family (Gustafson *et al.*, 1989 and references therein). In lettuce infectious yellows crinivirus (family *Closteroviridae*), RNA 1 (in which the 5'-gene codes for the replicase) contains a 97-nt 5'-UTR which is devoid of the uORFs, whereas RNA 2 (which may serve as an mRNA for the 5K and 62K nonstructural proteins) has a 326-nt 5'-UTR containing five uORFs (Klaassen *et al.*, 1995). The regulatory potential of the uORFs in translation of these plant virus RNAs awaits experimental evaluation.

ACCESS TO INTERNAL GENES

The common strategies used by many RNA plant viruses to make their genes available for the ribosomes are the genome partitioning and the use of subgenomic mRNAs. An apparent evolutionary advantage of the genome partitioning is that this renders more genes accessible for the translating ribosomes, and allows separate control of the gene expression by the *cis*-elements influencing the replication and translation of individual RNA components. In view of its impact on translation, subgenomization is similar to genome partitioning, and a possible evolutionary scenario may be envisaged where the genome splitting occurs through recombinational acquisition of the replicative *cis*-elements by a sgRNA, thus conferring on it the ability to replicate autonomously. In the steps to follow, the sequence portions that are no longer necessary (the autonomized part in the parental RNA and the subgenomic promoter in the newly formed RNA component) may be deleted. Examples of viral RNA recombination mechanisms which might be instrumental in this scenario are available. Replication of citrus tristeza closterovirus *in vivo* is accompanied by formation of autonomously replicating defective RNAs (D-RNAs), some of which contain long 5'- and 3'-terminal genomic sequences including the integral ORFs (Mawassi *et al.*, 1995). In the genomes of tobraviruses, RNA 1 contains genes for the replicase, the MP, and a small cysteine-rich protein, whereas RNA 2 contains the CP gene and, depending on the virus isolate, may also carry a complete or partial copy of the cysteine-rich protein (Angenent *et al.*, 1989). Notably, despite the 5'-proximal position of the CP gene in tobravirus RNA 2, its expression requires formation of a sgRNA (Figure 1A; Goulden *et al.*, 1990). The genetically unstable tobravirus system could thus illustrate a recent evolutionary transition from a monopartite ('tobamovirus-like') to a divided genome driven by RNA recombination.

To express the internal genes, some plant viruses also utilize non-orthodox mechanisms that violate the rules of eukaryotic translation at its three basic steps: initiation, elongation, and termination. Ribosomes may get direct access to an internal gene in viral RNA (internal initiation), or bypass the first AUG in the virus RNA to initiate the protein synthesis at a downstream AUG (leaky scanning). Likewise, in certain cases the ribosome may bypass a stop codon to continue translation of a downstream ORF, either by switching the reading frame (translational frameshifting), or by suppressing the stop codon and proceeding with translation in the same frame (readthrough of leaky stop codons).

Divided vs. Monopartite Genomes

The genomes of bromo-, cucumo-, ilar-, idaeo-, tobra-, diantho-, hordei-, furo-, crini-, bymo-, como-, and nepoviruses, are split among two or three RNA components. Notably, the members of a group of closely related viruses may have either monopartite or divided genomes. The respective examples are genera *Potyvirus* and *Bymovirus* in the family *Potyviridae*, and *Closterovirus* and *Crinivirus* in the family *Closteroviridae*.

The genome splitting may concern even the most conserved replicative gene module. Thus in the tripartite genomes of hordeiviruses, the conserved domains of methyltransferase (MT) and helicase (HEL), on the one hand, and POL, on the other, are expressed as distinct translation products of the genomic RNAs α and γ, respectively (Figure 1B; Gustafson *et al.*, 1989; Savenkov *et al.*, 1998). In the closely related tobamo-, tobra-, and furoviruses, these domains are found in a single gene product (Figure 1A, C; Chapter 3). A similar division of the replicative domains among the products of RNA 1 and 2 is also seen in the tripartite bromo-, cucumo-, and ilarviruses (Ahlquist, 1992). However, in the bipartite genome of a closely related idaeovirus the MT, HEL and POL domains are encoded in the monocistronic RNA 1 (Figure 1B; Ziegler *et al.*, 1992).

The viruses bearing the TGB of MP genes also provide examples of complicated genome division patterns. The monopartite genomes of potex- and carlaviruses encompass the genes for replicase, TGB proteins, and CP (Figure 1C; Chapter 3). In beet necrotic yellow vein benyvirus (BNYVV) and peanut clump furovirus (PCV), this gene array is split among the genomic RNA 1 (coding for the replicase) and RNA 2 (coding for the CP and TGB proteins) (Herzog *et al.*, 1994, and references therein). Further, in the tripartite genomes of the furoviruses, potato mop top virus (PMTV) and beet soil-borne virus (BSBV), the genetic content of the 'RNA 2' is split so that the TGB and CP genes are assigned to two individual genomic RNAs (Figure 1C; Kashiwazaki *et al.*, 1995; Koenig *et al.*, 1997).

Transcriptional Control: Subgenomic RNAs

Many plant viruses express the internal genes via sgRNAs that are transcribed from the full-sized minus RNA strand and represent the 5'-terminally truncated 3'-coterminal copies of the genomic RNA (Buck, 1996). As a rule, each sgRNA

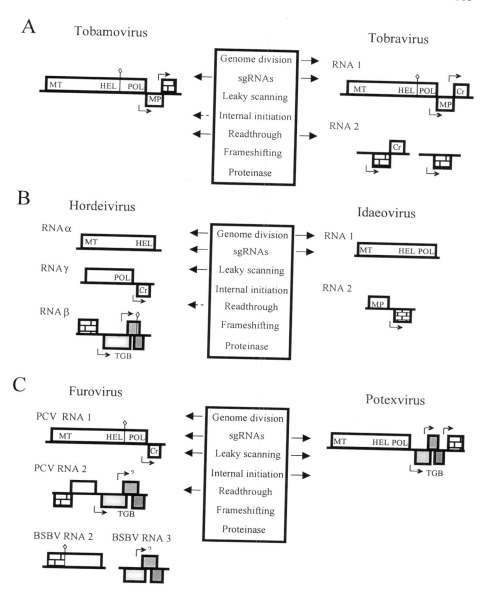

Figure 1. Pairwise comparisons of the expression strategies in alpha-like plant viruses. The basic expression mechanisms are shown in the central box, arrow points to the group that uses given mechanism; dashed arrows mark mechanisms used by a single member but not by the others in the group, or mechanisms of obscure significance. MT, methyltransferase; HEL, helicase; POL, polymerase; MP, movement protein; TGB, Triple Gene Block proteins; Cr, cysteine-rich protein; 'brick wall' fill-in, coat protein. Open diamonds, the readthrough sites; bent arrows, sgRNA start and direction.

species serves for translation of just one viral gene exposed at its 5' end. However, some sgRNAs of luteoviruses (Tacke *et al.*, 1990; Dinesh-Kumar *et al.*, 1992), tombusviruses (Rochon and Johnson, 1991), and hordeiviruses (Zhou and Jackson, 1996) direct the expression of two genes nested in different reading frames. The sgRNAs are utilized both by the viruses with monopartite and with divided genomes (Figure 1A, B, C). Their number varies in different virus groups, from just one in bromoviruses and dianthoviruses to six to nine in closteroviruses (Buck, 1996; Agranovsky, 1996).

An important phenomenon associated with the gene expression from sets of viral genomic and sgRNAs is the temporal regulation of protein synthesis. The synthesis of different viral mRNAs (and proteins) follows distinct time patterns. Thus, among the proteins encoded in the BSMV genome, the POL and the TGB proteins are expressed transiently at the early infection phases, the MT-HEL is expressed early but declines only gradually, and the CP and the regulatory cysteine-rich protein are expressed stably throughout the course of the infection (Donald *et al.*, 1993). Likewise, the TMV proteins involved in replication and movement are expressed transiently at the early phases, whereas the CP is expressed at the late phases of infection (reviewed in Dawson, 1992). Apparently, the subgenomic promoters are the key elements in determining the temporal regulation of the TMV gene expression. Particularly, insertion of the MP gene (normally controlled by an early promoter) under the CP gene sg promoter results in delays in the MP gene expression and the virus cell-to-cell movement (Lehto *et al.*, 1990).

The 5'-termini of the sgRNAs of a number of plant viruses have been mapped, including those of bromo-, cucumo-, hordei-, tymo-, tobra-, tombus-, carmo-, luteo-, diantho-, and closteroviruses (reviewed in French and Ahquist, 1988; Dinesh-Kumar *et al.*, 1992; Agranovsky, 1996; Buck, 1996). This allowed one to identify the *cis*-acting sequences directing the sgRNA synthesis (sg promoters) and to dissect these experimentally. The sg promoter elements have been thoroughly characterized for the *Bromoviridae* family members. In BMV RNA 3, the promoter sequence governing the synthesis of the CP sgRNA consists of ca. 70 nt grouped into four core and modulator domains (French and Ahlquist, 1988; Marsh *et al.*, 1988). It is worth mentioning that the sg promoters of BMV and the related animal alphaviruses show stretches of homologous sequences (Marsh *et al.*, 1988). Moreover, the BMV replicase can recognize the alphavirus sg promoter *in vitro* and initiate accurate transcription, albeit with low efficiency (Siegel *et al.*, 1997). The sequences at the 5' termini of the CP sgRNAs of several viruses having closely related replicases, namely bromo-, cucumo-, tobamo-, and closteroviruses, also show some similarity, thus suggesting parallel evolution of the sg promoter-binding domains in virus replicases and the recognition signals in the genomic RNAs (Agranovsky *et al.*, 1994b).

Recently, a unique mechanism of RNA-mediated *trans*-activation of sgRNA transcription was described for the red clover necrotic mosaic dianthovirus (RCNMV), whose genome is divided among RNA 1 (coding for the replicase and the CP) and RNA 2 (coding for the MP). The synthesis of the RCNMV CP sgRNA was shown to require base pairing between the sg promoter in the RNA 1 (-)strand and a 34-nt region in RNA 2 (Sit *et al.*, 1998).

Non-orthodox Mechanisms of Initiation

Leaky Scanning

The leaky scanning mechanism is essentially based on discrimination between two or more initiation sites having unequal contexts (reviewed in Fütterer and Hohn, 1996). As a result, the 40S ribosome subunit may bypass the first AUG in an mRNA and initiate translation at a downstream AUG. Several variants of this mechanism can be distinguished, with initiation at two sites on one ORF, on overlapping ORFs, and on consecutive ORFs (Fütterer and Hohn, 1996).

Leaky scanning accounts for the synthesis of the overlapping 95- and 105-kDa proteins encoded in a single ORF of the cowpea mosaic comovirus M-RNA (Verver *et al.*, 1991). Translation of the 5'-proximal gene in the BSMV TGB may be initiated from two alternative in-frame AUG codons to yield two functionally active forms of the βb protein (Petty and Jackson, 1990). In addition, leaky scanning is employed to express the partially overlapping middle and 3'-proximal genes of the BSMV TGB (Zhou and Jackson, 1996). In the genomic RNA 2 of PCV, the stop codon in the 5'-proximal CP gene and the initiating codon of the downstream 39K gene overlap by two bases (...**AUGA**..., stop codon underlined, start codon in bold; Manohar *et al.*, 1993). It was found that about one third of the translating ribosomes bypass the AUG in the CP gene (which is in a poor context) to initiate translation at the AUG of the 39K gene (Herzog *et al.*, 1995). It has been reported that upon translation of the southern bean mosaic sobemovirus RNA, a suboptimal context of the AUG in the 5'-most ORF 1 allows a part of the scanning ribosomes to reach the downstream ORF 2 coding for the replicase polyprotein (Sivakumaran and Hacker, 1998).

In the genomes of luteoviruses, the ORF coding for the 17-kDa movement protein (p17) is nested, in a different reading frame, within the CP ORF, both genes being expressed from one and the same sgRNA species (Dinesh-Kumar and Miller, 1993). As found for the barley yellow dwarf luteovirus (BYDV), the 5'-proximal AUG, that of the CP gene, is rather inefficient since it has a suboptimal context and is masked by the RNA secondary structure elements. As a consequence, most of the scanning 40S ribosomal subunits bypass this codon to initiate translation at the downstream AUG for the p17 gene. A feedback element in this leaky scanning mechanism is that a ribosome, once it has initiated translation at the second AUG, provokes initiation at the first AUG, perhaps by melting the RNA secondary structure and by stalling the following ribosomal subunits (Dinesh-Kumar and Miller, 1993). Coordinated translation of the two nested ORFs coding for the proteins involved in the virus movement, was also revealed in a tombusvirus (cucumber necrosis virus) (Rochon and Johnston, 1991).

Internal Ribosome Entry Sites (IRES Elements)

The mechanism of cap-independent internal initiation of translation was first described for animal picornaviruses. The *cis*-signals mediating direct access of the

40S ribosomal subunits to the initiation codon represent an elaborate set of primary and secondary RNA structure elements in the picornavirus genomic leaders (reviewed in Belsham and Sonnenberg, 1996). The IRES elements were also found in the genomes of plant picorna-like viruses, namely poty- and comoviruses, although their 5'-leaders are not as long and complex as those of picornaviruses (reviewed in Fütterer and Hohn, 1996). Also, in contrast to picornaviruses, the use of IRES by plant viruses appears to be optional. Indeed, the comovirus M-RNA translation may be initiated either by internal initiation or by leaky scanning (Verver *et al.*, 1991). Recently, the IRES elements have been implicated in expression of the CP genes of the alpha-like plant viruses, namely potato virus X (Hefferon *et al.*, 1997) and a crucifer-infecting tobamovirus (cr-TMV; Ivanov *et al.*, 1997). In both potexviruses and tobamoviruses, the CP gene is normally expressed through monocistronic sgRNAs. It has been supposed that the internal initiation mechanism allows direct translation of the CP gene in the genomic RNA at the early stages of infection, thus providing limited amounts of the CP necessary for the cell-to-cell and/or long-distance transport of the virus infection (Hefferon *et al.*, 1997; Ivanov *et al.*, 1997). Notably, the IRES in the crTMV genome is not conserved in the related tobamoviruses, implying that this element is a recent evolutionary acquisition (Ivanov *et al.*, 1997).

Non-orthodox Mechanisms of Elongation and Termination

Readthrough of Leaky Stop Codons

The readthrough of a leaky terminating codon is used for POL expression in disparate (+)RNA virus groups, including plant tobamo-, tobra-, furo-, carmo- and tombusviruses, and some animal alphaviruses. The readthrough is also employed for the expression of the *gag-pol* fusion protein in some retroviruses. The readthrough mechanism was first demonstrated for TMV by Pelham (1978). In the genome of TMV, the UAG stop codon separates the replicase gene into portions coding for MT plus HEL and POL, respectively (Figure 1A). In translation, 95% and 5% of the translating ribosomes terminate at the stop codon or suppress it, which results in the synthesis of the respective major 120K and the minor 183K products. It was found that two suppressor $tRNA^{Tyr}$ species present in the cytoplasm of tobacco cells are able to mediate the readthrough by offset pairing with the UGA stop codon, which is thus decoded as a tyrosine triplet (Zerfass and Beier, 1992a). The pseudouridine residue in the $tRNA^{Tyr}$ GΨA anticodons is a host determinant of efficient UAG suppression (Zerfass and Beier, 1992a). The TMV-specific determinant of the readthrough is the sequence of two triplets immediately downstream of the UAG stop codon in the viral RNA; the consensus UAG CAR YYA (where R and Y are a pyrimidine and a purine residue, respectively) was found to favor the suppression *in vivo* (Skuzeski *et al.*, 1991; Zerfass and Beier, 1992a). A similar, yet differing in some details, readthrough mechanism accounts for the synthesis of the closely related MT-HEL-POL-containing replicases encoded in the RNA 1 of tobacco rattle tobravirus (TRV; Zerfass and Beier, 1992b) and furoviruses SBWMV and PCV

(Figure 1A, C; Shirako and Wilson, 1993; Herzog *et al.*, 1994). In the case of TRV, two tRNA^{Trp} species with methylated CmCA anticodons (naturally occurring in tobacco cytoplasm and chloroplasts) were found to be necessary for the suppression (Zerfass and Beier, 1992b). The replicases of carmoviruses and tombusviruses lack the MT and HEL, and contain the POL domain only distantly related to that of the alpha-like virus superfamily (Koonin and Dolja, 1993). The POL expression in these viruses also depends on the readthrough of a leaky stop codon located far upstream of the POL domain (Scholthof *et al.*, 1995).

In furoviruses (except PCV), the suppression of the stop codons in the CP gene mediates the synthesis of the C-terminally extended CP forms (Schmitt *et al.*, 1992; Shirako and Wilson, 1993). SBWMV thus provides the so far only example of a virus having two readthrough signals residing on each of the genomic RNA components. A similar strategy for the expression of conventional and extended CP is employed by luteoviruses (Figure 2A; Bahner *et al.*, 1990; Tacke *et al.*, 1990). Notably, efficient suppression of the leaky stop in the BYDV CP gene requires a composite signal consisting of several CCX XXX repeats found 3' to the leaky UAG stop, and a distal sequence element located 697 to 758 residues downstream of the stop codon (Brown *et al.*, 1996). It was found that a few copies of the aberrant CP are incorporated into the isometric particles of luteoviruses (Filichkin *et al.*, 1994) and the rod-like particles of furoviruses (Haeberle *et al.*, 1994), where the protruding readthrough domains ensure interactions with their respective vectors, aphids and fungal zoospores. Apparently, the readthrough mechanisms have evolved independently in the unrelated CP genes of luteo- and furoviruses, exemplifying functional convergence.

Recently, it has been found that the βd protein encoded in the middle gene of the BSMV TGB, is expressed in two forms (of 14 and 23 kDa), of which the larger is a readthrough product of the amber stop codon (Zhou and Jackson, 1996). It cannot be excluded that other virus non-replicative and non-CP proteins may be expressed in a similar fashion.

Ribosomal Frameshifting

In elongation, the frequency of spontaneous ribosome switching to another reading frame (frameshifting) is quite low, being on the order of 10^{-4} (Lindsey and Gallant, 1993). However, the presence of specific signals, including the so-called 'slippery sequences', in the viral RNA templates increases the incidence of frameshifting manifold, up to $2-3 \times 10^{-1}$. The mechanism of leftward (or -1) frameshifting postulates 'simultaneous slippage' of two tRNAs bound to a slippery sequence with the consensus X XXY YYZ, to decode it as XXX YYY (Jacks *et al.*, 1988). In many cases, the reading frame switching at the slippery site is stimulated by a nearby pseudoknotted secondary structure (ten Dam *et al.*, 1990) which possibly forces the out-of-frame triplet recognition by impeding the progress of ribosomes along the template (Tu *et al.*, 1992). Like the readthrough signals described in the previous section, the -1 frameshifting signals exist in the mRNAs of viruses belonging to different classes, including the (+)RNA viruses of plants (luteo-, diantho-, and sobemoviruses) and animals (corona-, arteri-, and astroviruses); the small double-stranded RNA viruses; and retroviruses (reviewed in Jacks *et al.*, 1988; Farabaugh, 1993; Agranovsky, 1996). In luteoviruses and dianthoviruses, the efficiency of -1 frameshifting leading to the synthesis of a fusion replicase, is as low as ~1 % (Brault and Miller, 1992; Prüfer *et al.*, 1992; Garcia *et al.*, 1993; Kim and Lommel, 1994). In the genome of cocksfoot mosaic sobemovirus (CfMV), the overlap region of ORFs 2 and 3 (coding for the proteinase and POL domains, respectively) contains a -1 frameshifting signal that directs the synthesis of the replicase-containing polyprotein with high efficiency, up to 30% (Mäkkinen *et al.*, 1995; Ryabov *et al.*, 1996). This signal is conserved in the related sobemoviruses. Surprisingly, however, the proteinase and POL domains in the sobemoviruses other than CfMV are encoded in one continuous ORF (Figure 2A; Mäkkinen *et al.*, 1995; Ryabov *et al.*, 1996). Hence, in the evolution of sobemoviruses only one member of the group has engaged the frameshifting mechanism for POL expression.

In the genomes of beet yellows virus (BYV) and other closteroviruses, ORF 1a (coding for MT and HEL) and ORF 1b (coding for POL) are in an unusual 0/+1 configuration (Figure 2B). It is suggested that the BYV POL is expressed as an ORF 1a/1b fusion product resulting from +1 ribosomal frameshifting (Agranovsky *et al.*, 1994a; Agranovsky, 1996). The rightward (or +1) frameshifting, which requires distinct signals, has been reported only for prokaryotes and yeast *Ty* retrotransposons (reviewed in Farabaugh, 1993).

It is possible that the plant virus genes other than replicase may also be expressed by the ribosomal frameshifting mechanism. Gramstadt *et al.* (1994) reported that the gene array coding for CP and 12K nucleic acid-binding protein of carlavirus potato virus M (PVM) may be expressed by -1 frameshifting to yield the CP-12K fusion (Figure 2B). In addition, the 12K gene may be expressed by the internal initiation mechanism (see above). Site-directed mutagenesis and *in vitro* translation identified an unusual -1 frameshifting signal including the UGA codon in the CP gene and the upstream sequence of four A residues; the secondary structure elements were found to be of no importance (Gramstadt *et al.*, 1994).

PROTEOLYTIC PROCESSING

In many viruses, the completion of gene translation does not exhaust the act of expression, as the newly synthesized protein chains may have to undergo post-translational processing by the virus-encoded proteinases into mature structural and nonstructural proteins. The two types of proteolytic enzymes encoded in plant and animal (+)RNA viruses are chymotrypsin-like serine or cysteine proteinases (S-PRO or C-PRO) and papain-like cysteine (thiol) proteinases (PCP), as identified by the (limited) similarity in their sequence and spatial fold to the archetype cell enzymes (Dougherty and Semler, 1993; Koonin and Dolja, 1993).

In viruses belonging to picorna-like superfamily, including plant poty-, nepo-, and comoviruses, the genome is expressed as a polyprotein precursor by contiguous translation of a single large ORF. In such systems, there is no transcriptional control, and the principal means of controlling the gene expression is the proteolytic

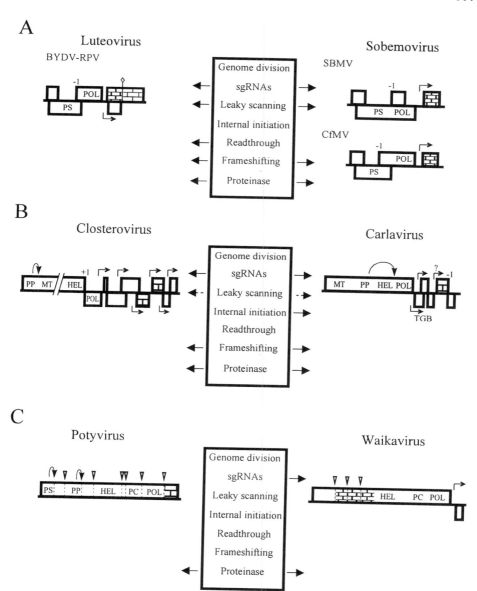

Figure 2. Pairwise comparisons of the expression strategies in sobemo-like (A), alpha-like (B), and picorna-like (C) plant viruses. In (B), the middle part of the closterovirus ORF 1a (coding for a non-conserved spacer) is not shown. PS and PC, chymotrypsin-like serine and cysteine proteinases, respectively; PP, papain-like proteinase; bent arrows, *cis*-cleavage sites; open arrowheads, *trans*-cleavage sites; -1 and +1, frameshifting sites. Other designations are as in Figure 1.

processing, subject to temporal regulation through a complex cascade of cleavage events and production of protein intermediates.

Chymotrypsin-like Serine or Cysteine Proteinases

The S-PRO domains contain a catalytic triad of the His, Asp, and Ser (of which the latter is the active residue) arranged in a conserved spatial structure. In picornavirus 3C proteinase and the related C-PRO of plant picorna-like viruses, Ser is substituted by Cys. The C-PRO domains play a key role in polyprotein processing in a number of (+)RNA viruses of animals (picorna-, flavi-, pesti-, and coronaviruses) and plants (poty-, bymo-, como-, and nepoviruses) (reviewed in Dougherty and Semler, 1993). The sequence data analysis also suggested the existence of the chymotrypsin-like PRO domains in the proteins of luteoviruses, sobemoviruses (Koonin and Dolja, 1993), sequiviruses (Turnbull-Ross *et al.*, 1993), and waikaviruses (Shen *et al.*, 1993) (Figure 2A, C).

The potyvirus polyproteins are cleaved *in cis* and *in trans* by the C-PRO domain located in the NIa protein (Figure 2C). In contrast, the S-PRO domain found at the N-terminus of the polyprotein is responsible for a single *in cis* cleavage and release of the potyvirus P1 protein (Figure 2C; Verchot *et al.*, 1991). The polyproteins encoded in bipartite genomes of como- and nepoviruses are processed exclusively by the 3C-like proteinases related to the NIa (Dessens and Lomonossoff, 1991; Peters *et al.*, 1992; reviewed in Dougherty and Semler, 1993). Possibly, in this case the absence of additional proteinase domains is compensated by the genome partitioning. A unique phenomenon described for the comovirus polyprotein processing is the involvement of the 32-kDa protein in the proteolysis by the 24-kDa C-PRO (encoded in the same genomic component, B-RNA). Albeit not a proteinase *per se*, the 32-kDa protein possibly serves as a cofactor of the C-PRO that influences the turnover of the M-RNA- and B-RNA-encoded polyproteins (Peters *et al.*, 1992). A notable exception in the picorna-like virus superfamily is the rice tungro spherical waikavirus, which apparently employs a C-PRO for the processing of a polyprotein encompassing the conserved structural and nonstructural proteins, and a sgRNA to express the unique 3'-proximal ORF(s) (Figure 2C; Shen *et al.*, 1993).

Papain-like Cysteine Proteinases

The active center of PCP contains a catalytic dyad of Cys and His residues, and its cleavage site has a consensus UG/X (where U is a hydrophobic residue, and X is Gly, Ala, or Val). The PCP was first identified in the helper component proteinase (HC-Pro) of a potyvirus (Oh and Carrington, 1989). Later, similar proteinase domains have been predicted (and, in some cases, confirmed experimentally) in the encoded proteins of more and more (+)RNA viruses, e.g., plant tymoviruses (Bransom and Dreher, 1994), closteroviruses (Agranovsky *et al.*, 1994a; Agranovsky, 1998), carlaviruses (Lawrence *et al.*, 1995), marafiviruses (Edwards *et al.*, 1997), capilloviruses (Koonin and Dolja, 1993) and BNYVV (Hehn *et al.*, 1997), as well as animal alpha-, arteri-, corona-, rubi-, and aphthoviruses

(Gorbalenya *et al.*, 1991; reviewed in Strauss and Strauss, 1994; Snijder and Meulenberg, 1998).

The potyvirus HC-Pro exerts a single *in cis* cleavage to detach its own C-terminus from the rest of the polyprotein (Figure 2C). In closteroviruses, one or two related PCP domains located at the N-terminus of the ORF 1a product, effect the release of the leader protein(s) from the rest of the polyprotein containing the MT and HEL domains (Figure 2B; reviewed in Agranovsky, 1996, 1998). In the polyprotein of BNYVV, the PCP is located between the HEL and POL domains (Hehn *et al.*, 1997), whereas in those of tymoviruses and carlaviruses it is found between the MT and HEL domains (Figure 2B; Bransom and Dreher, 1994; Lawrence *et al.*, 1995). Regardless of its position, the PCP of tymo-, carla-, and benyviruses mediates the splitting of the MT-HEL-containing and POL-containing fragments.

Hence, in all the reported cases, the PCPs of plant viruses possess only the *cis*-activity and cleave at a single site downstream of the proteinase active center. It should be noted, however, that the alphavirus PCP directs multiple *in cis* and *in trans* cleavages, being thus responsible for the entire processing of the nsP1-4 precursor of nonstructural proteins (Strauss and Strauss, 1994). The leader PCP in foot-and-mouth disease aphthovirus initially mediates the *in cis* cleavage to release itself from the rest of the polyprotein. However, the mature leader PCP participates in the shut-off of the cell cap-dependent translation by cleaving the p220 protein in eIF-4F complex (Devaney *et al.*, 1988). With respect to striking similarities of animal and plant viruses in picorna-like and alpha-like supergroups, such multiple activities of plant virus proteinases cannot be ruled out. The recent immunological analysis of the BYV replicative proteins *in vivo* indicated that, in addition to the processing by the leader PCP, the 1a polyprotein is further cleaved (by a yet unidentified proteinase activity) into the large fragments with MT and HEL domains (T. Erokhina, R. Zinovkin, M. Vitushkina, W. Jelkmann and A. Agranovsky, manuscript submitted).

CONCLUSIONS

Apparently, the expression mechanisms developed and 'tuned' in the process of evolution serve to optimally use the information in the compact genetic systems such as RNA viruses. The examples considered in this chapter show how the expression strategies of (+)RNA plant viruses meet a key challenge of the infection: the necessity to express the proteins at certain times and in fixed amounts. Also, expression of the compressed viral genomes illustrates the economical use of the coding sequence to get 'two yields from one plot'. This trend is most clearly seen in translation of overlapping genes in viral RNAs, and in the synthesis of the N- and C-terminally extended versions of a protein.

The two basic strategies to regulate the viral replicase synthesis rely on (i) the use of specific signals (leaky stop codons or frameshifting signals) that separate the sequence encoding POL from the upstream gene (to allow downregulation of the replicase expression), and (ii) the controlled expression of the replicative domains as distinct products resulting from polyprotein processing, or from translation of individual genomic RNAs. Expression of the viral replicase as a set of distinct components might serve for coordination of different enzymatic functions in RNA replication and transcription, namely unwinding of duplexes, asymmetrical synthesis of (+) and (-) strands, transcription of sgRNAs, and RNA capping.

The expression modes in quite disparate groups of (+)RNA viruses may be strikingly similar, e.g., in animal corona-like viruses and plant closteroviruses, which implies convergent evolution of the genome layout and expression (Agranovsky et al., 1994a). However, in many cases the viral expression strategies are quite variable even among the related (+)RNA viruses belonging to the same taxa. The examples are the existence of monopartite and divided genomes; the expression of the POL domains by either frameshifting, readthrough, or translation of a contiguous ORF; and the absence or presence of the proteinase domains, along with different localization of proteinase(s) and their cleavage sites in the encoded polyproteins. These examples, the list of which may be extended, clearly indicate that the specific expression mechanisms may be poorly conserved traits (see also Koonin, 1991b). In other words, distinct gene expression patterns in some virus groups arose from rather recent events in their evolution.

REFERENCES

Agranovsky, A. A. (1996). The principles of molecular organization, expression and evolution of closteroviruses: Over the barriers. *Adv.Virus Res.* 47, 119-158.

Agranovsky, A. A. (1998). Closterovirus papain-like cysteine endopeptidases. *In* "Handbook of Proteolytic Enzymes" (A. J. Barrett, N. D. Rawlings and J. F. Woessner, Eds.). pp. 700-702. Academic Press, New York.

Agranovsky, A. A., Boyko, V. P., Karasev, A. V., Lunina, N. A., Koonin, E. V., and Dolja, V. V. (1991). Nucleotide sequence of the 3'-terminal half of beet yellows closterovirus RNA genome: Unique arrangement of eight virus genes. *J. Gen. Virol.* 72, 15-23.

Agranovsky, A. A., Koonin, E. V., Boyko, V. P., Maiss, E., Frötschl, R., Lunina, N. A., and Atabekov, J. G. (1994a). Beet yellows closterovirus: complete genome structure and identification of a leader papain-like thiol proteinase. *Virology* 198, 311-324.

Agranovsky, A. A., Koenig, R., Maiss, E., Boyko, V. P., Casper, R., and Atabekov, J. G. (1994b). Expression of the beet yellows closterovirus capsid protein and p24, a capsid protein homologue, *in vitro* and *in vivo. J. Gen.Virol.* 75, 1431-1439.

Ahlquist, P. (1992). Bromovirus RNA replication and transcription. *Curr. Opin. Gen. Dev.* 2, 71-76.

Angenent, G. C., Posthumus, E., Brederode, F.T., and Bol, J. F. (1989). Genome structure of tobacco rattle virus strain PLB: Further evidence on the occurrence of RNA recombination among tobraviruses. *Virology* 171, 271-274.

Bahner, I., Lamb, J., Mayo, M. A., and Hay, R. T. (1990). Expression of the genome of potato leafroll virus: readthrough of the coat protein gene termination codon in vivo. *J. Gen. Virol.* 71, 2251-2256.

Belsham, G. J., and Sonenberg, N. (1996). RNA protein interactions in regulation of picornavirus RNA translation. *Microbiol. Rev.* 60, 499-511.

Bransom, K. L., and Dreher, T. W. (1994). Identification of the essential cysteine and histidine residues of the turnip yellow mosaic virus protease. *Virology* 198, 148-154.

Brault, V., and Miller, W. A. (1992). Translational frameshifting mediated by a viral sequence in plant cells. *Proc. Natl. Acad. Sci. USA* 89, 2262-2266.

Brown, C. M., Dinesh-Kumar, S. P., and Miller, W. A. (1996). Local and distant sequences are required for efficient readthrough of the barley yellow dwarf virus PAV coat protein gene stop codon. *J. Virol.* 70, 5884-5892.

Browning, K. S., Lax, S. R., Humphreys, J., Ravel, J. M., Jobling, S. A., and Gehrke, L. (1988). Evidence that the 5'-untranslated leader of mRNA affects the requirement for wheat germ initiation factors 4A, 4F, and 4G. *J. Biol. Chem.* 263, 9630-9634.

Buck, K. W. (1996). Comparison of the replication of positive-stranded RNA viruses of plants and animals. *Adv. Virus Res.* 47, 159-251.

Carrington, J. C., and Freed, D. D. (1990). Cap-independent enhancement of translation by a plant potyvirus 5' nontranslated region. *J. Virol.* 64, 1590-1597.

Dawson, W. O. (1992). Tobamovirus-plant interactions. *Virology* 186, 359-367.

De Jong, W., Mise, K., Chu, A., and Ahlquist, P. (1997). Effects of coat protein mutations and reduced movement protein expression on infection spread by cowpea chlorotic mottle virus and its hybrid derivatives. *Virology* 232, 167-173.

Dessens, J. T., and Lomonossoff, G. P. (1991). Mutational analysis of the putative catalytic triad of the cowpea mosaic virus 24K protease. *Virology* 184, 738-746.

Devaney, M. A., Vakharia, V. N., Lloyd, R. E., Ehrenfeld, E., and Grugman, M. J. (1988). Leader protein of foot-and-mouth disease virus is required for cleavage of the p220 component of the cap-binding protein complex. *J. Virol.* 62, 4407-4409.

Dinesh-Kumar, S. P., Brault, V., and Miller, W. A. (1992). Precise mapping and *in vitro* translation of a trifunctional subgenomic RNA of barley yellow dwarf virus. *Virology* 187, 711-722.

Dinesh-Kumar, S. P. and Miller, W. A. (1993). Control of start codon choice on a plant viral RNA encoding overlapping genes. *Plant Cell* 5, 679-692.

Dolja, V. V., Karasev, A. V., and Koonin, E. V. (1994). Molecular biology and evolution of closteroviruses: Sophisticated build-up of large RNA genomes. *Annu. Rev. Phytopathol.* 32, 261-285.

Donald, R. G. K., Zhou, H., and Jackson, A. O. (1993). Serological analysis of barley stripe mosaic virus-encoded proteins in infected barley. *Virology* 195, 659-668.

Donze, O., and Spahr, P. F. (1992). Role of the open reading frames of Rous sarcoma virus leader RNA in translation and genome packaging. *EMBO J.* 11, 3747-3757.

Dougherty, W. G., and Semler, B. L. (1993). Expression of virus-encoded proteinases: Functional and structural similarities with cellular enzymes. *Microbiol. Rev.* 57, 781-822.

Edwards, M. C., Zhang, Z., and Weiland, J. J. (1997). Oat blue dwarf marafivirus resembles the tymoviruses in sequence, genome organization, and expression strategy. *Virology* 232, 217-229.

Farabaugh, P. J. (1993). Alternative readings of the genetic code. *Cell* 74, 591-596.

Filichkin, S. A., Lister, R. M., McGrath, P. F., and Young, M. J. (1994). *In vivo* expression and mutational analysis of the barley yellow dwarf virus readthrough gene. *Virology* 205, 290-299.

French, R., and Ahlquist, P. (1988). Characterization and engineering of sequences controlling *in vivo* synthesis of brome mosaic virus subgenomic RNA. *J. Virol.* 62, 2411-2420.

Fütterer, J., and Hohn, T. (1996). Translation in plants - rules and exceptions. *Plant Mol. Biol.* 32, 159-189.

Gallie, D. R. (1991). The cap and poly(A) tail function synergistically to regulate mRNA translational efficiency. *Genes Devel.* 5, 2108-2116.

Gallie, D. R. (1996). Translational control of cellular and viral mRNAs. *Plant Mol. Biol.* 32, 145-158.

Gallie, D. R., Sleat, D. E., Watts, J. W., Turner, P. C., and Wilson, T. M. A. (1987). A comparison of eukaryotic viral 5'- leader sequences as enhancers of mRNA expression *in vivo*. *Nucleic Acids Res.* 15, 8693-8711.

Gallie, D. R., Tanguay, R., and Leathers, V. (1995). The tobacco etch viral 5' leader and poly(A) tail are functionally synergistic regulators of translation. *Gene* 165, 233.

Garcia, A., van Duin, J., and Pleij, C. (1993). Differential response to frameshift signals in eukaryotic and prokaryotic translational systems. *Nucleic Acids Res.* 21, 401-406.

Geballe, A. P., and Morris, D. R. (1994). Initiation codons within 5'-leaders of mRNAs as regulators of translation. *Trends Biochem. Sci.* 19, 159-164.

Gibbs, A. (1987). A molecular evolution of viruses: 'trees', 'clocks', and 'modules'. *J. Cell Sci.* (Suppl.) 7, 319-337.

Gorbalenya, A. E., Koonin, E. V., and Lai, M. M. C. (1991). Putative papain-related proteinases of positive-strand RNA viruses. *FEBS Lett.* 288, 201-205.

Goulden, M. G., Lomonossoff, G. P., Davies, J. W., and Wood, K. R. (1990). The complete nucleotide sequence of PEBV RNA 2 reveals the presence of a novel open reading frame and provides insights into the structure of tobraviral subgenomic promoters. *Nucleic Acids Res.* 18, 4507-4512.

Gramstadt, A., Prüfer, D., and Rohde, W. (1994). The nucleic acid-binding zinc finger protein of potato virus M is translated by internal initiation as well as by ribosomal frameshifting involving a stop codon and a novel mechanism of P-site slippage. *Nucleic Acids Res.* 22, 3911-3917.

Gustafson, G. D., and Armour, S. L. (1986). The complete nucleotide sequence of RNAβ from the Type strain of barley stripe mosaic virus. *Nucleic Acids Res.* 14, 3895-3909.

Gustafson, G. D., Armour, S. L., Gamboa, G. C., Burgett, S. G., and Shepherd, J. W. (1989). Nucleotide sequence of barley stripe mosaic virus RNAα: RNAα encodes a single polypeptide with homology to corresponding proteins from other viruses and to the BSMV βb protein. *Virology* 170, 370-377.

Haeberle, A.-M., Stussi-Garaud, C., Schmitt, C, Garaud, J.-C., Richards, K. E., Guilley, H. and Jonard, G. (1994). Detection by immunogold labelling of P75 readthrough protein near an extremity of beet necrotic yellow vein virus particles. *Arch. Virol.* 134, 195-203.

Hefferon, K. L., Khalilian, H., Xu, H., and AbouHaidar, M. G. (1997). Expression of the coat protein of potato virus X from a dicistronic mRNA in transgenic potato plants. *J. Gen. Virol.* 78, 3051-3059.

Hehn, A., Fritsch, C., Richards, K. E., Guilley, H., and Jonard, G. (1997). Evidence for *in vitro* and *in vivo* autocatalytic processing of the primary translation product of beet necrotic yellow vein virus RNA 1 by a papain-like proteinase. *Arch. Virol.* 142, 1051-1058.

Herzog, E., Guilley, H., Manohar, S. K., Dollet, M., Richards, K. E., Fritsch, C., and Jonard, G. (1994). Complete nucleotide sequence of peanut clump virus RNA 1 and relationships with other fungus-transmitted rod-shaped viruses. *J. Gen. Virol.* 75, 3147-3155.

Herzog, E., Guilley, H., and Fritsch, C. (1995). Translation of the second gene of peanut clump virus RNA 2 occurs by leaky scanning *in vitro*. *Virology* 208, 215-225.

Ivanov, P. A., Karpova, O. V., Skulachev, M. V., Tomashevskaya, O. L., Rodionova, N. P., Dorokhov, Yu. L., and Atabekov, J. G. (1997). A tobamovirus genome that contains an internal ribosome entry site functional *in vitro*. *Virology* 232, 32-43.

Jacks, T., Madhani, H. D., Masiarz, F. R., and Varmus, H. E. (1988). Signals for ribosomal frameshifting in the Rous sarcoma virus *gag-pol* region. *Cell* 55, 447-458.

Jackson, A. O., Petty, I. T. D., Jones, R. W., Edwards, M. C., and French, R. (1991). Analysis of barley stripe mosaic virus pathogenicity. *Semin. Virol.* 2, 107-119.

Jelkmann, W., Maiss, E., and Martin, R. R. (1992). The nucleotide sequence and genome organization of strawberry mild yellow edge-associated potexvirus. *J. Gen. Virol.* 73, 475-479.

Jobling, S. A., and Gehrke, L. (1987). Enhanced translation of chimeric messenger RNAs containing a plant viral untranslated leader sequence. *Nature* 325, 622-625.

Kamer, G., and Argos, P. (1984). Primary structural comparison of RNA-dependent polymerases from plant, animal and bacterial viruses. *Nucleic Acids Res.* 12, 7269-7283.

Kanyuka, K., Vishnichenko, V., Levay, K., Kondrikov, D., Ryabov, E., and Zavriev, S. K. (1992). Nucleotide sequence of shallot virus X RNA reveals a 5'-proximal cistron closely related to those of potexviruses and a unique arrangement of the 3'-proximal cistrons. *J. Gen. Virol.* 73, 2553-2560.

Kashiwazaki, S., Scott, K. P., Reavy, B., and Harrison, B. D. (1995). Sequence analysis and gene content of potato mop-top virus RNA 3: Further evidence of heterogeneity in the genome organization of furoviruses. *Virology* 206, 701-706.

Kim, K H., and Lommel, S. A. (1994). Identification and analysis of the site of -1 ribosomal frameshifting in red clover necrotic mosaic virus. *Virology* 200, 574-582.

Klaassen, V. A., Boeshore, M., Koonin, E. V., and Falk, B. W. (1995). Genome structure and phylogenetic analysis of lettuce infectious yellows virus, a whitefly-transmitted, bipartite closterovirus. *Virology* 208, 99-110.

Koenig, R., Commandeur, U., Loss, S., Beier, C., Kauffmann, A., and Lesemann, D. E. (1997). Beet soil-borne virus RNA 2: Similarities and dissimilarities to the coat protein gene-carrying RNAs of other furoviruses. *J. Gen. Virol.* 78, 469-477.

Koonin, E. V. (1991a). The phylogeny of RNA-dependent RNA-polymerases of positive-strand RNA viruses. *J. Gen. Virol.* 72, 2197-2206.

Koonin, E. V. (1991b). Genome replication/expression strategies of positive-strand RNA viruses: A simple version of a combinatorial classification and prediction of new strategies. *Virus Genes* 5, 273-282.

Koonin, E. V., and Dolja, V. V. (1993). Evolution and taxonomy of positive-strand RNA viruses: Implications of comparative analysis of amino acid sequences. *CRC Crit. Rev. Biochem. Mol. Biol.* 28, 375-430..

Kozak, M. (1986). Point mutations define a sequence flanking the AUG initiator codon that modulates translation by eukaryotic ribosomes. *Cell* 44, 283-292.

Kozak, M. (1991). Structural features in eukaryotic mRNAs that modulate the initiation of translation. *J. Biol. Chem.* 266, 19867-19870.

Lawrence, D. M., Rozanov, M. N., and Hillman, B. I. (1995). Autocatalytic processing of the 223-kDa protein of blueberry scorch carlavirus by a papain-like proteinase. *Virology* 207, 127-135.

Lehto, K., and Dawson, W. O. (1990). Changing the start codon context of the 30K gene of tobacco mosaic virus from "weak" to "strong" does not increase expression. *Virology* 174, 169-176.

Lehto, K., Grantham, G. L., and Dawson, W. O. (1990). Insertion of sequences containing the coat protein subgenomic RNA promoter and leader in front of the tobacco mosaic virus 30K ORF delays its expression and causes defective cell-to-cell movement. *Virology* 174, 145-157.

Lindsey, D., and Gallant, J. (1993). On the directional specificity of ribosome frameshifting at a "hungry" codon. *Proc. Natl. Acad. Sci. USA* 90, 5469-5473.

Lovett, P. S., and Rogers, E. J. (1996). Ribosome regulation by the nascent peptide. *Microbiol. Rev.* 60, 366-385.

Lütcke, H. A., Chow, K. C., Mickel, F. S., Moss, K. A., Kern, H. F., and Scheele, G. A. (1987). Selection of AUG codons differs in plants and animals. *EMBO J.* 6, 43-48.

Mäkkinen, K., Naess, V., Tamm, T., Truve, E., Aaspôlu, A., and Saarma, M. (1995). The putative replicase of the cocksfoot mottle sobemovirus is translated as a part of the polyprotein by -1 ribosomal frameshift. *Virology* 207, 566-571.

Manohar, S. K., Guilley, H., Dollet, M., Richards, K., and Jonard, G. (1993). Nucleotide sequence and genome organization of peanut clump virus RNA 2 and partial characterization of deleted forms. *Virology* 195, 33-41.

Marsh, L. E., Dreher, T. W., and Hall, T. C. (1988). Mutational analysis of the core and modulator sequences of the BMV RNA3 subgenomic promoter. *Nucleic Acids Res.* 16, 981-995.

Mawassi, M., Karasev, A. V., Mietkiewska, E., Gafny, R., Lee, R. F., Dawson, W. O., and Bar-Joseph, M. (1995). Defective RNA molecules associated with citrus tristeza virus. *Virology* 208, 383-387.

Memelink, J., van der Vlugt, C. I. M., Linthorst, H. J. M., Derks, A. F. L. M., Asjes, C. J., and Bol, J. F. (1990). Homologies between the genomes of a carlavirus (lily simptomless virus) and a potexvirus (lily virus X) from lily plants. *J. Gen. Virol.* 71, 917-924.

Morozov, S. Yu., Dolja, V. V., and Atabekov, J. G. (1989). Probable reassortment of genomic elements among elongated RNA-containing viruses. *J. Mol. Evol.* 29, 52-62.

Morozov, S. Yu., and Rupasov, V. V. (1985). On the possibility of a common origin of the genes encoding the RNA polymerases of bacterial, plant, and animal positive-strand RNA viruses. *Biol. Nauki* 10, 19-24 (in Russian).

Oh, C., and Carrington, J. C. (1989). Identification of essential residues in potyvirus proteinase HC-Pro by site-directed mutagenesis. *Virology* 173, 692-699.

Pelham, H. R. B. (1978). Leaky UAG termination codon in tobacco mosaic virus RNA. *Nature* 272, 469-471.

Peters, S. A., Voorhost, W. G. B., Wery, B., Wellink, J., and van Kammen, A. (1992). A regulatory role for the 32K protein in proteolytic processing of cowpea mosaic virus polyproteins. *Virology* 191, 81-89.

Petty, I. T. D., and Jackson, A. O. (1990). Two forms of the major barley stripe mosaic virus nonstructural protein are synthesized *in vivo* from alternative initiation codons. *Virology* 177, 829-832.

Prüfer, D., Tacke, E., Schimtz, J., Kull, B., Kaufmann, A., and Rohde, W. (1992). Ribosomal frameshifting in plants: a novel signal directs the -1 frameshift in the synthesis of the putative viral replicase of potato leafroll luteovirus. *EMBO J.* 11, 1111-1117.

Rochon, D. M., and Johnston, J. C. (1991). Infectious transcripts from cloned cucumber necrosis virus cDNA: Evidence for a bifunctional subgenomic RNA. *Virology* 181, 656-665.

Rohde, W., Gramstadt, A., Schimtz, J., Tacke, E., and Prüfer, D. (1994). Plant viruses as model systems for the study of non-canonical translation mechanisms in higher plants. *J. Gen. Virol.* 75, 2141-2149.

Ryabov, E. V., Krutov, A. A., Novikov, V. K., Zheleznikova, O. V., Morozov, S. Yu., and Zavriev, S. K. (1996). Nucleotide sequence of RNA from the sobemovirus found in infected cocksfoot shows a luteovirus-like arrangement of the putative replicase and protease genes. *Phytopathology* 86, 391-397.

Sachs, A. B., Sarnow, P., and Hentze, M. W. (1997). Starting at the beginning, middle, and end: Translation initiation in eukaryotes. *Cell* 89, 831-838.

Savenkov, E. I., Solovyev, A. G., and Morozov, S. Yu. (1998). Genome sequences of poa semilatent and lychnis ringspot hordeiviruses. *Arch. Virol.* 219, 9-18.

Schmitt, C., Balmori, E., Jonard, G., Richards, K. E., and Guilley, H. (1992). *In vitro* mutagenesis of biologically active transcripts of beet necrotic yellow vein virus RNA 2: Evidence that a domain of the 75-kDa readthrough protein is important for efficient virus assembly. *Proc. Natl. Acad. Sci. USA* 89, 5715-5719.

Schmitz, J., Prüfer, D., Rohde, W., and Tacke, E. (1996). Non-canonical translation mechanisms in plants: Efficient *in vitro* and *in planta* initiation at AUU codons of the tobacco mosaic virus translational enhancer. *Nucleic Acids Res.* 24, 257-263.

Scholthof, K. G., Scholthof, H. B., and Jackson, A. O. (1995). The tomato bushy stunt virus replicase proteins are coordinately expressed and membrane associated. *Virology* 208, 365-369.

Shen, P., Kaniewska, M., Smith, C., and Beachy, R. N. (1993). Nucleotide sequence and genomic organization of rice tungro spherical virus. *Virology* 193, 621-630.

Shirako, Y. (1998). Non-AUG translation initiation in a plant RNA virus: A forty-amino-acid extension is added to the N terminus of the soil-borne wheat mosaic virus capsid protein. *J. Virol.* 72, 1677-1682.

Shirako, Y., and Wilson, T. M. A. (1993). Complete nucleotide sequence and organization of the bipartite RNA genome of soil-borne wheat mosaic virus. *Virology* 195, 16-32.

Siegel, R. W., Adkins, S., and Kao, C. C. (1997). Sequence-specific recognition of a subgenomic RNA promoter by a viral RNA polymerase. *Proc. Natl. Acad. Sci. USA* 94, 11238-11243.

Sit, T. L., Vaewhongs, A. A., and Lommel, S. A. (1998). RNA-mediated *trans*-activation of transcription from a viral RNA. *Science* 281, 829-832.

Sivakumaran, K., and Hacker, D. L. (1998). The 105-kDa polyprotein of southern bean mosaic virus is translated by scanning ribosomes. *Virology* 246, 34-44.

Skuzeski, J. M., Nichols, L. M., Gesteland, R. F., and Atkins, J. F. (1991). The signal for a leaky UAG stop codon in several plant viruses includes the two downstream codons. *J. Mol. Biol.* 218, 365-373.

Snijder, E. J., and Meulenberg, J. M. (1998). The molecular biology of arteriviruses. *J. Gen. Virol.* 79, 961-979.

Steinhauer, D. A., and Holland, J. J. (1987). Rapid evolution of RNA viruses. *Annu. Rev. Biochem.* 41, 409-433.

Strauss, J. H., and Strauss, E. G. (1994). The alphaviruses: Gene expression, replication and evolution. *Microbiol. Rev.* 58, 491-562.

Tacke, E., Prüfer, D., Salamini, F., and Rohde, W. (1990). Characterization of a potato leafroll luteovirus subgenomic RNA: Differential expression by internal translation initiation and UAG suppression. *J. Gen. Virol.* 71, 2265-2272.

ten Dam, E. B., Pleij, C. W. and Bosch, L. (1990). RNA pseudoknots: Translational frameshifting and readthrough on viral RNAs. *Virus Genes,* 4, 121-136.

Timmer, R. T., Benkowski, L. A., Schodin, D., Lax, S. R., Metz, A. M., Ravel, J. M., and Browning, K. S. (1993). The 5' and 3' untranslated regions of satellite tobacco necrosis virus RNA affect translational efficiency and dependence on a 5' cap structure. *J. Biol.Chem.* 268, 9504-9510.

Tomashevskaya, O. L., Solovyev, A. G., Karpova, O. V., Fedorkin, O. N., Rodionova, N. P., Morozov, S. Yu., and Atabekov, J. G. (1993). Effects of sequence elements in the potato virus X RNA 5' non-translated αβ-leader on its translation enhancing activity. *J. Gen. Virol.* 74, 2717-2724.

Tu, C., Tzeng, T.-H. and Bruenn, J. A. (1992). Ribosomal movement impeded at a pseudoknot required for frameshifting. *Proc. Natl. Acad. Sci. USA* 89, 8636-8640.

Turnbull-Ross, A. D., Mayo, M. A., Reavy, B., and Murant, A. F. (1993). Sequence analysis of the parsnip yellow fleck virus polyprotein: Evidence of affinities with picornaviruses. *J. Gen. Virol.* 74, 555-561.

Verchot, J., Koonin, E. V., and Carrington, J. C. (1991). The 35-kDa protein from the N-terminus of the potyviral polyprotein functions as a third virus-encoded proteinase. *Virology* 185, 527-535.

Verver, J., LeGall, O., van Kammen, A., and Wellink, J. (1991). The sequence between nucleotides 161 and 512 of cowpea mosaic virus M RNA is able to support internal initiation of translation *in vitro*. *J. Gen. Virol.* 72, 2339-2345.

Wang, S., and Miller, W. A. (1995). A sequence located 4.5 to 5 kilobases from the 5' end of the barley yellow dwarf virus (PAV) genome strongly stimulates translation of uncapped mRNA. *J. Biol.Chem.* 270, 13446-13452.

Zaccomer, B., Haenni, A. L., and Macaya, G. (1995). Remarkable variety of plant virus RNA genomes. *J. Gen. Virol.* 76, 231-247.

Zelenina, D. A., Kulaeva, O. I., Smirnyagina, E. V., Solovyev, A. G., Miroshnichenko, N. A., Fedorkin, O. N., Rodionova, N. P., Morozov, S. Yu., and Atabekov, J. G. (1992). Translational enhancing properties of the 5'-leader of potato virus X RNA. *FEBS Lett.* 296, 267-270.

Zerfass, K., and Beier, H. (1992a). Pseudouridine in the anticodon $G\Psi A$ of plant cytoplasmic tRNA^{Tyr} is required for UAG and UAA suppression in the TMV-specific context. *Nucleic Acids Res.* 20, 5911-5918.

Zerfass, K., and Beier, H. (1992b). The leaky UGA termination codon of tobacco rattle virus RNA is suppressed by tobacco chloroplast and cytoplasmic tRNA^{Trp} with CmCA anticodon. *EMBO J.* 11, 4167-4173.

Zhou, H., and Jackson, A. O. (1996). Expression of the barley stripe mosaic virus RNAβ "Triple Gene Block". *Virology* 216, 367-379.

Ziegler, A., Natsuaki, T., Mayo, M. A., Jolly, C. A., and Murant, A. F. (1992). The nucleotide sequence of RNA-1 of raspberry bushy dwarf virus. *J. Gen. Virol.* 73, 3213-3218.

Zimmern, D. (1988). Evolution of RNA viruses. *In* "RNA Genetics" (J. J. Holland, E.R. Domingo, and P. Ahlquist, Eds.). pp. 211-240. CRC Press, Boca Raton, Florida.

5 MOLECULAR BASIS OF GENETIC VARIABILITY IN RNA VIRUSES

Jozef J. Bujarski

INTRODUCTION

The variability of genome organization, so essential for RNA virus evolution, among more than 900 plant RNA viruses (56 genera) is enormous (Gibbs *et al.*, 1997). In fact, the potential for variation of the RNA genome is so large, that a term of quasispecies was proposed to reflect the nature of RNA virus populations (Domingo *et al.,* 1995, Eigen, 1993, 1996; Holland *et al.,* 1992; Moya and Garcia-Arenal, 1995). This concept predicts that a single virus isolate is not a single RNA sequence but rather a mixture of mutant sequences averaging around a consensus sequence. The biological selection acts upon the quasispecies to allow variants with greatly improved fitness to arise and predominate in the population. When the host or other characteristics of the environment change, such shifts in selection pressure can be easily overcome by changing the predominant sequences of the RNA genome. The biological implications of the quasispecies nature of plant RNA viruses are profoundly instrumental for rapid viral adaptation to new environments.

GENETIC MECHANISMS OF VARIABILITY

Mutation

Any change in nucleotide sequence of the viral RNA genome is considered a mutation. Mutations include point mutations, deletions, and insertions. Mutations can occur in natural infections and can also be induced experimentally. Mutagen treatments (involving special chemicals, gamma radiation and UV light) are used to increase the frequency of mutants in the population that is subsequently screened with appropriate selection techniques. To maximize the probability of isolating the most desired mutants carrying single nucleotide alterations, mutants are induced with the lowest possible mutagen dose. The advent of molecular biology techniques, especially polymerase chain reaction (PCR) and reverse genetics (Boyer and Haenni 1994), allows the use of useful site-directed mutagenesis protocols where specially designed oligonucleotide primers serve to target the mutagenizing events.

One of the most desirable traits for analysis of gene function, that could arise during mutations, is a temperature-sensitive (*ts*) trait in which viral mutants grow abnormally at the nonpermissive temperature. Such *ts* mutants arise when base changes lead to amino acid substitution that results in defective function at the nonpermissive temperature. It seems that most of the genes are susceptible to *ts* mutations which can also occur in a noncoding region by modifying the function that this region controls (Morse, 1994).

Point Mutations

Point mutations arise mainly due to low fidelity of RNA genome replication, most likely due to the lack of proofreading activities of RNA replicating enzymes [viral RNA-dependent RNA polymerases (RdRp) or replicases]. The rate of mutation has been estimated in the order of 10^{-4} per nucleotide per round of replication cycle (Smith and Inglis, 1987; Steinhauer and Holland, 1987). However, a much lower rate, in the order of 10^{-7}, was observed for particular mutations in poliovirus (Parvin *et al.*, 1986) and Sindbis virus (Durbin and Stollar, 1986). Among the two types of point mutations possible, transitions are the most common. They comprise the substitution of one pyrimidine by the other, or one purine by the other. G-A substitutions are the most common errors made by replicase enzyme because they require that C is substituted by U in nascent RNA strand, which can make a pair (weaker though) with the G residue in the original template (Levin, 1997). In transversions, which are less common, a purine residue is replaced by a pyrimidine residue or vice versa. Essentially, in transversions an A-U pair is substituted by a U-A or C-G pair.

Another possible mechanism causing point mutations in RNA genomes involves the so called RNA editing which is a process of posttranscriptional base modification. This enzymatic mechanism, described for several animal RNA viruses, waits for its discovery in plant viruses. Given that plant cells contain the enzymatic machinery required for RNA editing, it is conceivable that this strategy may also be employed by plant RNA viruses.

The lack of error-checking capabilities causes that the mutation rates in RNA viruses are many orders of magnitude greater than those of their host cells. This leads to rapid evolution of RNA viruses. However, the so called mutation frequency, which is the actual rate of misincorporations detected (established) in virus population (Roossinck, 1997), varies extensively among different virus systems. For instance, tobamoviruses do not show high variability (Rodriguez-Cerezo *et al.*, 1991) but cucumber mosaic cucumovirus (CMV) shows a high degree of variability. Between two poleroviruses, potato leafroll virus (PLRV) has little variation whereas beet western yellows virus has much greater variation. Interestingly, the degree of mutation frequency (variability) seems to correlate directly with variability in host range. A speculation is that the greater mutation frequency allows the virus to adapt to a higher number of different hosts (Roossinck 1997). Of course, natural selection and competition between variants are the driving forces for the survival of the fittest mutant and these mechanisms greatly contribute to the observed populations of variants.

Differences in mutation frequency can be observed not only between different viruses, but also inside the genomic RNA of a given virus. For instance, hot spots of genetic change were found in the RNA of satellite tobacco mosaic virus (Kurath and Dodds 1994) and in satellite RNA of CMV (Kurath and Palukaitis, 1989). It is speculated that polymerase stoppage by upstream secondary structures is responsible for the observed hot spots.

Insertions and Deletions

Insertions can be of larger RNA fragments or of only few nontemplated nucleotides. Usually, larger RNA fragments are inserted due to RNA recombination. Insertions of a few nontemplated nucleotides can occur due to errors in RNA replication at template regions that are more difficult for copying by the replicase enzyme. In fact, observations from several systems suggest that the addition of nontemplated nucleotides is a general phenomenon of viral and nonviral polymerases, including DNA-dependent RNA polymerases and RdRps (Bertholet *et al.* 1987). Polymerase stuttering ('slippage'), defined as reiterative copying of one or a few adjacent nucleotides, is one possible mechanism of nontemplated addition and has been described for several viral and nonviral polymerases, including DNA-dependent RNA polymerases and RdRps (Bertholet *et al.,* 1987; Cascone *et al.,* 1990; Jacques *et al.,* 1994; Nagy and Bujarski, 1995). A role of RNA duplex structure in promoting polymerase stuttering has been demonstrated for the polyadenylation of influenza virus RNA (Luo *et al.,* 1991). Other mechanisms such as a terminal transferase-like activity or some form of RNA editing (Cattaneo, 1991) might be responsible for addition of nontemplated nucleotides.

Deletions are quite often present in defective-interferring (DI) RNAs and numerous examples are discussed later. Much like insertions, the size of deleted fragments can vary from a few nucleotides to large portions of the RNA molecule. The latter truncation can remove up to several genes from the parental genome. In order to be competitive during infection, these deletions cannot involve signal regions that control RNA replication and RNA stability in a particular host.

Frameshift Mutations

Ribosomal frameshift mutations occur (during translation) when deletions or insertions change the reading of the regular genetic triplet code. In nature, ribosomal frameshifting is a method of regulation of protein expression. For instance, in all luteoviruses the polymerase polypeptide is translated from a minus 1 frameshift inside the first open reading frame (ORF). At a specific signal sequence a small fraction of the ribosomes is able to march back one base and to resume translation in a new reading frame (Miller and Rasochova, 1997). The shifty heptanucleotide sequence includes the stop codon and essentially the whole nucleotide context allows the ribosomes to bypass the stop codon. The structure of the shifty site includes a consensus sequence followed by a secondary structure region (Miller and Rasochova, 1997).

Mutant Stability

Most of the mutants (except deletion mutations) can undergo reverse mutations. Due to the errors during RNA replication following by selection, the RNA sequence can revert to the initial [wild-type (wt)] arrangement. The correction of the original mutation is called "back mutation". Alternatively, second site "compensatory mutations" can occur at physically distant location(s) from the original mutation. This can lead to suppression, which is the inhibition of a mutant phenotype by a second suppressor mutation. Such genetic suppression results in a wt phenotype despite its genetic mutation, resulting in formation of the so called pseudorevertants. Suppression mutation can occur in a different virus gene and even in a host gene. Since suppression may allow viruses to overcome negative effects of sequence alterations, it may be positively selected. (Morse, 1994).

The stability of individual mutations differs significantly between each other. This depends on the nature of the molecular alteration in the RNA genome (e.g. deletion mutation cannot be restored by reversion processes but only by recombination) and selection pressure that subjects the mutations into competition with other variants during evolution.

Effects of Mutations on Host-Virus Interactions

At the molecular level, mutations can cause changes in viral replication, viral movement, virus accumulation, stability of the assemblied virions, vector transmission, etc. This quite often links to changes in the observed phenotypic characteristics of plant viruses and as such may have profound implications for adaptation and evolution of these viruses. As an example, effects of mutations in coat proteins and in movement proteins are discussed. There are several reports indicating the requirement of capsid protein (CP) sequences for elicitation of the hypersensitive response (HR) by plant virus infections. The CP mutations affected the HR of *Nicotiana sylvestris* to tobacco mosaic virus (TMV) (Knorr and Dawson, 1988; Saito *et al*., 1987) and in tobacco to alfalfa mosaic virus (AlMV) (Neeleman *et al*., 1991). In contrast, CP is not required for eliciting local lesions on leaves of *Chenopodium amaranticolor* inoculatd by turnip crinkle virus (TCV) (Hacker *et al.,* 1992) and on leaves of Xanthi-nc tobacco plants inoculated by TMV (Dawson *et al.,* 1988). For brome mosaic bromovirus (BMV), mutational analysis of CP gene revealed effects on elicitation of local lesions in *C. hybridum* (Flasinski *et al*., 1995).

Similar to symptom elicitation, the involvement of CP in cell-to-cell transport differs among different virus groups. For instance, tobamoviruses and bromoviruses can move from cell-to-cell without the CP gene (Allison *et al.,* 1990; Dawson *et al*., 1988; Flasinski *et al*., 1995) while CMV, tobacco etch virus (TEV) and potato virus X (PVX) require CP for cell-to-cell movement (Suzuki *et al*., 1991; Chapman *et al*., 1992; Boccard and Baulcombe, 1993; Bujarski *et al*., 1994; Dolja *et al*., 1994, 1995; Taliansky and Garcia-Arenal, 1995). The requirement for CP in long distance movement through sieve elements of the phloem tissue has been shown for a variety of plant viruses (Sacher and Ahlquist, 1989; Allison *et al.,*

1990; Hull, 1991; Maule, 1991; Chapman *et al.*, 1992; Hilf and Dawson, 1993; Flasinski *et al.*, 1995).

Virus spread can be affected by mutations in movement protein (MP) ORFs of tomato aspermy cucumovirus strain V (V-TAV) (Moreno *et al.*, 1997), TMV (Deom and He, 1997), and BMV (Rao and Grantham, 1995). Specifically, for V-TAV, introduction of a stop codon in MP gene abolished cell-to-cell movement and thus infectivity despite increased RNA replication levels in protoplasts. This mutation, found in about 70% of V-TAV RNA populations was complemented by a movement-competent variant, ensuring virus infectivity. This shows that complementation may play an important role in the determination of genetic features in RNA genome populations. For TMV, second-site reversion mutations (two additional amino acid substitutions in a central portion of the MP protein) restored movement functions in the virus, suggesting that distal portions of the protein may interact in the active MP conformation. In the case of BMV, a genetic determinant for uniform vein chlorosis in systemically infected *N. benthamiana* was mapped due to a single transitional mutation that changes Val-266 to Ile-266 in the movement protein. The mutation had no detectable effect on accumulation of the virus and is specific for *N. benthamiana*, as it did not change the symptoms in barley plants. Mutations in the MP gene were also found to alter the host range of such DNA viruses as geminiviruses (Ingham and Lazarowitz, 1993) or interfere with virus movement for cauliflower mosaic virus (Thomas *et al.*, 1993).

However, not every mutation leads to phenotypic changes. Some mutations can be translationally silent (those at wobble positions in the codon triplets) or the amino acid changes can be neutral towards a particular function.

RNA Recombination

Several recent review describe the mechanisms of recombination in several viral systems (Lai, 1992; Pilipenko *et al.*, 1995; Simon and Bujarski, 1994; Nagy and Simon, 1997; MacFarlane, 1997). In general, RNA-RNA recombination events can occur among different viral RNAs (derived from either different viruses or from the same virus) and involve either homologous or nonhomologous crossover events. The former occurs between two nearly identical RNAs (or within nearly identical RNA regions) while the latter occurs between unrelated RNAs (or dissimilar regions). The evidence for the role of recombination in the evolution of plant viruses is overwhelmingly visible in the phylogenetic relationships among the variety of virus groups (Dolja and Carrington, 1992; Gibbs, 1995; Morozov *et al.*, 1989), and is based on the comparative genome organizations of RNA viruses (Zaccomer *et al.*, 1995). It can be studied in several developed experimental systems (confer, *Seminars in Virology*, Vol. 7/6, 1996 and 8/2, 1997).

The various experimental systems with which RNA recombination could be studied in RNA viruses include BMV, TCV, tombusviruses, and *Closteroviridae* from plant viruses and coronaviruses, picornaviruses, alphaviruses, nodaviruses, bacteriophage Qβ, vesicular stomatitis virus, and retroviruses from animal viruses (confer, *Seminars in Virology* 7/6, 1996; and 8/2, 1997). One of the most useful experimental systems in plant RNA viruses are bromoviruses. Previous studies showed that a partially debilitating BMV RNA3 mutant (designated M4) carrying a deletion in the 3' RNA replication promoter was repaired *in vivo* by exchanges with the sequences of other BMV RNA components (Bujarski and Kaesberg, 1986). By using this BMV system it was found that BMV RNA constructs supporting frequent recombination events can be designed;   local hybridization is important for nonhomologous recombination;   there are sequences supporting efficient homologous crossovers; and both 1a and 2a proteins of BMV replicase complex participate in RNA recombination.

Much work has been conducted in experimentally inducing recombinations in various plant viruses. Nonhomologous recombination was induced in AlMV where a *ts* mutation in RNA3 acquired a 5'terminal fragment from RNA1 during infection (Huissman *et al.,* 1989). In TCV, nonhomologous recombination occured between satellite RNAs (Cascone *et al.,* 1993), between satellite and genomic RNAs (Zhang *et al.,* 1991), and between DI and satellite RNAs (Cascone *et al.,* 1993). Recombination here was suggested to take place during plus strand synthesis. Recombination occured between two types of defective RNAs of tomato bushy stunt tombusvirus (TBSV) (White and Morris, 1994).

Natural Sequence Rearrangement

Two complete nucleotide sequences of the genomic RNAs of a large number of plant RNA viruses belonging to different virus groups have been obtained. This made possible comparisons of viral families at the molecular level which revealed the relatedness of various RNA viruses. In turn, comparative sequence data have led to a greater understanding of various aspects of the RNA virus evolution. Those aspects include genome organization of RNA viruses (Strauss and Strauss, 1988; Zaccomer *et al.*, 1995; Chapter 3), amino acid sequence similarities in viral proteins (Goldbach, 1986; Hodgman and Zimmern, 1987; Zaccomer *et al.*, 1995), and the differentiation of RNA genomes (Holland *et al.,* 1982; Smith and Inglis, 1987; Steinhauer and Holland, 1987; Zaccomer *et al.,* 1995). All these studies document RNA recombination as one of the major driving forces of RNA virus evolution. Natural sequence rearrangements were found in the following plant RNA viruses: AlMV (Huisman *et al.*, 1989), beet necrotic yellow vein furovirus (BNYVV) (Bouzoubaa *et al.*, 1991), bromoviruses (see below), hordeiviruses (Edwards *et al.*, 1992), luteoviruses (Mayo and Jolly, 1991), nepoviruses (Rott *et al.*, 1991), tobamoviruses (Shirako and Brakke, 1984; Dawson *et al.,* 1989), tobraviruses (Robinson *et al.*, 1987), tombusviruses (Hillman *et al.*, 1987), and TCV (Cascone *et al.*, 1990). This by itself confirms that RNA recombination events are common among various groups of plant viruses.

Recombination Between Viral and Host RNAs

Recombination between viral and plant host RNAs occurs. Several potato leafroll virus (PLRV) isolates contain sequences homologous to an exon of tobacco chloroplast (Mayo and Jolly, 1991). Acquisition of chloroplast sequences during RNA recombination has also been observed in BMV by us.

That plant RNA viruses are able to recombine with host mRNAs was confirmed by using and widely reported in transgenic plants expressing viral RNA sequences (Chapter 12). Majority of such recombinations occur at high selection pressure. Inefficient crossovers were observed between the transgenic CP mRNA and the mutated genomic RNA of potato virus Y potyvirus (PVY) that lacked the 3' noncoding region (3' UTR) (Jakab *et al.*, 1997). This supports the earlier observations that the 3' UTR plays a significant role in RNA recombination. Homologous recombination events between the transcripts of CP-defective plum pox potyvirus (PPV) and the functional PPV CP mRNA expressed in transgenic *N. benthamiana* plants were observed (Maiss *et al.*, 1997). The crossover events were not efficient, most likely due to complementation by functional CP. This demonstrates that complementation might reduce selection pressure on the accumulation of recombinants.

Role of RNA Structure

Mechanisms of recombination, based on results obtained with BMV and TCV systems, are discussed below.

During the initial recombination experiments with BMV, the lack of major recombination hotspots, the relatively low recombination frequency, and the existence of strong selection pressure favoring viable 3' recombinants have been observed (Nagy and Bujarski, 1992; Rao *et al.*, 1990; Bujarski and Dzianott, 1991). This was overcome by constructing an RNA3-based recombination vector (designated PN0 RNA3) (Nagy and Bujarski, 1993). Insertion of recombinationally active sequences into PN0 induced recombination with high frequency and the system was utilized in studying the role of RNA sequences in recombination. To assess the involvement of short base-paired regions between the RNA recombination substrates in selection of crossover sites (Nagy and Bujarski, 1992; Bujarski and Dzianott, 1991), an RNA1-derived 141-40 nucleotide sequence was inserted into the cloning site of PN0 in complementary orientation (Nagy and Bujarski, 1993). This induced RNA1/RNA3 recombinants with high frequency with junction sites clustered (targeted) within the heteroduplexed region. These studies supported evidence for heteroduplex-based nonhomologous recombination.

A series of PN0 derivatives with RNA1-derived complementary inserts revealed that longer than 30 nt heteroduplexed regions can support recombination at high frequency. Also, analysis of junction sites revealed that they were clustered at one side of the heteroduplex region. A model (Fig. 1) which explains such characteristics proposes that RdRp has difficulties passing through the heteroduplex region. (Nagy and Bujarski, 1993). According to the model, the formation of heteroduplexes can facilitate crossovers due to bringing the RNA substrates into a close proximity within the cells and/or structure or stability of heteroduplex regions which slows down or stalls the approaching replicase. The replicase may by-pass (loop out) the heteroduplex in a processive manner by sliding under this region and

A. Replication mode

I. Internal hairpin   II. Heteroduplex

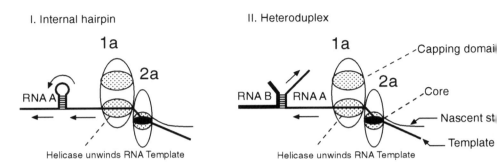

B. Recombination mode

I. Internal hairpin   II. Heteroduplex

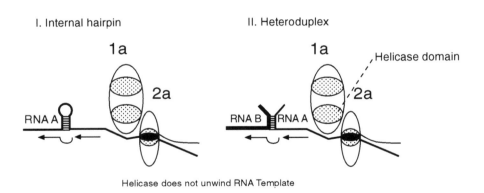

Figure 1. A model of the mechanism of heteroduplex-mediated recombination in BMV. Proteins 1a and 2a (represented by ellipses), which contain functional domains (represented by smaller dotted ellipses) of nucleotidyl transferase, helicase and core RNA polymerase, interact with each other and with host factors (represented by a smaller solid circle). The figure shows the two modes the replicase enzyme may operate near double-stranded (ds) (either intramolecular hairpins or intermolecular heteroduplex) structures: replication mode and recombination mode. During replication mode, the RNA (represented by a solid line) interacts with both helicase and core domains. Helicase unwinds ds regions and the original RNA template is copied through. During recombination mode, the conformation of the complex is different so that the RNA interacts only with the 2a core domain. In this case, undissociated ds regions "slide" through the RNA polymerase active site. The arrows indicate the direction of replicase migration.

switching to the new acceptor RNA (Bujarski and Nagy, 1994; Nagy and Bujarski, 1993).

The analyses revealed several important differences between homologous and nonhomologous recombinants, including the selection of crossover sites (regions), the precision of the crossover events, and the frequency of their generation (Bujarski and Nagy, 1992, 1994). To study the requirements of homologous recombination, an RNA3-based recombination vector (designated PN100) was constructed (Nagy and Bujarski, 1993) and a 60 nt RNA2-derived region (designated region R) was inserted into PN100 RNA3 in direct orientation. The resulting PN-H26 RNA3 construct recombined efficiently with wt RNA2, with the majority of crossovers occurring precisely within the common regions (Nagy and Bujarski, 1995). Mismatch mutations were introduced into region R in RNA3 to provide markers for localization of junction sites. Surprisingly, there was no correlation between predicted secondary structures in region R and the distribution of junction sites, arguing that in contrast to nonhomologous recombination the secondary structures at the hotspot regions play a limited role in homologous recombination.

Imprecise homologous recombinants having mismatched, deleted, inserted or nontemplated nucleotides within the crossover region were also isolated (Nagy and Bujarski, 1995). Since they were concentrated within the AU-rich stretches, it suggested that the AU-rich stretches can cause either misalignment between the RNA templates, replicase stuttering (nontemplated addition of several nucleotides), nucleotide misincorporation or their different combinations (Nagy and Bujarski, 1995b, 1996). This model was tested by increasing the length of the AU-rich stretches and by using artificial sequences of different A/U content. This revealed that an efficient homologous recombination hot spot should contain a stretch of GC-rich nucleotides followed by an AU-rich stretch (Nagy and Bujarski, 1996, 1997). It has been postulated that these sequences act as recombination activators: the downstream AU-rich region is the "core" region, and the upstream GC-rich region is the "enhancer" region. The AU-rich sequences in the donor RNA may cause replicase pausing and/or the release of the enzyme while the GC-rich upstream portion may facilitate the interactions with the complementary region of the acceptor RNA. The AU-rich sequence in the acceptor RNA may facilitate the opening of the partially double-stranded (ds) replication intermediate (RI) *via* local bubble structure formation (Nagy and Bujarski, 1997, 1998a,b).

Turnip crinkle virus is a small single component RNA virus that is associated with a number of subviral RNAs, such as satellite RNA D and chimeric RNA C. The latter is composed of sequences similar to RNA D at its 5' end and from TCV genomic RNA at its 3' side, and as such is a natural RNA recombinant (Simon and Nagy, 1996). High frequency recombination was observed *in vivo* between RNAs C and D where the recombination junctions were clustered near the 3' end of RNA D. A secondary structure element involving a longer hairpin in the central portion of RNA C was demonstrated to be responsible for recombination events (Cascone *et al.*, 1993). Based on these observations a template switching model was proposed where viral replicase uses the nascent plus strand of RNA D to prime RNA elongation on the acceptor minus-strand RNA C template and that the hairpin is responsible for re-attachment of the replicase complex during re-initiation of RNA synthesis (Simon and Nagy, 1996).

An *in vitro* system utilizing a partially purified RdRp TCV replicase and a chimeric RNA template containing an *in vivo* recombinationally-active region from RNA D joined to another hot-spot region from RNA C was developed (Nagy *et al.,* 1998b). This study demonstrated the important roles for a priming stem sequence in the RNA C portion and the TCV RdRp binding hairpin, also from the RNA3 portion. Both competition experiments and the use of mutations destabilizing the hairpin structure affected the RNA copying in the *in vitro* extension assays (Nagy *et al.*, 1998b), suggesting that the observed phenomenon might reflect the late steps of the *in vivo* RNA recombination, i. e. strand transfer and primer elongation. Later on, the analysis of the role of the hairpin and the priming stem and flanking sequences in *in vitro* recombination has been extended (Nagy *et al.,* 1998b; Nagy and Simon, 1998 a and b). In particular, the *in vitro* extension reaction depended on the proper sequence context of the hairpin and confirmed that the TCV RdRp recognizes the secondary and/or tertiary structure of the hairpin. The nucleotide sequence of the hairpin was found less important in this process. For the priming stem, it was found that its sequence can influence the site of initiation while its stability can affect the efficiency of the *in vitro* extension reaction. It is likely that the observed effects reflect those functioning during *in vivo* recognition of the negative strand initiation promoter and the subgenomic promoter, and are analogous to the observed *in vivo* recombination events.

Role of Replicase Proteins

The results described above have allowed for creation of efficient recombination systems with which the role of replicase proteins could be tested. BMV genomic RNAs with mutations within the corresponding ORFs of 1a and 2a replicase proteins were utilized. Regular and *ts* 1a protein mutants within the putative helicase region were tested (Nagy *et al.*, 1995). A characteristic 5' shift in crossover sites towards energetically less stable portions of the RNA1-RNA3 heteroduplex at the elevated (but still permissible) temperature was observed only for the temperature sensitive mutant under a suboptimal temperature. This suggests that the helicase domain of 1a participates in heteroduplex-mediated crossover events.

A single amino acid mutation within the core domain of the polymerase, protein 2a, (designated DR7) inhibited the frequency of nonhomologous recombination below the level of detection (Figlerowicz *et al.*, 1997). In contrast, the frequency of homologous events was comparable in infections with DR7 and wt RNA2. Also, two viable mutations (designated MF-II and MF-V) in the N-terminal portion of the 2a protein, the region that is known to interact with the C-terminal domain of 1a (Figlerowicz *et al.*, 1998) were found to reduce the heteroduplex-mediated nonhomologous recombination. For homologous recombination, MF-II generated the recombinants at a high frequency but MF-V reduced markedly the homologous recombination frequency. Thus, the nature and characteristics of both homologous and nonhomologous recombinants were influenced by mutations in the polymerase gene of BMV. This gives a crucial support for replicase-driven template-switching (though different) mechanisms operating in nonhomologous and in homologous systems. The term "recombinosome" was proposed to describe a complex between the recombining RNAs, the replicase, and other putative factors involved in recombination (Figlerowicz *et al.*, 1997). The stability of the recombinosome (the time required for crossover events before the disassembly of the recombinosome takes place) is likely to be influenced by both the RNA components and the replicase in a different way for both recombination types.

In conclusion, these data show that BMV replicase proteins do participate in RNA recombination. The role of such proteins in RNA recombination for other viruses remains to be established. Indirectly so, alterations in these proteins may affect the occurrence of crossovers and consequently the evolutionary traits in RNA viruses. Also, this opens up the possibility to engineer recombinant RNA viruses which will recombine at reduced levels, thus increasing the stability as well as safety of viral expression vectors and vaccines.

Pseudorecombination

Pseudorecombination occurs when RNA components of different strains of a virus with a divided genome are mixed together and form the so called reassortants. Exchanging segments between viral species has been proposed as a mechanism introducing variation during mixed infections. Among animal RNA viruses the most classical example is provided by influenza virus where different pathogenic strains were proven to emerge due to reassortment combinations.

Several plant RNA virus genera have divided RNA genomes. The occurrence of reassortants might have an important impact on evolving new plant viral species in such cases when reassortant confers a selectively advantageous broader host range. The occurrence of reassortment in nature has been demonstrated for tricornaviruses by isolation of a natural reassortant that contains an RNA3 component from CMV and RNAs 1 and 2 from peanut stunt cucumovirus. Also, reassortants between tobacco rattle tobravirus and pea early browning tobravirus have been described (Robinson *et al.*, 1987).

Pseudorecombinant exchanges have been used for genetic mapping of functional genes and to study viral gene function. For instance, exchanges between two BMV strains revealed that the information that determines their host range was located within the RNA3 component (De Jong and Ahlquist, 1991). Also, the use of nitrous acid mutants of two cowpea chlorotic mottle virus (CCMV) strains indicated that systemic symptoms on cowpea were genetically directed by RNA3 (Kuhn and Wyatt, 1979). However, reassortants between BMV and CCMV (strain T) demonstrated that host specificity determinants are encoded in all three genomic RNAs (Allison *et al.*, 1988; Bancroft and Lane, 1973). To map RNA3 sequences that specify the reactions of bromoviruses with their hosts, hybrid BMV/CCMV RNAs were constructed by precise exchanges of the BMV and CCMV 3a gene (Mise *et al.*, 1993). This revealed that bromovirus MPs must be specifically adapted for successful bromovirus infection. In AlMV, construction of pseudorecombinants between two virus strains revealed that symptom differences observed on tobacco plants mapped to RNA3. The use of chimeric RNA3 molecules further mapped some of the essential AlMV sequences to the coat protein open reading frame (Neeleman *et al.*, 1991).

In order to investigate host-related genetic information in bromoviruses, viable CCMV pseudorecombinants were obtained by exchanging individual genomic RNA components among S, D and N strains. This demonstrated that the RNA3-encoded genetic information is involved in the induction of anti-CCMV resistance and symptom formation in soybean. However, RNA1 and RNA2 components also contributed to the interactions of CCMV with its hosts (Shang and Bujarski, 1993).

Defective Interfering (DI) RNAs

Infections by RNA viruses are frequently accompanied by DI RNAs that are derived from the helper virus genome (Graves *et al.*, 1996). As DI RNAs can interfere with helper virus accumulation, they play an important role in controling disease severity and the overall evolutionary fitness of RNA viruses. While DI RNAs have been reported for virtually all the animal RNA viruses (Holland, 1991), they have not been extensively studied in plant RNA viruses (Roux *et al.*, 1991). In positive-strand RNA viruses of plants (Fig. 2), the DI RNA was first observed in TBSV (Hillman *et al.*, 1987) and the molecular characterization of several cloned TBSV DI RNAs has revealed a consistent pattern of rearranged TBSV sequences flanked by unmodified terminal regions. Similar DI RNA molecules have been described for cucumber necrosis tombusvirus, cymbidium ringspot tombusvirus, and for TCV (Li *et al.*, 1989). While in all tombusviruses the DI RNAs had an attenuating effect on infection (Burgyan *et al.*, 1989; Hillman *et al.*, 1987), the TCV DI RNAs tended to increase the severity of symptoms (Li *et al.*, 1989).

Sequencing of DI RNAs associated with tombusviruses has revealed a mosaic type of DI RNAs (Fig. 3). Besides being heavily rearranged, these DI RNA molecules have the following features: they are relatively small (~300-800 nt), do not code for a protein or code short oligopeptides, and are not efficiently encapsidated. Analysis of TBSV DI RNAs suggests that in some cases the base-pairing between an incomplete replicase-associated nascent strand and the acceptor template can mediate the selection of the rearrangement sites (White and Morris, 1994). The authors propose a stepwise deletion model to describe the temporal order of events leading to the formation of the mosaic TBSV DI RNAs.

Besides mosaic type, single deletion type DI RNAs were isolated from infections with several viruses including BNYVV (Bouzoubaa *et al.*, 1991), soil-borne wheat mosaic furovirus (SBWMV) (Chen *et al.*, 1994), peanut clump furovirus (PCV) (Manohar *et al.*, 1993), clover yellow mosaic potexvirus (CYMV; White *et al.*, 1991), sonchus yellow net rhabdovirus (SYNRV; Ismail and Milner, 1988), and tomato spotted wilt tospovirus (TSWV) (Resende *et al.*, 1991), the latter

Figure 2. Types of defective RNAs in plant RNA viruses. Single-deletion type defective RNAs found in broad bean mottle virus (BBMV) and in CMV infections are shown in the top two pannels. A double-deletion type DI RNA of a tombusvirus is shown at the bottom. In each pannel the top line shows the parental genomic RNA while the bottom line shows the actual defective RNA molecule.

Figure 2. Types of defective RNAs in plant RNA viruses. Single-deletion type defective RNAs found in broad bean mottle virus (BBMV) and in CMV infections are shown in the top two pannels. A double-deletion type DI RNA of a tombusvirus is shown at the bottom. In each pannel the top line shows the parental genomic RNA while the bottom line shows the actual defective RNA molecule. A complicated, mosaic-type DI RNA of Sindbis virus, an animal virus, is also shown.

BBMV

CMV

SINDBIS VIRUS

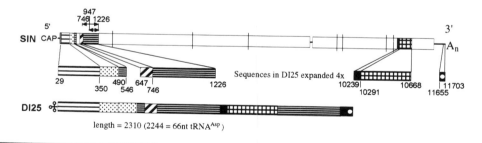

TOMBUS VIRUS

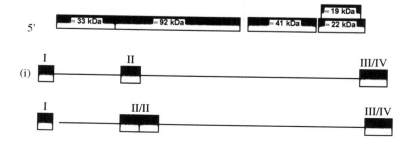

two being negative and ambisense viruses, respectively. In CYMV, the defective RNAs originated from single in-frame deletions which involved more than 80% of the viral genome and affected all the ORFs. In the quadripartite BNYVV furovirus, spontaneous deletions were observed in the coding regions of RNA3 and RNA4 components. In SBWMV, defective RNAs were derived from a shorter RNA2 segment. These data reveal that single deletion type DI RNAs are generated with the preservation of the ORF downstream of deletion, that they are relatively large RNA molecules, and that they are efficiently encapsidated.

For broad bean mottle bromovirus (BBMV), the naturally existing DI RNAs were derived by single deletions in the RNA2 component (Romero *et al.*, 1993; Pogany *et al.*, 1994, 1997). The accumulation and encapsidation of BBMV DI RNAs are specific to host, emphasizing the importance of host in generation of modified RNA genomes during viral adaptation and evolution. Such features as the possibility of introducing the *in vitro* transcribed parental DI RNA constructs into host plants by co-inoculation with RNAs of supporting strain, the *de novo* generation of DI RNAs, a great deal of tolerance for modifications within internal parts of the RNA2-derived molecules, and multi-partite nature of BBMV genome, all make the BBMV system well suitable for testing the mechanism of DI RNA generation and accumulation. Two secondary structure-mediated models for BBMV DI RNAs have been proposed (Fig. 3). In the first one, role of homology between remote portions of RNA molecule is stressed out. In this case the replicase pauses at a signal sequence, falls off the original RNA template, and the nascent strand reassociates at a homologous region further upstream so RNA synthesis can be resumed. In another model, local complementary regions bring remote parts of RNA2 together, facilitating crossover events. The RNA polymerase, after approaching such secondary structures, skips the region and continues on other side of double-stranded area. Essentially, this model is analogous to the heteroduplex-mediated model of recombination in BMV (Fig. 2; Bujarski *et al.*, 1994; Simon and Bujarski, 1994).

Similar to BBMV, single deletion type DI RNAs are also found associated with cucumoviruses in which a single deletion in RNA3 (instead of RNA2 for BBMV) component formed the defective RNA molecule (Graves and Roossinck, 1995).

Artificial defective RNA2-derived molecules have been constructed in BMV (Marsh *et al.*, 1991). There were three types of RNA2 constructs: replicating interfering, nonreplicating interfering, and nonreplicating non-interfering, which further reflects variability among viral RNAs but also the usefullness of defective RNAs in functional analysis of the viral genome.

Overall, it is obvious that RNA viruses are able to adapt their genomes rapidly to new environments, even by deleting large parts of the genomic RNA molecules that are not required under particular conditions. Apparently, the entire genomic RNAs are under continuous selection pressure during viral infection (Graves *et al.*, 1996).

CONCLUSIONS

Plant viruses have a variety of mechanisms available, including point mutations, deletions, insertions, frameshift mutations, and RNA recombination, that contribute

A. Homology-driven

B. Complementarity-driven

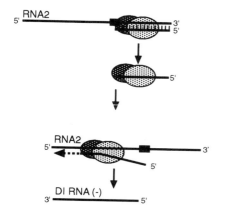

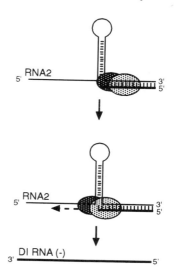

Figure 3. Two copy-choice models of the mechanism leading to formation of deletions during replication of BBMV RNA. A. During homology-driven mechanism, viral replicase (represented by two filled ellipses) pauses at a signal sequence (represented by a filled square) during synthesis of minus strands and falls off the original RNA template. Then the nascent strand reattaches at a homologous site further upstream and the synthesis of minus strand resumes. B. Acoording to a complementarity-driven mechanism two remote parts of the RNA template are brought together through complementary sequences, the polymerase skips this region, e.g., due to difficulties in unwinding, and continues at the other side of this double-stranded region.

to high variability of their RNA genomes. Because these sequence changes alter genetic properties of viral RNAs, they play important roles in virus evolution. Their relative importance during virus evolution becomes apparent when kinds of differences that distinguish different viral taxa are examined. Namely, the most closely related virus isolates differ mostly by point mutations (often silent); here point mutations and reassortment are important but recombination and *de novo* gene generation are much less important. Also, within the species virus isolates usually have only a small number of point mutations. However, at the more distant levels of relationship between viruses, one can observe the contribution of other mechanisms. Namely, at the genus (group) level, mutation is less important but recombination and *de novo* gene generation are contributing more; more frequent occurrence of non-silent mutations and insertions/deletions, quite often involving new genes, could be observed. This indicates that the evolution within the same genus of viruses has been linked and that they co-diverged. However, gene and genome comparisons show that this is not true between most families, suggesting that recombination is the most important factor of evolution at the family level (Gibbs *et al.*, 1997).

The high adaptability and capability of rapid change has practical implications for plant viruses. For example, many strategies of pathogen-derived resistance engineered, for instance, in transgenic plants, are highly specific (Beachy, 1993; Lomonossoff, 1995). One can predict that due to the enormous plasticity of viral genomes, these strategies may not function for a long time in the crop plants. The Red Queen theory (Clay and Kover, 1996) predicts that in case of host-pathogen interactions, the pathogens and the hosts constantly struggle with each other and have to undergo constant dynamic changes in order to achieve a safe, dynamic stalemate in their relationship. Thus, the use of specific and genetically non-adaptable strategies, that just provide new selection pressure, may not be efficient to prevent highly adaptable RNA virus infections. As M. Roossinck points out in her recent review (Roossinck, 1997) "it may be more practical to consider ways to reduce virus-induced disease symptoms without necessarily affecting virus load". Also, virus evolution may lead to expansion of the host range and this leads to new virus emergence, as exemplified by the emergence of tospoviruses in plant kingdom due to evolving from animal bunyaviruses and to expansion of their thrips vectors (Goldbach and Peters, 1994). The increasing intervention of human activities upon environment changes the selection pressure that will likely lead to emergence of new RNA virus species.

ACKNOWLEDGEMENTS

I thank Drs. Nathalie Rauffer-Bruyere, Arnaud Bruyere, and Marek Figlerowicz as well as Mr. Marcus Wantroba for helpful discussions and critical comments during the course of writing this review. Work in my laboratory is supported by grants from the United States Department of Agriculture NRI Competitive Grants Program (96-39210-3842), the National Institutes of Health (3R01-A126769), the National Science Foundation (MCB-9630794), and by Plant Molecular Biology Center at Northern Illinois University.

REFERENCES

Allison, R. C., Thompson, C., and Ahlquist, P. (1990). Regeneration of a functional RNA virus genome by recombination between deletion mutants and requirement for cowpea chlorotic mottle virus 3a and coat genes for systemic infection. *Proc. Natl. Acad. Sci. USA* 87, 1820-1824.

Allison, R., Janda, M., and Ahlquist, P. (1988). Infectious *in vitro* transcripts from cowpea chlorotic mottle virus cDNA clones and exchange of individual components with brome mosaic virus. *J. Virol.* 62, 3581-3588.

Bancroft, J. B., and Lane, L. C. (1973). Genetic analysis of cowpea chlorotic mottle virus and brome mosaic virus. *J. Gen. Virol.* 19, 381-389.

Beachy, R. N. (1993). Transgenic resistance to plant viruses. *Semin. Virol.* 4, 327-331.

Bertholet, C. E., van Meir, B., Heggeler-Bordier, T., and Wittek, R. (1987). Vaccinia virus produces late mRNAs by discontinuous synthesis. *Cell* 50, 153-162.

Boccard, F., and Baulcombe, D. C. (1993). Mutational analysis of *cis*-acting sequences and gene function in RNA3 of cucumber mosaic virus. *Virology* 193, 563-578.

Bouzoubaa, S., Niesbach-Klosgen, U., Jupin, I., Guilley, H., Richards, K. and Jonard, G. (1991). Shortened forms of beet necrotic yellow vien virus RNA3 and -4: Internal deletions and subgenomisc RNA. *J.Gen. Virol.* 72, 259-266

Boyer, J. C., and Haenni, A. L. (1994). Infectious transcripts and cDNA clones of RNA viruses. *Virology* 198, 415-426.

Bujarski, J. J., and Kaesberg, P. (1986). Genetic recombination in a multipartite plant virus. *Nature* 321, 528-531.

Bujarski, J. J., and Dzianott, A. M. (1991). Generation and analysis of nonhomologous RNA-RNA recombinants in brome mosaic virus: Sequence complementarities at crossover sites. *J. Virol.* 65, 4153-4159.

Bujarski, J. J., Nagy, P. D., and Flasinski, S. (1994). Molecular studies of genetic RNA-RNA recombination in brome mosaic virus. *Adv. Vir. Res.* 43, 275-302.

Bujarski, J. J., and Nagy, P. D. (1994). Genetic RNA-RNA recombination in positive-stranded RNA viruses of plants. *In* "Homologous Recombination in Plants" (J. Paszkowski, Ed.), pp. 1-24. Kluwer Academic Publisher, Dordrecht.

Burgyan, J. Grieco, F., and Russo, M. (1989). A defective interfering RNA molecule in cymbidium ringspot virus infections. *J. Gen. Virol.* 70, 235-239.

Cascone, P. J., Carpenter, C. D., Li, X. H., and Simon A. E.. (1990). Recombination between satellite RNAs of turnip crinkle virus. *EMBO J.* 9, 1709-1715.

Cascone, P. J., Haydar, T. F., and Simon, A. E. (1993). Sequences and structures required for recombination between virus-associated RNAs. *Science* 260, 801-805.

Cattaneo, R. (1991). Different types of messenger RNA editing. *Annu. Rev. Genet.* 25, 71-88.

Chapman, S. G., Hills, J., Watts, A., and Baulcombe, D. (1992). Mutational analysis of the coat protein gene of potato virus X: Effects on virion morphology and viral pathogenicity. *Virology* 191, 2 23-230.

Chen, J., MacFarlane, S. S., and Wilson, T. M. A. (1994). Detection and sequence analysis of a spontaneous deletion mutant of soil-borne wheat mosaic virus RNA2 associated with increased symptom severity. *Virology* 202, 921-929.

Clay, K., and Kover, P. X. (1996). The red queen hypothesis and plant/pathogen interactions. *Annu. Rev. Phytopathol..* 34, 29-50.

Dawson, W. O., Bubrick, P., and Grantham, G. L. (1988). Modifications of tobacco mosaic virus Coat protein gene affecting replication, movement and symptomatology. *Phytopathology* 78, 783-789.

Dawson, W. O., Lewandowski, D. J., Hilf, M. E., Bubrick, P., Raffo, A. J., Shaw, J. J., Grantham, G. L., and Desjardins, P. R. (1989). A tobacco mosaic virus-hybrid expresses and loses an added gene. *Virology* 172, 285-292.

De Jong, W., and Ahlquist, P. (1991). Bromovirus host specificity and systemic infection. *Semin. Virol.* 2, 97-105.

Deom, C. M., and He, X. Z. (1997). Second-site reversion of a dysfunctional mutation in a conserved region of the tobacco mosaic tobamovirus movement protein. *Virology* 232, 13-18.

Dolja, V. V., and Carrington, J. C. (1992). Evolution of postive-strand RNA viruses. *Semin. Virol.* 3, 315-26

Dolja, V. V., Haldeman, R., Robertson, N. L., Dougherty, W. G., and Carrington, J. (1994). Distinct functions of capsid protein in assembly and movement of tobacco etch potyvirus in plants. *EMBO J.* 13, 1482-1491.

Dolja ,V. V., Haldeman-Cahill, R., Montgomery, A. E., Vandenbosch, K. A., and Carrington, J. C. (1995). Capsid protein determinants involved in cell-to-cell and long distance movement of tobacco etch potyvirus. *Virology* 206, 1007-1016.

Domingo, E., Holland, J., Biebricher, C., and Eigen, M. (1995). Quasi-species: the concept and the word. *In* "Molecular Basis of Virus Evolution" (A. J. Gibbs, C. H. Calisher, and F. Garcia-Arenal, Eds.), pp. 181-91. Cambridge University Press, Cambridge.

Durbin , R. K., and Stollar, V. (1986). Sequence analysis of the E2 gene of a hyperglycosylated, host restricted mutant of Sindbis virus and estimation of mutation rate from frequency of revertants. *Virology* 154, 135-43

Edwards, M. C., Petty, I. T. D., and Jackson, A. O. (1992). RNA recombination in the genome of barley stripe mosaic virus. *Virology* 189, 389-392.

Eigen, M. (1993). Viral quasispecies. *Sci. Am.* July, 42-49.

Eigen, M. (1996). On the nature of virus quasispecies. *Trends Microbiol.* 4, 16-17.

Figlerowicz , M., Nagy, P. D., and Bujarski, J. J. (1997). A mutation in the putative RNA polymerase gene inhibits nonhomologous, but not homologous, genetic recombination in an RNA virus. *Proc. Natl. Acad. Sci. USA* 94, 2073-2078.

Figlerowicz , M., Nagy, P. D., Tang, N., Kao, C. C., and Bujarski, J. J. (1998) . Mutations in the N-terminus of the brome mosaic virus polymerase affect genetic RNA-RNA recombination. *J. Virol.,* in press.

Flasinski, S., Dzianott, A., Pratt, S., and Bujarski, J. J. (1995). Mutational analysis of the coat protein gene of brome mosaic virus: Effects on replication and movement in barley and in *Chenopodium hybridum. Mol. Plant-Microbe Interact.* 8, 23-31.

Gibbs, A. (1987). Molecular evolution of viruses; "trees", "clocks", and "modules". *J. Cell Sci. Suppl* .7, 319-337.

Gibbs, M. (1995). The luteovirus super-group: Rampart recombination and persistent partnerships. *In* "Molecular Basis of Virus Evolution", pp. 351-368. Cambridge University Press, Cambridge.

Gibbs, M., Armrstrong, J., Weiller, G. F., and Gibbs, A. J. (1997). Virus evolution; the past, a window on the future? *In* " Virus-Resistant Transgenic Plants: Potential Ecological Impact, (M. Tepfer and E. Balazs, Eds.). pp. 1-17. Springer -Verlag, Berlin.

Goldbach, R. (19860. Molecular evolution of plant RNA viruses. *Annu. Rev. Phytopathol.* 24, 289-310.

Goldbach, R, and Peters, D. (1994). Possible causes of the emergence of tospovirus diseases. *Semin. Virol.* 5, 113-120.

Graves, M. G., and Roosinck, M. J. (1995). Characterization of defective RNAs derived from RNA 3 of the Fny strain of cucumber mosaic cucumovirus. *J. Virol.* 69, 4746-4751.

Graves, M. V., Pogany, J., and Romero, J. (1996). Defective interfering RNAs and defective viruses associated with multipartite RNA viruses of plants. *Semin. Virol.* 7, 399-408.

Hacker, D. L., Petty, I. T. D., Wei, N., and Morris, T. J. (1992). Turnip crinkle virus genes required for RNA replication and virus movement. *Virology* 186, 1-8.

Hillman, B. I., Carrington, J. C., and Morris, T. J. (1987). A defective interfering RNA that contains a mosaic of a plant virus genome. *Cell* 51, 427-433.

Hodgman , T. C., and Zimmern, D. (19870. Evolution of RNA viruses. *In* "RNA Genetics" (J. J. Holland, E. Domingo, and P. Ahlquist, Eds). CRC Press, Boca Raton, Fla.

Holland, J., Spindler, K., Horodyski, F., Grabau, E., Nichol, S., and Van de Pol, S. (1982). Rapid evolution of RNA genomes. *Science* 215, 1577-1585.

Hilf, M. E., and Dawson, W. O. (1993). The tobamovirus capsid protein functions as a host-specific determinant of long-distance movement. *Virology* 193, 106-114.

Hillman, B. I., Carrington, J. C., and Morris, T. J. (1987). A defective interfering RNA that contains a mosaic of a plant virus genome. *Cell* 51, 427-433.

Holland, J. J. (1991). Generation and replication of defective viral genomes. *In* "Virology" (B.N. Fields, Ed.), 2nd Edition. Raven Press, New York.

Holland, J. J., DeLaTorre, J. C., and Steinhauer, D. A. (1992). RNA virus populations as quasispecies. *In* "Genetic Diversity of RNA Viruses" (J. J. Holland, Ed.), pp. 1-20. Springer Verlag, Berlin.

Huisman, M. J., Cornelissen, B. J. C., Groenendijk, C. F. M., Bol, J. F., and van Vloten-Doting, L. (1989). Alfalfa mosaic virus temperature-sensitive mutants. V. The nucleotide sequence of TBS 7 RNA 3 shows limited nucleotide changes and evidence for heterologous recombination. *Virology* 171, 409-416.

Hull, R. (1991). The movement of viruses within plants. *Semin. Virol.* 2, 89-95.

Ingham , D.J., and Lazarowitz, S. (1993). A single mutation in the BR1 movement protein alters the host range of the squash leaf curl geminivirus. *Virology* 196, 694-702.

Ismail, I. D., and Milner, J. J. (1988). Isolation of defective interfering particles of sonchus yellow net virus from chronicallyy infcted plants. *J. Gen. Virol.* 69, 999-1006.

Jacques, J. P., Hausmann, A., and Kolakofsky, D. (1994). Paramyxovirus mRNA editing leads to G deletion as well as insertions *EMBO J,* 13, 5496-5503.

Jakab, G., Vaistij, F. E., Depz, E., and Malone, P. (1997). Transgenic plants expressing viral sequences create a favourable environment for recombination between viral sequences. *In* "Virus-Resistant Transgenic Plants: Potential Ecological Impact". (M. Tepfer, and E. Balaz, Eds.), pp. 45-51. Springer Verlag, Berlin.

Knorr, D. A., and Dawson, W. O. (1988). A point mutation in the tobacco mosaic virus capsid protein gene induces hypersensitivity in *Nicotiana sylvestris. Proc. Natl. Acad. Sci. USA* 85, 170-174.

Kuhn, C. W., and Wyatt, S. D. (1979). A variant of cowpea chlorotic mottle virus obtained by passage through beans. *Phytopathology* 69, 621-624.

Kurath, G., and Dodds, J. A. (1994). Satellite tobacco mosaic virus sequence variants with only five nucleotide differences can interfere with each other in a cross-protection-like phenomenon in plants. V*irlolgy* 202, 1065-1969.

Kurath, G., and Palukaitis, P. (1989). RNA sequence heterogeneity iin natural populations of three satellite RNAs of cucumber mosaic viurs. *Virology* 173, 231-240.

Lai, M. M. C. (1992). RNA recombination in animal and plant viruses. *Microbiol. Rev.* 56, 61-79.

Levin, B. (1997). Genes VI. Oxford University Press.

Li , X. H., Heaton, L. A., Morris, T. J., and Simon, A. E. (1989). Turnip crinkle virus defective interfering RNAs intensify viral symptoms and are generated *de novo*. *Proc. Natl. Acad. Sci. USA* 86, 9173-9177.

Lomonossoff, G. P. (1995). Pathogen-derived resistance to plant viruses. *Annu. Rev. Phytopathol.* 33, 323 -343.

MacFarlane, S. A. (1997). Natural recombination among plant virus genomes: Evidence from tobraviruses. *Semin. Virol.* 8, 25-31.

Mais, E., Varrelmann, M., DiFonzo, C., and Raccah, B. (1997). Risk assessment of transgenic plants expressing the coat protein gene of plum pox potyvirus (PPV). *In* "Virus-Resistant Transgenic Plants: Potential Ecological Impact". (M. Tepfer, and E. Balazs, Eds.). pp. 85-93. Springer Verlag, Berlin.

Marsh , L. E., Pogue, G. P., Connell, J. P., and Hall, T. C. (1991). Artificial defective interfering RNAs derived from brome mosaic virus. *J. Gen. Virol.* 72, 1787-1792.

Maule, A. J. (1991). Virus movement in infected plants. *Critical Rev. Plant Sci.* 9, 457-473.

Mayo, M. A.., and Jolly, C. A. (1991). The 5' terminal sequence of potato leafroll virus RNA: evidence for recombination between virus and host RNA. *J. Gen. Virol.* 72, 2591-2595.

Miller, W. A., and Rasochova, L. (1997). Barley yellow dwarf viruses. *Annu. Rev. Phytopathol.* , 35, 167-190.

Mise, K., Allison, R. F., Janda, M., and Ahlquist, P. (1993). Bromovirus movement protein genes play a crucial role in host specificity. *J. Virol.* 67, 2815 -2825.

Manohar, S. K.., Guilley, H., Dollet, M., Richards, K., and Jonard, G. (1993). Nucleotide sequence and genetic organization of peanut clump virus RNA2 and partial characterization of deleted forms. *Virology* 195, 33-41.

Moreno, I. M., Malpica, J. M., Rodriguez-Cerezo, E., and Garcia-Arenal, F. (1997). A mutation in tomato aspermy cucumovirus that abolishes cell-to-cell movement is maintained to high levels in the viral RNA population by complementation. *J. Virol.* 71, 9157-9162.

Morozov , S. Y., Dolja, V. V., and Atabekov, J. G. (1989). Probable reassortment of genomic elements among elongated RNA-containing plant viruses. *J. Mol. Evol.* 29, 52-62

Morse, S. S. (Ed.). (1994). " The Evolutionary Biology of Viruses". Raven Press, New York.

Moya, A., and Garcia-Arenal, F. (1995). Population genetics of viruses: An introduction. *In* "Molecular Basis of Virus Evolution". (A. J. Gibbs, C. H. Calisher, C. H., and Garcia-Arenal, F., Eds.) pp. 213-223. Cambridge Univ. Press, Cambridge.

Nagy, P. D., and Bujarski, J. J. (1992). Genetic recombination in brome mosaic virus: Effect of sequence and replication of RNA on accumulation of recombinants. *J. Virol.* 66, 6824-6828.

Nagy, P. D., and Bujarski, J. J. (1993). Targeting the site of RNA-RNA recombination in brome mosaic virus with antisense sequences. *Proc Natl Acad Sci USA* 90, 6390-6394.

Nagy, P. D., and Bujarski, J. J. (1995). Efficient system of homologous RNA recombination in brome mosaic virus: Sequence and structure requirements and accuracy of crossovers. *J. Virol.* 69, 131-140.

Nagy, P. D., and Bujarski, J. J. (1996). Homologous RNA recombination in brome mosaic virus AU-rich sequences descrease the accuracy of crossovers. *J. Virol.* 70, 415-426.

Nagy, P. D., and Bujarski, J. J. (1997). Engineering of homologous recombination hotspots with AU-rich sequences in brome mosaic virus. *J. Virol.* 71, 1294-1306.

Nagy, P. D., and Simon, A. E. (1997). New insight into the mechanisms of RNA recombination. *Virology*, 235, 1-9.

Nagy, P. D., and Simon, A. E. (1998a). *In vitro* characterization of late steps of RNA recombination in turnip crinkle virus. I. Role of the motif1-hairpin structure. *Virology*, in press.

Nagy, P. D., and Simon, A. E. (1998b). *In vitro* characterization of late steps of RNA recombination in turnip crinkle virus. II. The role of the priming stem and flanking sequences. *Virology*, in press.

Nagy, P. D., Dzianott, A., Ahlquist, P., and Bujarski, J. J. (1995). Mutations in the helicase-like domain of protein 1a alter the sites of RNA-RNA recombination in brome mosaic virus. *J. Virol.* 69, 2547 - 2556.

Nagy, P. D., Ogiela, C., and Bujarski, J. J. (1998a). Mapping sequences active in homologous RNA recombination in brome mosaic virus: prediction of recombination hot-spots. *Virology*, in press.

Nagy, P. D., Zhang, C., and Simon, A. E. (1998b). Disecting RNA recombination *in vitro*: Role of RNA sequences and the viral replicase. *EMBO J.* 17, 2392-2403.

Neeleman, L., van der Kuyl, A., and Bol, J. F. (1991). Role of alfalfa mosaic virus coat protein gene in symptom formation. *Virology* 181, 687-693.

Parvin, J. D., Moscona, A., Pan, W. T., Leider, J. M., and Palese, P. (1986). Measurement of the mutation rates of animal viruses: influenza A virus and poliovirus type I. *J. Virol.* 59, 377-383.

Pilipenko, E. V., Gmyl, A. P., and Agol, V. I. (1995). A model for rearrangements in RNA genomes. *Nucleic Acids Res.* 23, 1870-1875

Pogany, J., Huang, Q., Romero, J., Nagy, P., and Bujarski, J. J. (1994). Infections transcripts from PCR-amplified broad bean mottle bromovirus cDNA clones and variable nature of leader regions in RNA3 segment. *J. Gen.Virol.* 75, 693-699.

Pogany, J., Romero, J., and Bujarski, J. J. (1997). Effect of 5' and 3' terminal sequences, overall length and coding capacity on the accumulation of defective-like RNAs associated with broad bean mottle virus. *Virology* 228, 236-243.

Rao, A. L. N., and Grantham, G. L. (1995). A spontaneous mutation in the movement protein gene of brome mosaic virus modulates symptom phenotype in *Nicotiana benthamiana*. *J. Virol.* 69, 2689-2691.

Rao, A. L. N., and Hall, T. C. (1993). Recombination and polymerase error facilitate resotration of infectivity in brome mosaic virus. *J. Virol.* 67, 969-979.

Rao, A. L. N., Sullivan, B. P., and Hall, T. C. (1990). Use of *Chenopodium hybridum* facilitataes isolation of brome mosaic virus RNA recombinants. *J. Gen. Virol.* 71, 1403-407.

Resende, R. O., de Haan, P., de Avila, A. C., Kitajima, E. W., Kormelink, R., Goldbach, R., and Peters, D. (1991). Generation of envelope and defective interfering RNA mutants of tomato spotted wilt virus by mechanical passage. *J. Gen. Virol.* 72, 2375-2383.

Robinson, D. J, Hamilton, W. D. O., Harrison, B. D., and Baulcombe, D. C. (1987). Two anomalous tobravirus isolates: Evidence for RNA recombination in nature. *J. Gen. Virol.* 68, 2551 - 2561

Rodriguez-Cerezo, E., Moya, E., and Garcia-Arenal, F. (1991). High genetic stability in natural populations of the plant RNA virus tobacco mild green mosaic virus. *J. Mol. Evol.* 32, 328-332.

Romero, J., Huang, Q., Pogany, J., and Bujarski, J. J. (1993). Characterization of defective interfering RNA components that increase symptom severity of broad bean mottle virus infections. *Virology* 194, 576-584.

Roossinck, M. (1997). Mechanisms of plant virus evolution. *Annu. Rev. Phytopathol.* 35, 191-209.

Rott, M. E ., Tremaine, J. H., and Rochon, D. M. (1991). Comparison of the 5' and 3' termini of tomato ringspot virus RNA1 and RNA2: evidence for RNA recombination. *Virology* 185, 468-472.

Roux, L., Simon, A. E., and Holland, J. J. (1991). Effects of defective-interfering viruses on virus replication and pathogenesis *in vitro* and *in vivo*. *Adv. Virus Res.* 40, 181-211.

Sacher, R., and Ahlquist, P. (1989). Effects of deletions in the N- terminal basic arm of brome mosaic virus coat protein on RNA packaging and systemic infection. *J. Virol.* 63, 4545-4552.

Saito , T., Meshi, T. , Takamatsu, N., and Okada, Y. (1987). Coat protein gene sequence of tobacco mosaic virus encodes a host response determinant. *Proc. Natl. Acad. Sci. USA* 84, 6074 - 6077.

Shang, H., and Bujarski, J. J. (1993). Systemic spread and symptom formation of cowpea chlorotic mottle virus in soybean and cowpea plants map to the RNA3 component. *Mol. Plant-Microbe Interact.* 6, 755-763.

Shirako, Y., and Brakke, M. K. (1984). Spontaneous deletion mutation of soil-borne wheat mosaic virus, II. *J. Gen. Virol.* 65, 855-858.

Simon, A. E., and Bujarski, J. J. (1994). RNA-RNA recombination and evolution in virus-infected plants. *Annu. Rev. Phytopathol.* 32, 337-362.

Simon, A. E., and Nagy, P. D. (1996). RNA recombination in turnip crinkle virus: Its role in formation of chimeric RNAs, multimers and 3'-end repair. *Semin. Virol.* 7, 373-379.

Smith, D. B., and Inglis, S. C. (1987). The mutation rate and variability of eukaryatic viruses: an analytical review. *J. Gen.Virol.* 68, 2729-2740.

Steinhauer, D. A., and Holland, J. J. (1987). Rapid evolution of RNA viruses. *Ann. Rev. Microbiol.* 41, 409-433.

Strauss, E. G., and Strauss, J. H. (1983). Replication strategies of the single-stranded RNA viruses of eucaryotes. *Curr. Top. Microbiol. Immunol.* 105, 2-98.

Strauss, J. H. , and Strauss, E. G. (1988). Evolution of RNA viruses. *Annu. Rev. Microbiol.* 42, 657- 683.

Suzuki, M., Kuwata, S., Kataoka, J., Masuta, M., Nitta, N., and Takanami, T. (1991). Functional analysis of deletion mutants of cucumber mosaic virus RNA 3 using an *in vitro* transcription system. *Virology* 8 3, 106-113.

Taliansky, M. E., and Garcia-Arenal, F. (1995). Role of cucumovirus capsid protein in long-distance movement within the infected plant. *J. Virol.* 69, 916-922.

Thomas, L. C., Perbal, C., and Maule, A. J. (1993). A mutation of cauliflower mosaic virus gene I interferes with virus movement but not virus replication. *Virology* 192, 415-421.

White, K. A., and Morris, T. J. (1994a). Nonhomologous RNA recombination in tombusviruses: Generation and evolution of defective interfering RNAs by stepwise deletions. *J. Virol.* 68, 14-24.

White, K. A., and Morris, T. J. (1994b). Recombination between defective tombusvirus RNAs generates functional hybrid genomes. *Proc Natl. Acad. Sci. USA.* 91, 3642-3646.

White, K. A., Bancroft, J. B., and Mackie, G. A. (1991). Defective RNAs of clover yellow mosaic virus encode nonstructural/coat protein fusion products. *Virology* 183, 479-486.

Zaccomer , B., Haenni, A-L., and Macaya, G. (1995). The remarkable variety of plant RNA virus genomes. *J. Gen. Virol.* 76, 231-247.

Zhang, C., Cascone, P. J., and Simon, A. E. (1991). Recombination between satellite and genomic RNAs of turnip crinkle virus. *Virology* 184, 791-794.

6 GENETIC VARIABILITY AND EVOLUTION

F. García-Arenal, A. Fraile, and J.M. Malpica

INTRODUCTION

Variation is an intrinsic property of living entities. New genetic variants are generated as organisms reproduce and their frequencies in populations may change with time. This temporal change of the genetic structure of the population of an organism is the process of evolution. The study of evolution has two main goals. One is to clarify the evolutionary history of organisms, the other is to understand the mechanisms of evolution. Much work has been done in the recent past and also reviewed to clarify the evolutionary history and the resulting taxonomic relationships of viruses, including plant viruses (Gibbs *et al.*, 1995; Gibbs *et al.*, 1997; Morse, 1994; Zanotto *et al.*, 1998). This will not be addressed in this chapter which instead is focused on the other main goal - mechanisms of evolution.

Evidence for genetic variation of plant viruses was reported as early as 1926 (Kunkel, 1947), and since then it has been much analysed under experimental conditions. This has allowed identification of the mechanisms involved in the generation of genetic variation, and understanding of their molecular basis. These subjects have been reviewed excellently by Roossinck (1997), and also by Jozef Bujarski in Chapter 5 of this book. More recently, interest in the quantitative analysis of variation in natural populations of plant viruses has contributed to understanding the relative importance of different mechanisms of evolution for different viruses. Here the focus is on the analysis of genetic variation, its quantification in populations of plant viruses, and on the factors and processes that determine the genetic structure of these populations and its temporal change.

VARIABILITY UNDER EXPERIMENTAL CONDITIONS

Genetic variation in plant viruses was first observed by the isolation of symptom mutants from areas in systemically infected plants that showed atypical symptoms (e.g. McKinney, 1935). Also, it was found that viral characters could be altered by serial passaging under different conditions. Most typical of these experiments are those in which a virus, adapted to a particular host, was repeatedly passaged through a different one, and this resulted in a change of virus traits called host adaptation (Yarwood, 1979). This was interpreted as selection by passaging into the new host of variants present, or newly generated, in the original virus population.

Host adaptation was first analysed at the molecular level for satellite tobacco necrosis virus (STNV). A stock adapted to tobacco was passaged through mung bean: it was found that adaptation to mung bean involved reproducible changes, by gradual selection of sequence variants from the tobacco-adapted population (Donis-Keller *et al.*, 1981). Similarly, symptom or temperature-resistant mutants were obtained by gradual temperature shift passaging (Van Vloten-Doting and Bol, 1988).

Results from passage experiments, in addition to demonstrating the capacity of RNA plant viruses for genetic variation, also showed that laboratory stocks of plant viruses consisted of a heterogeneous mixture of variants. Biological cloning by single-lesion passage did not eliminate heterogeneity (Garcia-Arenal *et al.*, 1984), as new variants could arise by mutation. The intrinsic heterogeneity of RNA virus isolates was further demonstrated by the analysis of populations obtained by the multiplication of inocula derived from biologically active cDNA clones. Aldaoud *et al.* (1989) thus showed that both temperature sensitive and necrotic lesion variants were found in a population derived from infectious transcripts of a cDNA full-length clone of tobacco mosaic virus (TMV). The frequency of lesion mutants was different according to the host plant species in which the virus multiplied, but random differences were also observed among individual plants of one species. Thus, mutation and founder effects can rapidly affect the genetic structure of the TMV population.

Fast accumulation of mutations, host adaptation effects, and founder effects have been also analysed in detail in populations derived from infectious RNA transcripts from cDNA clones of the satellite RNA of cucumber mosaic virus (CMV-satRNA) (Kurath and Palukaitis, 1989, 1990; Palukaitis and Roossinck, 1996), and of satellite tobacco mosaic virus (STMV) (Kurath and Dodds, 1995). Thus, available data indicate that mutants arise easily upon amplification of molecularly cloned inocula. Accumulation of mutations in hotspots may be a reproducible phenomenon (Kurath and Palukaitis, 1990; Kurath and Dodds, 1995). No estimation of error rates for RNA-dependent RNA polymerases of plant viruses has been reported, since the only published experiment (Kearney *et al.*, 1993) was quite inconclusive, but data are compatible with high error rates as reported for RNA polymerases of bacterial and animal RNA viruses (in the order of 10^{-4}, Smith and Inglis, 1987). Although most of the work discussed above was done with RNA viruses and/or their satellites, heterogeneous populations and host passage effects have been also reported for viroids (e.g. Fagoaga *et al.*, 1995; Visvader and Symon, 1985) and for dsDNA virus (Al-Kaff and Covey, 1994; Daubert and Routh, 1990; Sanger *et al.*, 1991). Thus, no significant differences appear to exist in the capacity to vary of plant viruses (and viroids) with RNA or DNA genomes.

Quantitative analyses of the heterogeneity of laboratory stocks of plant viruses were also done at early dates. Kunkel, in 1940 (quoted in Van Vloten-Doting and Bol, 1988) found that one out of every 200 lesions induced by TMV in tobacco was a symptom mutant. Gierer and Mundry (1958) obtained similar values of 0.2-0.3% symptom mutants, also for TMV. Data from sequence analyses of cDNA clones of different RNA viruses indicate polymorphism in about one per 1000 nucleotides (Van Vloten-Doting and Bol, 1988), but the effect of possible errors introduced during cDNA synthesis should be considered. Ribonuclease T1 fragment analysis of biologically cloned, or uncloned, populations of tobacco mild green mosaic virus (TMGMV) showed that single lesion passage reduced population heterogeneity, estimated as nucleotide diversity per site (probability that a randomly chosen nucleotide is different between two randomly chosen sequences). Values were 2.5 and 1.02 x10^{-3} nucleotide substitutions per site for the uncloned and cloned population, respectively (Rodríguez-Cerezo and García-Arenal, 1989). Interestingly, these values were lower than those estimated from similar analyses with bacterial or animal RNA viruses. The heterogeneity of an uncloned isolate of STMV, expressed as heterozygosity (Nei, 1987, p. 177), can be estimated from ribonuclease protection assay (RPA) data (Kurath *et al.*, 1993) to be of 0.686 (our calculations), much higher than for its helper virus TMGMV (data from Rodríguez-Cerezo and García-Arenal, 1989, for uncloned TMGMV would yield a heterozygosity index of 0.358). This indicates that constraints to variation are bigger for the helper than for the satellite virus.

All reported data indicate that a laboratory stock of a plant virus isolate contains a major genotype in addition of a minor pool of variants kept at lower frequencies because of selective pressures. This genetic structure had been previously reported for bacterial or animal RNA viruses, and the quasispecies theoretical model was proposed to describe it (Domingo and Holland, 1997; Eigen and Biebricher, 1988). The quasispecies concept has deeply permeated the literature on population analyses of viruses, including plant viruses. Nevertheless, it has been questioned that the diversity of RNA genomes is so different to that of DNA genomes as to need special concepts to describe it. The large mutability of RNA genomes does not necessarily produce more diverse populations (see below), although diversity might evolve more quickly (Smith and Inglis, 1987). Also, the contribution of the quasispecies concept to explain features of virus biology has been questioned from different grounds (Smith *et al.*, 1997).

VARIABILITY UNDER NATURAL CONDITIONS

As indicated by data from experimental conditions, high potential to vary need not result in high variation rates for virus populations. Genetic stability was already shown by the similarity of stocks of plant viruses derived from a single source, and repeatedly passaged over the years in different laboratories, such as "strains" *vulgare* and U1 of TMV (Dawson *et al.*, 1986; Goelet *et al.*, 1982). With the advent of technology for the determination of nucleotide sequences, data have accumulated which show that genetic stability is the rule, rather than the exception, for RNA plant viruses. For different genera, sequence similarity among isolates of the same virus is well above 95% (Fraile *et al.*, 1995; Ward *et al.*, 1995), and most nucleotide changes are silent. Even when methods of genome comparison other than sequence determination indicate different genetic types in field isolates, sequence analysis shows little genetic divergence (below 5%) among types. This is the case for viruses with different genetic material, such as beet necrotic yellow vein virus (BNYVV) (ssRNA, Kruse *et al.*, 1994), rice dwarf virus (RDV) (dsRNA, Murao *et al.*, 1994) or cotton leaf curl virus (CLCuV) (ssDNA, Sanz *et al.*, 1998). This may not be the case of viruses that have evolved to differentiate into well defined subgroups, as for instance CMV (Palukaitis *et al.*, 1992), kennedya yellow mosaic virus (KYMV) (Skotnicki *et al.*, 1996) or several potyviruses (Aleman-Verdaguer *et al.*, 1997; Blanco-Urgoiti *et al.*, 1996). Still, in those cases, divergence within a subgroup is also small. Further examples of genetic stability come from the few reported instances in which viruses isolated in the same area over different time periods were compared. High nucleotide sequence similarities (above 97%) were found for wound tumor virus (WTV) sampled with a 50 year interval (Hillman *et al.*, 1991) or TMGMV sampled over a 100 year period (Fraile *et al.*, 1997b). The most striking instance of stability reported is that of isolates of turnip yellow mosaic virus (TYMV) representing populations from Europe and Australia, that differ by 3-5% in nucleotide sequence (mostly synonymous differences), even if it is likely that the Australian population was derived from the European population before the last glaciation, which ended 13-14,000 years ago (Blok *et al.*, 1987; Keese *et al.*, 1989).

High mutability may not result in high diversity of a virus because of many different factors, including the direction and strength of selection, and founder effects associated to population bottlenecks, both related to the virus life history. To identify and understand these factors it is necessary to analyse the genetic structure of the natural populations of the virus. This kind of analysis has been done with relatively few systems, illustrating a variety of possible scenarios. Gibbs and his colleagues have analysed the genetic structure of two tymoviruses that infect wild plants in Australia. TYMV isolates infecting *Cardamine lilacina*, (a wild plant endemic in the mountains of SE Australia, where it grows in isolated patches 2-3 km apart) were characterised by RNase protection analysis (RPA) of probes that represent different genomic regions (Skotnicki *et al.*, 1993). The number of RPA patterns respective to the number of analysed isolates, was different for each probe. Thus, genetic heterogeneity differs for different genomic regions, suggesting different selection pressures. Data from all probes was used to define haplotypes (i.e. genetic types), that were closely related (about 98% similar, Keese *et al.*, 1989) in sequence. The most frequent haplotype represented 13% of the population, and the four next more frequent ones, 6% of the population each; all other types were represented by one or few isolates; a quite classical distribution of genetic types. Genetic types did not associate with the location of the isolated patches in which the host plants grow, indicating a single, undifferentiated population. Interestingly, random amplified polymorphic DNA (RAPD) analysis of the DNA from individual *C. lilacina* plants also gave evidence of a single population (Nolan *et al.*, 1996), so that the genetic structure of the host plant and its virus parasite is the same. A different genetic structure was found for KYMV. This virus is found along the eastern coast of Australia infecting species of *Desmodium* in the north, and *Kennedya rubicunda* in the south. RPA analyses divide the isolates into three well-defined groups, those isolated from *Desmodium* in the north, and two isolated from *Kennedya* in the south. Within these main types, the RPA analysis defined many haplotypes, and the *Desmodium* population alone was more variable than the

KYMV population from *Cardamine* (Skotnicki *et al.*, 1996). Thus, the KYMV population appears to be differentiated into at least three separate subpopulations that are evolving independently.

García-Arenal and colleagues have analysed the population structure of two tobamoviruses. The genomes of pepper mild mottle virus (PMMV) isolated in Europe over a seven year period were compared by RNase T1 oligonucleotide fragment mapping (Rodríguez-Cerezo *et al.*, 1989). The results showed a very stable population, that maintained its diversity through time, and consisted of a main prevailing type (representing 30% of the isolates) from which closely related variants arose without replacing it. They also studied TMGMV in populations of *Nicotiana glauca*, a wild perennial that has colonised, over the past two centuries, most Mediterranean regions of the world from its origin in central South America (Fraile *et al.*, 1996; Moya *et al.*, 1993; Rodríguez-Cerezo *et al.*, 1991). It was found that TMGMV populations from different regions of the world were closely similar. For each population, diversity was not related to the size of the area from which the isolates were collected. It was smaller for the Australian and Spanish populations, and twice as big for the Californian and Cretan ones, that showed diversity values similar to that of the whole world population. This suggests an upper threshold for TMGMV diversity. Diversity values also indicated population differentiation, which was largest between the less diverse Australian and Spanish populations. This could be due in part to a founder effect, which is supported by data on the genetic stability of the Australian TMGMV population over the last 100 years (Fraile *et al.*, 1997b), and by considerations on effective population sizes (Moya *et al.*, 1993). The population diversity of STMV in *N. glauca* in California can be estimated from the RPA data in Kurath *et al.* (1993), giving a heterozygosity index of 0.919 (our calculations). Interestingly this value does not significantly differ from that obtained for the Californian population of the helper virus TMGMV, analysed in the same sample of *N. glauca* plants by RPA (heterozygosity was 0.888, Fraile *et al.*, 1996). This is at odds with the higher diversity of laboratory stocks of STMV as compared with TMGMV (see above). Thus, constraints to divergence in the field and in the glass-house may be different, and these results indicate the need to be cautious when extrapolating data from experimental conditions to explain data from natural populations.

The genetic structure of wheat streak mosaic virus (WSMV) population was analysed by comparing restriction fragment length polymorphism (RFLP) patterns of a reverse transcriptase-polymerase chain reaction (RT-PCR) fragment encompassing the CP ORF plus 3' non-coding region (McNeil *et al.*, 1996). Thirty two haplotypes were found for the 461 analysed isolates. One haplotype (type 4) represented 46% of the isolates, the next two more frequent types represented 21% and 14% of the population, respectively. Analysis of the distribution of types within fields, and within sampled areas, indicated that there was no genetic differentiation according to geography. On the other hand, the subpopulations sampled in 1994 and in 1995 differed significantly in the proportions of the four main haplotypes. Nevertheless, there was almost no genetic differentiation between years (our analysis of the data). Type 4 was the major haplotype both in 1994 and in 1995. It also corresponded to an isolate sampled in the same area in 1981. Thus, there is a main genetic type that is not displaced by new ones.

RFLP analysis has been also used to characterise the population structure of two geminiviruses. Stenger and McMahon (1997) have analysed the beet curly top virus (BCTV) population in sugar beet in the Western USA. The population was built of 38 haplotypes belonging to two strains, CHF and Worland (19 types each). Type, and strain, distribution in the different sampled locations showed a clear genetic differentiation according to geography. At odds with all the cases before, the frequency of the different types in the population was rather similar. On the opposite, for the population of CLCuV infecting cotton plants in the Punjab province of Pakistan, haplotypes were not differentiated according to geographic location or year of sample, indicating a single, undifferentiated population. As with most cases before, haplotype frequency varied largely, the population being built of few major, and several minor, types (Sanz *et al.*, 1998).

In all the above-discussed cases, virus population is a single, undifferentiated population, or is genetically differentiated according to host plant, geography or time. A different situation has been described for CMV isolates sampled from different host plants, locations and years in Spain, and analysed by RPA of probes representing ORFs in the three genomic RNAs (Fraile *et al.*, 1997a). There was no correlation between haplotype and host plant species, place, or year of isolation, but the frequency composition for these types was significantly different for subpopulations sampled at one certain place and year. This random fluctuation in genetic structure corresponds to a metapopulation with local extinction and recolonization, and strong founder effects. A metapopulation structure has also been described for the CMV vectors *Myzus persicae* and *Aphis gossypii* (Martínez-Torres *et al.*, 1998), and seems to be the rule for other, non-viral, plant microparasites (Thompson and Burdon, 1992). Interestingly, when the population of the CMV-satRNA has been analysed, it corresponds to a single, non-differentiated population (Aranda *et al.*, 1993; Grieco *et al.*, 1997). The genetic structure, and dynamics, of populations of CMV and its satRNA are uncoupled indicating that satRNA spreads epidemically, as a hyperparasite, on CMV population (Alonso-Prados *et al.*, 1998).

Thus, available data indicate a wide range of genetic structures for plant viruses. So far, no correlation can be established between genetic structure and properties of the virus such as type of genetic material or life cycle.

A precise quantification of genetic diversity, as nucleotide diversity per site, of the above-discussed virus populations cannot be done in all cases. Genetic distances between isolates can be estimated from RFLP or RNase T1 fragment analyses (Nei, 1987, pp. 106-107), and of course from nucleotide sequence data. Genetic distances cannot be estimated from RPA data unless the method has been previously calibrated with a set of known sequences (Aranda *et al.*, 1995). Table 1 shows the populational diversities for a set of plant viruses. Values in Table 1 indicate low genetic diversity for the analysed viruses. They also indicate that there is no correlation between population diversity and the type of genetic material (RNA or DNA) of the virus analysed. The low diversities also reported for laboratory stocks (see above) could suggest that high genetic stability is a general property of plant viruses, but the number of analysed systems is too small to draw general conclusions. Data from a wider range of animal RNA viruses yield similar low values (Moya and García-Arenal, 1995) or much higher ones (Ina and Gojobori, 1994; Ina *et al.*, 1994; Lanciotti *et al.*, 1994) depending on the virus. The data in Table 1 also show a bigger diversity for the population of CMV-satRNA than for the other five viruses, which could be related to differences in functional constraints for helper virus and satellite, as also suggested for TMGMV and STMV (see above).

Table 1. Intrapopulation nucleotide diversities in some plant viruses [a]

Virus	Diversity Value	Estimated From
PMMV [b]	0.018	RNase T1 fragment analysis
TMGMV [b]		
Spain	0.022	RNase T1 fragment analysis
Spain	0.020	Sequence analysis
Australia	0.022	Sequence analysis
World	0.057	Sequence analysis
WSMV [c]	0.031	RFLP analysis
CMV-sat RNA [b]	0.064	RPA analysis
BCTV [c]		
CHF	0.026	RFLP analysis
Worland	0.021	RFLP analysis
CLCuV [b]		
CP	0.019	RFLP analysis
AC1	0.024	RFLP analysis

a) Diversities expressed as nucleotide diversity per site (Nei 1987, p. 276)
b) Data as in Rodríguez-Cerezo *et al.* (1989) for PMMV; Rodríguez-Cerezo *et al.* (1991) and Fraile *et al.* (1996) for TMGMV; Aranda *et al.* (1993) for CMV-sat RNA; Sanz *et al.* (1998) for CLCuV.
c) Our estimates from data in McNeil *et al.* (1996) for WSMV and in Stenger and McMahon (1997) for BCTV

FACTORS DETERMINING GENETIC STRUCTURE OF VIRUS POPULATIONS

As stated above, from the scanty data available, the genetic diversity and structure of viral populations does not depend so much on the input of such variation (such as the frequency of mutation or recombination), as on a range of factors related to the virus life history. Knowledge of these factors, and of how do they determine genetic structure of viral populations, is key to understand virus evolution.

Founder Effects

The founder effects are associated with population bottlenecks during colonisation of new hosts (i.e. during transmission) or new areas. Population bottlenecks introduce a random element in the genetic composition of populations, and their role in decreasing diversity within a population and in increasing diversity between populations (i.e. in promoting genetic differentiation into subpopulations) has been well documented for different organisms. For viruses, the work done with vesicular stomatitis virus (VSV, an animal rhabdovirus) should be singled out (Duarte *et al.*, 1992; Novella *et al.*, 1995). Though similar detailed work has not been done with plant viruses, random changes in the main genotype, observed during passage experiments with different viruses (Aldaoud *et al.*, 1989; Kurath and Dodds, 1995; Kurath and Palukaitis, 1989) can be explained by founder effects. Founder effects may also contribute to small intrapopulation diversity, and to geographic subdivision of populations, in TMGMV (Moya *et al.*, 1993; Fraile *et al.*,1996), and to the metapopulation structure of CMV in Spain (Fraile *et al.*, 1997a).

Selection

It is more often invoked in virus literature to explain evolutionary processes. The effect of selection in the evolution of organisms is directional, as opposed to the random effects of bottlenecks. Selection also results in a decrease of diversity within a population, and may also cause an increase of diversity between populations, so that the effects of bottlenecks can be (and often are) misinterpreted as due to selection. Selection pressures can be associated with different elements in the virus life cycle. A first group of pressures can be associated to structural features of the virus. For instance, amino acids that have a role in assembly and stabilisation of nucelocapsids are conserved among different tobamoviruses (Altschuh *et al.*, 1987). Also, selection might have a role in the maintenance of structures of viral genomes important for their replication, such as the 3' non-coding regions of ssRNA genomes (e.g. Bacher *et al.*, 1994; Kurath *et al.*, 1993), or the intergenic region of the geminiviruses (Argüello-Astorga *et al.*, 1994; Sanz *et al.*, 1998). The maintenance of a functional structure would be the primary selection factor for subviral, non-coding replicating RNAs, as CMV-satRNA (Aranda *et al.*, 1997; Fraile and Garcia-Arenal, 1991) or viroids (Elena *et al.*, 1991; Kofalvi *et al.*, 1997).

A second group of selection factors can be associated with the host plant. Host plant adaptation has been already discussed. Even if results of some reported passage experiments could be explained by random founder effects, consistent selection of different variants in different host plants has been well documented for a number of cases (e.g. Donis-Keller *et al.*, 1981; Kaper *et al.*, 1988; Kurath and Palukaitis, 1990; Moriones *et al.*, 1991). Data from natural populations that can be explained by host-associated selection are population differentiation according to host plant species, as reported for KYMV (Skotnicki *et al.*, 1996) or hop stunt viroid (HSVd) (Kofalvi *et al.*, 1997) and, of course, the well known phenomenon of overcoming of resistance genes (Harrison, 1981). Host-associated selection may have had a major role in the evolution of virus genera with host-specific species. It might also be that co-evolution of a group of viruses and their host plants had occurred, as has been proposed, and supported on different grounds, for the *Tobamovirus* (Lartey et al., 1996; Gibbs, 1999).

In addition with its host plant, a virus must interact efficiently with its vector to spread successfully in populations. Vector associated selection has not been as often documented as host-associated selection. Virus stocks maintained in the laboratory by non-vector transmission may lose the ability to be vector transmitted. This was first reported for WTV (Reddy and Black, 1977), and has also been shown in several other instances (Pirone and Blanc, 1996). This is evidence of vector selection of a virus-encoded function. More direct evidence is the specific selection from mixtures of cucumovirus variants by aphids (Fraile et al., 1997a; Perry and Francki, 1992), or of reovirus variants by leafhoppers (Suga et al., 1995; Uyeda et al., 1995).

Sequence analyses shows that in most cases selection pressures result in negative, or purifying selection, i.e. in the elimination from the population of the less fit variants. An indication of negative selection is the low frequency of insertions/deletions in coding sequences, particularly if the ORF is interrupted. The degree of negative selection in genes can be estimated by comparing the ratio between nucleotide diversities at nonsynonymous versus synonymous positions, which indicates the amount of nucleic acid variation resulting in variation of the encoded protein. Table 2 shows some of these ratios for different genes of plant viruses. The ratios vary largely according to the gene and the virus. It is no surprise that different virus-encoded functions are differently constrained. What is to be stressed about the data in Table 2 is that the d_{NS}/d_S ratios fall within the range reported for DNA-encoded genes of cellular organisms (Nei, 1987, Table 5.4).

Table 2. Nucleotide diversities in some virus genes

Virus [a]	Gene	d_{NS}	d_S	d_{NS}/d_S
TMGMV	183k	0.012	0.052	0.230
CMV	1a	0.006	0.301	0.020
	2a	0.007	0.222	0.032
	Coat protein	0.004	0.086	0.050
PVA	Coat protein	0.014	0.073	0.206
PVY	Coat protein	0.027	0.238	0.115
CLCuV	AC1	0.076	0.505	0.150
	Coat protein	0.021	0.556	0.048
CaMV	Coat protein	0.016	0.143	0.115

a) All data estimated according to the method described by Pamilo and Bianchi (1993) and Li (1993). Data are from: Fraile et al. (1996) for TMGMV; Fraile et al. (1997a) for CMV; Ooi et al. (1997) for tobacco leaf curl virus (TLCV); Sanz et al. (1998) for CLCuV. For potato virus A (PVA) (21 sequences), potato virus Y (PVY) (39 sequences) and cauliflower mosaic virus (CaMV) (7 sequences), data were estimated from sequences in the EMBL databank.

Thus, virus-encoded proteins are under similar levels of constraint as those of their hosts, a point highly relevant to understand virus evolution.

Variation due to genetic exchange, either by reassortment of genomic segments or by recombination, may be also reduced by negative selection. As for mutants, the fate in the population of a new type generated by genetic exchange will depend on its fitness. The increase of fitness due to genetic exchange has been documented (Fernandez-Cuartero *et al.*, 1994), and the occasional appearance of more fit reassortant or recombinant viruses may have an important role in virus evolution (e.g. Lartey *et al.*, 1996; Revers *et al.*, 1996; White et al., 1995). Experiments with cucumovirus (Fraile *et al.*, 1996; Perry and Francki, 1992), with tospovirus (Qiu *et al.*, 1998) and with reovirus (Uyeda *et al.*, 1995) showed that reassortants between different genetic types were not random, i.e. certain genetic combinations were preferred, and there was evidence for host and vector selection of reassortants. Under natural conditions, reassortants of CMV also showed preferential association of segments from different genetic types. Moreover, reassortants were infrequent in the population, and did not become fixed, indicating selection against them (Fraile *et al.*, 1996). Recombinants also seem to be selected against in CMV natural populations (Fraile *et al.*, 1996). This may not be the case for other viruses, as shown in natural populations of geminiviruses (Zhou *et al.*, 1997, 1998), and as comparison of different isolates of CaMV (Chenault and Melcher, 1994) and PVY (Revers *et al.*, 1996) suggests. A phenomenon mechanistically related to recombination is the appearance of deletion mutants that can multiply as defective interfering (DI) nucleic acids. DI RNAs have been shown to appear in passage experiments for many plant and animal viruses, and they have been shown to affect the genetic structure of the helper virus population (Kirkwood and Bangham, 1994; Roux *et al.*, 1991). The role of DIs in natural populations of viruses remains to be established. In the only case in which a deliberate search was done in natural populations of an RNA virus, DIs were not found (Celix *et al.*, 1997). Conversely DI DNAs were found frequently in natural populations of BCTV (Stenger, 1995), although their effect, if any, on the genetic structure of BCTV has not been established.

The strength of selection may be lesser in non-coding genomic regions or non-coding RNAs such as satellite RNAs and viroids. Indeed, it has been shown that mutation accumulation, and recombination, is less constrained in populations of CMV-satRNA (Aranda *et al.*, 1997) than in those of the helper virus CMV. Mutation accumulation and recombination, insertions and deletions are also frequent in viroids (Daros and Flores, 1995; Visvader and Symon, 1985).

Selection can also be positive selection, i.e. by favouring new mutants that are more fit than the parent isolate. Positive selection could be acting in the Spanish population of TMGMV (Fraile *et al.*, 1996; Moya *et al.*, 1993).

Complementation

It is the third important factor in virus evolution as less fit or deleterious variants can be maintained in a population through this mechanism. Thus, complementation modifies the effect of selection. In spite of its importance, complementation is often overlooked and, to our knowledge, its role in viral populations has been analysed in only one case. In a laboratory stock of tomato aspermy virus (TAV), Moreno *et al.* (1997) have analysed complementation of a mutant in RNA3 that abolishes cell-to-cell movement but increases its replication efficiency. Data on the relative fitness of the movement defective and competent variants allowed the estimation of a lower limit for the degree of complementation, i.e. for the probability that the non-functional individual will perform the function relative to the functional one. This value was of 0.13, and it led to a frequency of the defective mutant in the population of 0.76, indicating the importance of complementation in shaping the genetic structure of virus populations.

All the above discussed factors may act on the populations of a virus. In nature mixed infections can be frequent, and interactions between different viruses may determine the population structure of them. A most studied type of interaction is that of helper viruses and satellites. The genetic structure of CMV-satRNA has been shown to be affected by the strain of the helper virus in passage experiments, though the reverse does not seem to occur (Palukaitis and Roossinck, 1995; Roossinck and Palukaitis, 1995). Analyses of field populations, though, do not show such an effect (Grieco *et al.*, 1997; Alonso-Prados *et al.*, 1998). An interesting interaction between the tobamoviruses TMV and TMGMV has been described recently (Fraile *et al.*, 1997b). Both TMV and TMGMV were isolated from Australian *N. glauca* herbarium specimens collected before 1950, but only TMGMV was isolated from more recent specimens. Gene sequences of these isolates showed that although the TMGMV population did not increase in genetic diversity with time, the TMV population did. It was also shown that although TMGMV accumulated to the same concentration in double or single infected *N. glauca* plants, TMV attained only one tenth the concentration in plants double-infected with TMGMV that in single infected plants. Thus, a likely cause for the disappearance of TMV from the *N. glauca* population is that a mutational meltdown occurred as a result of the process known as Muller's ratchet (Lynch *et al.*, 1993). The drop of TMV concentration in doubly infected plants may go below a threshold needed to ensure the transmission to a new plant of the more fit genotype. Then, the TMV population will progressively be dominated by less fit genomes, and will succumb.

Other factors in the complex ecology of viruses may have a role in virus evolution, as reviewed by Gibbs (1983). To our knowledge, no analysis of their role in the evolution of plant virus populations has been reported in recent years. Also, it should be pointed out that epidemiological factors, not analysed for plant viruses, may also have an important role in the evolution of virus populations (May, 1995).

CONCLUSIONS

Since early work in the 1920s, evidence has accumulated for the high potential to vary of plant viruses and viroids. Passage experiments, and the analyses of diversity in laboratory and field populations, indicate that they are usually built of many different variants. The qualitative description of variants in virus populations has often led to the conclusion that populations were very diverse. Nevertheless, the genetic diversity of a population does not depend only on the number of variants present, but on the frequency distribution of these variants, and on the relatedness (i.e. genetic distance) among them. In the few reported analyses of virus populations that allow the estimation of intrapopulation diversities, estimated values are low, showing genetic stability rather than the high diversity that the number of identified variants could suggest. Abundant, less direct data also point to high genetic stability for plant virus populations. Diversity in populations of plant RNA viruses is not greater than in populations of plant DNA viruses.

Different genetic structures have been reported for different plant viruses, as regards genetic differentiation (or lack of it) according to spatial or temporal patterns, or according to the taxonomy of the host plant. The genetic structure may be determined by evolutionary factors related to the virus life cycle, but so far the identification of these factors is largely speculative: more systems need to be analysed, and in deeper detail, to understand the relationship between the virus life cycle and its genetic structure. Selection is the evolutionary process more often invoked in the literature to explain the genetic structure and evolution of virus populations. While data showing selection (mostly negative selection) are reasonably abundant, in few instances the selection factor can be identified. Again, more work is needed on this important aspect of virus evolution. The strength of selection may result in similar levels of constraint for viral and for their host and vector proteins. Virus proteins have to interact with those of their hosts and vectors, what may result in the selection acting on these also constricting genetic divergence of virus-encoded proteins. Caution should be taken not to misinterpret as due to selection phenomena due to genetic drift associated with population bottlenecks in the virus life cycle. In many instances, the severe bottlenecks associated with transmission (Pirone and Blanc, 1996) between plants (and hence, between areas) could be the primary factor in virus evolution. A third main factor, not usually considered in the analysis of virus evolution, is complementation, as it results in the modification of selection, and may have a big impact in the evolution of viruses.

More work is needed on the description of the genetic structure of virus populations, as this should allow the generality of the trends described here to be tested. In addition, efforts should be directed to a much needed estimation of basic parameters in virus evolution. These include mutation and recombination rates, population numbers, and epidemiological parameters as the size of bottlenecks during transmission and the number of successful infections per individual host, all of which will influence the structure and evolution of virus populations.

Although the genetic analysis of virus populations, and their evolution has progressed considerably in the last fifteen years, much more work is needed to understand the factors that determine virus evolution in nature, so that predictions could be made. In addition to the intrinsic, basic interest of such work, it should be possible to manage host plant resistance, either conventional or transgenic, for the control of diseases induced by plant viruses.

ACKNOWLEDGMENTS

We want to thank Adrian Gibbs; Australian National University, Canberra, for access to unpublished data and for critical reading of the manuscript.

REFERENCES

Al-Kaff, N., and Covey, S. N. (1994).Variation in biological properties of cauliflower mosaic virus clones. *J Gen Virol* 75, 3137-3145.

Aldaoud, R., Dawson, W. O., and Jones, G. E. (1989). Rapid, random evolution of the genetic structure of replicating tobacco mosaic virus populations. *Intervirology* 30, 227-233.

Aleman-Verdaguer, M. E., Goudou-Urbino, C., Dubern, I., and Beachy, R. N. (1997). Analysis of the sequence diversity of P1, HC, P3, NIb and CP genomic regions of several yam mosaic potyvirus isolates: implications for the intraspecies molecular diversity of potyviruses. *J Gen Virol* 78, 1253-1264.

Alonso-Prados, J. L., Aranda, M. A., Malpica, J. M., García-Arenal, F., and Fraile, A. (1998). Satellite RNA of cucumber mosaic cucumovirus spreads epidemically in natural populations of its helper virus. *Phytopathology* 88, 520-524.

Altschuh, D., Lesk, A. M., Bloomer, A. C., and Klug, A. (1987). Correlation of co-ordinated amino acid substitutions with function in viruses related to tobacco mosaic virus. *J. Mol. Biol.* 193, 693-707.

Aranda, M. A., Fraile, A., and García-Arenal, F. (1993). Genetic variability and evolution of the satellite RNA of cucumber mosaic virus during natural epidemics. *J. Virol.* 67, 5896-5901.

Aranda, M. A., Fraile, A., García-Arenal, F., and Malpica, J. M. (1995). Experimental evaluation of the ribonuclease protection assay method for the assessment of genetic heterogeneity in populations of RNA viruses. *Arch. Virol.* 140, 373-1383.

Aranda, M. A., Fraile, A., Dopazo, J., Malpica, J. M., and García-Arenal, F. (1997). Contribution of mutation and RNA recombination to the evolution of a plant pathogenic RNA. *J. Mol. Evol.* 44, 81-88.

Argüello-Astorga, G., Herrera-Estrella, L., and Rivera-Bustamante, R. (1994). Experimental and theoretical definition of geminivirus origin of replication. *Plant Mol. Biol.* 26, 553-556.

Bacher, J. W., Warkentin, D., Ramsdell, D., and Hancock, F. (1994). Selection versus recombination: What is maintaining identity in the 3' termini of blueberry leaf mottle nepovirus RNA1 and RNA2? *J. Gen. Virol.* 75, 2133-2137.

Blanco-Urgoiti, B., Sánchez, F., Dopazo, J., and Ponz, F. (1996). A strain-type clustering of potato virus Y based on the genetic distance between isolates calculated by RFLP analysis of the amplified coat protein gene. *Arch. Virol.* 141, 2425-2442.

Blok, J., Mackenzie, A., Guy, P., and Gibbs, A. J. (1987). Nucleotide sequence comparisons of turnip yellow mosaic virus isolates from Australia and Europe. *Arch. Virol.* 97, 283-295.

Celix, A., Rodríguez-Cerezo, E.., and García-Arenal, F. (1998). New satellite RNAs, but no DI RNAs, are found in natural populations of tomato bushy stunt tombusvirus. *Virology* 239, 277-284.

Chenault, K. D., and Melcher, U. (1994). Phylogenetic relationships reveal recombination among isolates of cauliflower mosaic virus. *J. Mol. Evol.* 39, 496-505.

Daros, J. A., and Flores, R. (1995). Characterization of multiple circular RNAs derived from a plant viroid-like RNA by sequence deletions and duplications. *RNA* 1, 734-744.

Daubert, S., and Routh, G. (1990). Point mutations in cauliflower mosaic virus gene VI confer host-specific symptoms changes. *Mol.. Plant-Microbe Interac.* 3, 341-345.

Dawson, W. O., Beck, D. L., Knorr, D. A., and Grantham, G. L. (1986). cDNA cloning of the complete genome of tobacco mosaic virus and production of infectious trancripts. *Proc. Natl. Acad. Sci. USA* 83, 1832-1836.

Domingo, E., and Holland, J. J. (1997). RNA virus mutations and fitness for survival. *Annu. Rev. Microbiol.* 51, 151-178.

Donis-Keller, H., Browning, K. S., and Clarck, J. M. (1981). Sequence heterogeneity in satellite tobacco necrosis virus RNA. *Virology* 110, 43-54.

Duarte, E. A., Clarke, D. K., Moya, A., Domingo, E., and Holland, J. J. (1992). Rapid fitness losses in mammalian RNA virusd clones due to Muller's ratchet. *Proc. Natl. Acad. Sci. USA* 89, 6015-6019.

Eigen, M., and .Biebricher, C. K. (1988). Sequence space and quasiespecies distribution. In "Variability of RNA Genomes" (E. Domingo, J. J. Holland, and P. Ahlquist, Eds.), Vol. III, pp. .211-245. CRC Press, Boca Raton, Florida.

Elena, S. F., Dopazo, J., Flores, R., Diener, T. O., and Moya, A. (1991). Phylogeny of viroids, viroid-like satellite RNAs, and the viroidlike domain of hepatitis δ virus RNA. *Proc. Natl. Acad. Sci. USA* 88, 631-5634.

Fagoaga, C., Semancik, J. S., and Durán-Vila, N. (1995). A citrus exocortis viroid variant from broad bean (*Vicia faba* L): Infectivity and pathogenesis. *J. Gen. Virol.* 76, 2271-2277.

Fernandez-Cuartero, B., Burgyan, J., Aranda, M. A., Salanki, K., Moriones, E., and García-Arenal, F. (1994). Increase in the relative fitness of a plant virus RNA associated with its recombinant nature.: *Virology* 203, 373 -377.

Fraile, A., and García-Arenal, F. (1991). Secondary structure as a constraint on the evolution of a plant viral satellite RNA. *J. Mol. Biol.* 221, 1065-1069.

Fraile, A., Aranda, M. A., and García-Arenal, F. (1995). Evolution of the tobamoviruses. In "Molecular Basis of Virus Evolution" (A. J. Gibbs, C. H. Calisher, and F. García-Arenal, Eds.), pp. 338 - 350. Cambridge University Press, Cambridge.

Fraile, A., Malpica, J. M., Aranda, M. A., Rodríguez-Cerezo, E., and García-Arenal, F. (1996). Genetic diversity in tobacco mild green mosaic tobamovirus infecting the wild plant *Nicotiana glauca*. *Virology* 223, 148 -155.

Fraile, A., Alonso-Prados, J. L., Aranda, M. A., Bernal, J. J., Malpica, J. M., and García-Arenal, F. (1997a). Genetic exchange by recombination or reassortment is infrequent in natural populations of a tripartite RNA plant virus. *J. Virol.* 71, 934 -940.

Fraile, A., Escriu, F., Aranda, M. A., Malpica, J. M., Gibbs, A. J., and García-Arenal, F. (1997b). A century of tobamovirus evolution in an Australian population of *Nicotiana glauca*. *J. Virol.* 71, 8316 - 8320.

García-Arenal, F., Palukaitis, P., and Zaitlin, M. (1984). Strains and mutants of tobacco mosaic virus are both found in virus derived from single-lesion-passaged inoculum. *Virology* 132, 131-137.

Gibbs, A. J. (1983). Virus ecology- "Struggle" of the genes. In "Encyclopaedia of Plant Physiology" (L. O. Lange, P. S Nobel, C. B Osmond, and H. Zieger, Eds.), Vol. XIIC, pp. 537-558. Springer-Verlag, Berlin;

Gibbs, A. J. (1999). Evolution and origins of tobamoviruses. *Trans. Royal Soc., London B*, in press.

Gibbs, A. J., Calisher, C. H., and García-Arenal. F., Eds. (1995). Molecular Basis of Virus Evolution. Cambridge University Press, Cambridge. 603 pp. "

Gibbs, M. J., Armstrong, J., Weiller, G. F, and Gibbs, A. J. (1997). Virus evolution; the past, a window on the future? In "Virus-resistant Transgenic Plants: Potential Ecological Impact" (M. Tepfer,, and E. Balázs) pp. 1-19. Springer-Verlag, Berlin

Gierer, A., and Mundry, K. W. (1958). Production of mutants of tobacco mosaic virus by chemical alteration of its ribonucleic acid *in vitro*. *Nature* 182, 1457-1458.

Goelet, P., Lomonossoff, G. P., Butler, P. J. G, Akam, M. E., Gait, M. J., and Karn, J. N. (1982). Nucleotide sequence of tobacco mosaic virus RNA. *Proc. Natl. Acad. Sci. USA* 79, 5818 -5822.

Grieco, F., Lanave, C., and Gallitelli, D. (1997) Evolutionary dynamics of cucumber mosaic virus satellite RNA during natural epidemics in Italy. *Virology* 229, 166 -174.

Harrison, B. D. (1981). Plant virus ecology: Ingredients, interactions, and environment influences. *Ann. Appl. Biol.* 99, 195 -209.

Hillman, B. I., Anzola, J. V., Halpern, B. T., Cavileer, T. D., and Nuss, D. L. (1991). First field isolation of wound tumor virus from a plant host: minimal sequence divergence from the type strain isolated from an insect vector. *Virology* 185, 896 -900.

Ina , Y., and Gojobori, T. (1994). Statistical analysis of nucleotide sequences of the hemagglutinin gene of human influenza A viruses. *Proc. Natl. Acad. Sci. USA* 91, 8388 - 8392.

Ina, Y., Mizokami, M., Ohoba, K., and Gojobori, T. (1994). Reduction of synonymous substitutions in the core protein gene of hepatitis C virus. *J. Mol. Evol.* 38, 50 -56.

Kaper, J. M., Tousignant, M. E., and Steen, M. T. (1988). Cucumber mosaic virus-associated RNA 5. XI. Comparison of 14 CARNA 5 variants relates ability to induce tomato necrosis to a conserved nucleotide sequence. *Virology* 163, 284 -292.

Kearney, C. M., Donson, J., and Jones, G. E., and Dawson, W. O. (1993). Low level genetic drift in foreign sequences replicating in an RNA virus in plants. *Virology* 192, 11-17.

Keese, P., Mackenzie, A., and Gibbs, A. (1989). Nucleotide sequence of the genome of an Australian isolate of turnip yellow mosaic tymovirus. *Virology* 172, 536 -546.

Kirkwood, T. B. L., and Bangham, C. R. M. (1994). Cycles, chaos, and evolution in virus cultures: a model of defective interfering particles. *Proc.. Natl. Acad. Sci. USA* 91, 8685 -8689.

Kofalvi, S. A., Marcos, J. F., Cañizares, M. C., Pallás, V., and Candresse, T. (1997). Hop stunt viroid (HSVd) sequence variants from *Prunus* species: evidence for recombination between HSVd isolates. *J. Gen. Virol.* 78, 3177 -3186.

Kruse, M., Koenig, P., Hoffmann, A., Kaufmann, A., Commandeur, U., Solovyev, A. G., Savenkof, I., and Burgermeister, W. (1994). Restriction fragment lenghth polymorphism analysis of reverse transcription-PCR products reveals the existence of two major strain groups of beet necrotic yellow vein virus. *J. Gen. Virol.* 75, 1835 -1842.

Kunkel, L. O. (1947). Variation in phytopathogenic viruses. *Annu. Rev. Microbiol.* 1, 85 -100.

Kurath, G., and Palukaitis, P. (1989). RNA sequence heterogeneity in natural populations of three satellite RNAs of cucumber mosaic virus. *Virology* 173, 231-240.

Kurath, G., and Palukaitis, P. (1990). Serial passage of infectious transcripts of a cucumber mosaic virus satellite RNA clone results in sequence heterogeneity. *Virology* 176, 8 -15.

Kurath, G., and Dodds, J. A. (1995). Mutation analyses of molecularly cloned satellite tobacco mosaic virus during serial passage in plants: Evidence for hotspots of genetic change. *RNA* 1, 491-500.

Kurath, G., Rey, M. E. C., and Dodds, A. (1992). Analysis of genetic heterogeneity within the type strain of satellite tobacco mosaic virus reveals several variants and a strong bias for G to A substitution mutations. *Virology* 189, 233 -244.

Kurath, G., Heick, J. A., and Dodds, J. A. (1993). RNase protection analyses show high genetic diversity among field isolates of satellite tobacco mosaic virus. *Virology* 194, 414 - 418.

Lanciotti, R. S., Lewis, J. G., Gubler, D. J., and Trent, D. W. (1994). Molecular evolution and epidemiology of dengue-3 viruses. *J. Gen. Virol.* 75, 65 -75.

Lartey, R. T., Voss, T. C., and Melcher, U. (1996). Tobamovirus evolution: gene overlaps, recombination, and taxonomic implications. *Mol. Biol. Evol.* 13, 1327 -1338.

Li, W.-H. (1993). Unbiased estimation of the rates of synonymous and non-synonymous substitution. *J. Mol. Evol.* 36, 96 -99.

Lynch, M. R., Burger, R., Butcher, D., and Gabriel, W. (1993). The mutational meltdown in asexual populations. *J. Hered.* 84, 339 -344.

Martinez-Torres, D., Carrio, R., Latorre, A., Simon, J. C., Hermoso, A., and Moya, A. (1998). Assessing the nucleotide diversity of three aphid species by RAPD. *J. Mol. Evol.* 10, 459 -477.

May, R. .M. (1995). The co-evolutionary dynamics of viruses and their hosts. *In* "Molecular Basis of Virus Evolution" (A. J. Gibbs, C. H. Calisher, and F. García-Arenal, Eds.), pp. 192-212. Cambridge University Press, Cambridge.

McKinney, H. H. (1935). Evidence of virus mutation in the common mosaic of tobacco. *J. Agr. Res.* 51, 951-981.

McNeil, J. E., French, R., Hein, G. L., Baezinger, P. S., Eskridge, K. M. (1996). Characterization of genetic variability among natural populations of wheat streak mosaic virus. *Phytopathology* 86, 1222 -1227.

Moreno, I., Malpica, J. M., Rodríguez-Cerezo, E., and García-Arenal, F. (1997). A mutation in tomato aspermy cucumovirus that abolishes cell-to-cell movement is maintained to high levels in the viral RNA population by complementation. *J. Virol.* 71, 9157 -9162.

Moriones, E., Fraile, A., and García-Arenal, F. (1991). Host–associated selection of sequence variants from a satellite RNA of cucumber mosaic virus. *Virology* 187, 465 - 468.

Morse, S. S. (1994). The Evolutionary Biology of Viruses. Raven Press, New York. 353 pp.

Moya, A., and García-Arenal, F. (1995). Population genetics of viruses: An introduction. *In* "Molecular Basis of Virus Evolution" (A. J. Gibbs, C. H. Calisher, and F. García-Arenal, Eds.), pp. 213 - 223. Cambridge University Press, Cambridge.

Moya, A., Rodríguez-Cerezo, E., and García-Arenal, F. (1993). Genetic structure of natural populations of the plant RNA virus tobacco mild green mosaic virus. *Mol. Biol. Evol.* 10, 449 - 456.

Murao, K., Suda, N., Uyeda, I., Isogami, M., Suga, H., Yamada, N., Kimura, I., and Shikata, E. (1994). Genomic heterogeneity of rice dwarf phytoreovirus field isolates and nucleotide sequences of variants of genome segment 12. *J. Gen. Virol.* 75, 1843 -1848.

Nei, M. (1987). Molecular Evolutionary Genetics. Columbia University Press, New York.

Nolan, M. F., Skotnicki, M. L., and Gibbs, A. J. (1996). RAPD variation in populations of *Cardamine lilacina* (Brassicaceae). *Aus. System. Bot.* 9, 291-299.

Novella, I. S., Elena, S. F., Moya, A., Domingo, E., and Holland, J. J. (1995). Size of genetic bottlenecks leading to virus fitness loss is determined by mean of initial population fitness. *J. Virol.* 69, 2869-2872.

Ooi, K., Ohshita, S., Ishii, I., and Yahara, T. (1997). Molecular phylogeny of geminivirus infecting wild plants in Japan. *J. Plant. Res.* 110, 247-257.

Palukaitis, P., and Roossinck, M. J. (1995). Variation of the hypervariable region of cucumber mosaic virus satellite RNAs is affected by the helper virus and the initial helper context. *Virology* 206, 765-768.

Palukaitis, P., and Roossinck, M. J. (1996). Spontaneous change of a benign satellite RNA of cucumber mosaic virus to a pathogenic variant. *Nat. Biotechnol.* 14, 1264-1268.

Palukaitis, P., Roossinck, M. J., Dietzgen, R. G., and Francki, R. I. B. (1992). Cucumber mosaic virus. *Adv. Virus Res.* 41, 281-348.

Pamilo, P., and Bianchi, N. O. (1993). Evolution of the Zfx and Zfy genes: rates and interdependence between the genes. *Mol. Biol. Evol.* 10, 271-281.

Perry, K. L., and Francki, R. I. B. (1992). Insect-mediated transmission of mixed and reassorted cucumovirus genomic RNAs. *J. Gen. Virol.* 73, 2105-2114.

Pirone, T. P., and Blanc, S. (1996). Helper-dependent vector transmission of plant viruses. *Annu. Rev. Phytopathol.* 34, 227-247.

Qiu, W. P., Geske, S. M., Hickey, C. M., and Moyer, J. W. (1998). Tomato spotted wilt *Tospovirus* genome reassortment and genome segment-specific adaptation. *Virology* 244, 186-194.

Reddy, D. V. R., and Black, L. M. (19770. Isolation and replication of mutant populations of wound tumour virions lacking certain genome segments. *Virology* 80, 336-346.

Revers, F., Legall, O., Candresse, T., Leromancer, M., and Dunez, J. (1996). Frequent occurrence of recombinant potyvirus isolates. *J. Gen. Virol.* 77, 1953-1965.

Rodríguez-Cerezo, E., and García-Arenal, F. (1989). Genetic heterogeneity of the RNA genome population of the plant virus U5-TMV. *Virology* 170, 418-423.

Rodríguez-Cerezo, E., Moya, A., and García-Arenal, F. (1989). Variability and evolution of the plant RNA virus pepper mild mottle virus. *J. Virol.* 63, 2198-2203.

Rodríguez-Cerezo, E., Elena, S. F., Moya, A., and García-Arenal, F. (1991). High genetic stability in natural populations of the plant RNA virus tobacco mild green mosaic virus. *J. Mol. Evol.* 32, 328-332.

Roossinck, M. J. (1997). Mechanisms of plant virus evolution. *Annu. Rev. Phytopathol.* 35, 191-209.

Roossinck, M. J., and Palukaitis, P. (1995). Genetic analysis of helper virus-specific selective amplification of cucumber mosaic virus satellite RNA. *J. Mol. Evol.* 40, 25-29.

Roux, L., Simon, A. E., Holland, J. J. (1991). Effects of defective interfering viruses on virus replication and pathogenesis *in vitro* and *in vivo*. *Adv. Virus Res.* 40, 181-211.

Sanger, M., Daubert, S., and Goodman, R. M. (1991). The regions of sequence variation in caulimovirus gene VI. *Virology* 182, 830-834.

Sanz, A. I., Fraile, A., Gallego, J. M., Malpica, J. M., and García-Arenal, F. (1998). Genetic structure and evolution of natural populations of cotton leaf curl geminivirus, a single-stranded DNA virus. Submitted.

Skotnicki, M. L., Mackenzie, A. M., Ding, S. W., Mo, J. Q., and Gibbs, A. J. (1993). RNA hybrid mismatch polymorphisms in Australian populations of turnip yellow mosaic tymovirus. *Arch. Virol.* 132, 83-99.

Skotnicki, M. L., Mackenzie, A. M., and Gibbs, A. J. (1996). Genetic variation in populations of kennedya yellow mosaic tymovirus. *Arch. Virol.* 141, 99-110.

Smith, D. B., and Inglis, S. C. (1987). The mutation rate and variability of eukaryotic viruses: An analytical review. *J. Gen. Virol.* 68, 2729-2740.

Smith, D. B., McAllister, J., Casino, C., and Simmonds, P. (1997). Virus "quasispecies": Making a mountain out of a molehill?. *J. Gen. Virol.* 78, 1511-1519.

Stenger, D. C. (1995). Genotypic variability and the occurrence of less than genome-length viral DNA forms in a field population of beet curly top geminivirus. *Phytopathology* 85, 1316-1322.

Stenger, D. C., and McMahon, C. L. (1997). Genotypic diversity of beet curly top virus populations in the Western United States. *Phytopathology* 87, 737-744.

Suga, H., Uyeda, I., Yan, J., Murao, K., Kimura, I., Tiongco, E. R., Cabautan, P., and Koganezawa, H. (1995). Heterogeneity of rice ragged stunt oryzavirus genome segment 9 and its segregation by insect vector transmission. *Arch. Virol.* 140, 1503 -1509.

Thompson, J. N., and Burdon, J. J. (1992). Gene-for-gene coevolution between plant and parasites. *Nature* 360, 121-125.

Uyeda, I., Ando, Y., Murao, K., and Kimura, I. (1995). High resolution genome typing and genomic reassortment events of rice dwarf phytoreovirus. *Virology* 212, 724 -727.

Van Vloten-Doting, L., and Bol, J. F. (1988). Variability, mutant selection, and mutant stability in plant RNA viruses. *In* "RNA Genetics" (J. J. Holland, E. Domingo, and P. Ahlquist, Eds.).,Vol. III. pp. 37-52. CRC Press, Boca Raton.

Visvader, J. E., and Symons, R. H. (1985). Eleven new sequence variants of citrus exocortis viroid and the correlation of sequence with pathogenicity. *Nucleic Acids Res.* 13, 2907-2920.

Ward, C. W., Weiller, G. F., Shukla, D. D., and Gibbs, A. (1995). Molecular systematics of the *Potyviridae*, the largest plant virus family. *In* "Molecular Basis of Virus Evolution" (A. J. Gibbs, C. H. Calisher,, and F. García-Arenal, Eds.), pp. 477 -500. Cambridge University Press, Cambridge.

White, P. S., Morales, F. J., and Roossinck, M. J. (1995). Interspecific reassortment in the evolution of a cucumovirus. *Virology* 207, 334 -337.

Yarwood, C. E. (1979). Host passage effects with plant viruses. *Adv. Virus Res.* 24, 169 -190.

Zanotto, P. M. A., and Gibbs, M. J., Gould, E. A., and Holmes, E. C. (1998). A reevaluation of the higher taxonomy of viruses based on RNA polymerases. *J. Virol.* 70, 6083 - 6096.

Zhou, X., Liu, Y., Calvert, L., Muñoz, C., Otim-Nape, G. W., Robinson, D. J., and Harrison, B. D. (1997). Evidence that DNA-A of a geminivirus associated with severe cassava mosaic disease in Uganda has arisen by interspecific recombination. *J. Gen. Virol.* 78, 2101-2111.

Zhou, X., Robinson, D. J., and Harrison, B. D. (1998). Four DNA-A variants among Pakistan isolates of cotton leaf curl virus and their affinities to DNA-A of geminivirus isolates from okra. *J. Gen. Virol.* 79, 915 - 923.

7 MOLECULAR BASIS OF VIRUS TRANSPORT IN PLANTS

Scott M. Leisner

INTRODUCTION

Virus movement in plants is one of the fastest growing areas of plant virology. Several excellent reviews on this topic have been published, particularly since 1990 (Atabekov and Dorokhov, 1984; Hull, 1989; Atabekov and Taliansky, 1990; Maule, 1991; Hull, 1991; Deom *et al.,*1992; Citovsky, 1993; Citovsky and Zambryski, 1993; Leisner and Turgeon, 1993; Lucas and Gilbertson, 1994; Carrington *et al.,* 1996; Gilbertson and Lucas, 1996; Mezitt and Lucas, 1996; Seron and Henni, 1996; Ghoshroy *et al.,* 1997; McLean *et al.,* 1997; and Nelson and van Bel, 1998). The emphasis here is on the genetic basis of the process and a detailed discussion of the various aspects of the movement proteins.

PATHWAYS OF VIRUS MOVEMENT

Spread of virus through a host is of two types; short distance or cell-to-cell and long-distance or vascular transport (Samuel, 1934; Hull, 1989). During mechanical inoculation of a leaf, virus is introduced into epidermal (ECs), and sometimes mesophyll (MCs) cells. As the virus propagates, it begins to spread to neighbouring cells (Fig. 1) as virus particles or as non-virion nucleoprotein complexes (Deom *et al.,* 1992), most likely by re-deploying a host short-distance transport pathway (Mezitt and Lucas, 1996). Virus spreads in spherical waves to neighbouring ECs and MCs until it reaches a vein. Invasion of a vein, movement through and exit from the vasculature, is long-distance movement (Leisner and Turgeon, 1993).

Viruses most likely enter the vascular system through minor (class IV and V) veins (Nelson and van Bel, 1998) which a virus is most likely to encounter after spreading from an infected cell. Once virus passes through the bundle sheath cell (BSC) layer and enters the vascular tissue proper, it then invades the phloem. Some viruses also travel through xylem, but this pathway is less well understood (Opalka *et al.,*1998).

Although viruses enter the vascular system through class IV and V veins, the cellular pathway of phloem invasion is still open to question. Virus may enter sieve

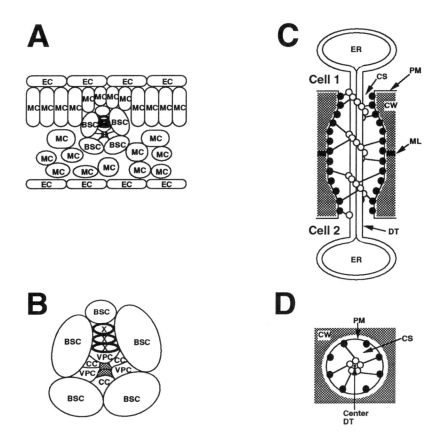

Figure 1. Pathways of virus movement in plants. (A) Schematic drawing of a leaf cross section, (B) enlargement of a minor vein, (C) schematic drawing of a plasmodesma, (D) cross section through a plasmodesma (Nelson and van Bel, 1998; Overall and Blackman, 1996). In (A) and (B); EC, epidermal cell; MC, mesophyll cell; BSC, bundle sheath cell; X, Xylem elements; VPC, Vascular parenchyma cell; CC companion cell; shaded cells in (B) sieve elements, in (C) and (D) cell wall; ER, endoplasmic reticulum; PM, plasma membrane; CW, cell wall; DT, desmotubule; CS, cytoplasmic sleeve; ML, middle lamella.

elements (SEs) via a symplastic pathway from epidermal cell (EC) to mesophyll cell (MC) to bundle sheath cell (BSC) to vascular parenchyma cell (VPC) to sieve element (SE) (Nelson and van Bel, 1998). Other possible routes are from EC to MC to BSC to VPC to companion cell (CC) to SE or from EC to MC to BSC to CC to SE. Viruses could also invade the phloem through vein endings or branches because MCs often directly abut SEs at these locations. The last route may explain why inoculated leaves contain a limited number of vascular invasion sites (Cohen *et al.,* 1957).

   Once in the SEs, virus travels through these long files of cells along with flow of photoassimilates (Leisner and Turgeon, 1993) which are transported from their locations of synthesis (source tissues) to sites of utilization (sink tissues). Young leaves import photoassimilates while mature leaves export them. This explains why the youngest leaves on a plant contain virus while older ones often remain uninfected. Many viruses probably spread through the phloem as particles, although a non-virion nucleoprotein complex, containing coat protein has also been proposed (Atabekov and Dorokhov, 1984). It is likely that long-distance movement of viral nucleic acids re-deploys a host transport system and is not completely passive (Nelson and van Bel, 1998).

After reaching sink tissues, virus probably exits SEs within class III and larger veins (Nelson and van Bel, 1998). Viruses likely travel from SEs to either CCs or VPCs, then to BSCs. Once there, viruses spread into MCs and ECs by cell-to-cell movement. Some viruses are limited to phloem SEs and CCs (and sometimes VPCs). This probably reflects differences in the connections between phloem and other cells.

Because plant cells are surrounded by a cell wall, virus movement must occur through cytoplasmic connections, called plasmodesmata (PD). In fact, the cytoplasm of almost all plant cells is continuous (called the symplast), making plants supracellular organisms (Lucas *et al.*, 1993a). Possible structure of simple or primary PD is shown in Fig. 1 (reviewed in Overall and Blackman, 1996). The plasma membrane is continuous between the two cells and running through the center of a plasmodesma is a tightly appressed cylinder of endoplasmic reticulum (ER) called the desmotubule (DT). Filaments that may contain actin and myosin connect protein located in the plasma membrane and the DT. Materials are thought to move from one cell to another through the cytoplasmic sleeve between the DT and the plasma membrane. The size exclusion limit (SEL) of the cytoplasmic sleeve of an average PD is about 1000 Daltons (1 kDa), permitting passage of solutes with a stokes radius of 1.5 nanometers. The PD SEL is regulated by callose deposition and by actin filaments (McLean *et al.*, 1997; Overall and Blackman, 1996).

PD are dynamic structures that can form, change morphology, and disappear. PD formation and disappearance can be regulated by the plant in response to virus attack (Shalla *et al.*, 1982). PD structure also changes during development. As a plant organ matures, the connections change from primary to secondary PD (Lucas and Gilbertson, 1994). Secondary PD often have appearance of two or more primary connections linked at their centres.

Virus movement between almost all types of cells within an infected leaf occurs through PD except the movement of a virus from one SE to another through sieve plate pores (Esau, 1964). Although derived from PD, sieve plate pores permit passage of much larger solutes than their progenitors but it is not clear how transport through these structures is regulated.

IDENTIFICATION OF VIRAL MOVEMENT PROTEIN GENES

If viruses move through PD, then they must encode a factor that facilitates transport through these structures. During the late 1970s, evidence began to accumulate that movement was controlled by a transport function encoded by the virus (reviewed in Atabekov and Dorokhov, 1984). During this time, several temperature-sensitive movement-defective tobacco mosaic virus (TMV) mutants were identified. The TMV Ls1 mutant was found to replicate well in infected cells, synthesize coat protein and encapsidate viral RNA, but it was unable to move from cell-to-cell at the restrictive temperature. The TMV open reading frame (ORF) encoding a 30 kDa protein had been proposed to be responsible for virus movement. Therefore, it was perhaps no surprise when a peptide fragment difference was discovered in the 30 kDa proteins of Ls1 and its wild type (L) parent. Sequencing of the 30 kDa protein genes of the L and Ls1 strains of TMV confirmed that a single base substitution resulted in a proline to serine change at position 153 and accounted for the peptide fragment difference. Site-directed mutagenesis of an infectious L TMV clone confirmed that the codon change was responsible for the movement properties of Ls1 (Meshi *et al.*, 1987). As a final proof, expression of the wild type TMV 30 kDa protein gene in transgenic plants was shown to complement the Ls1 mutation (Deom *et al.*, 1987). Therefore, the TMV 30 kDa protein controls virus cell-to-cell movement and is the movement protein (MP).

CHARACTERISTICS OF PLANT VIRAL MOVEMENT PROTEINS

Like the tobamoviruses, the genomes of the alfamo-, bromo-, caulimo-, como-, cucumo-, diantho-, ilar-, nepo-, tobra-, and tombusviruses, all encode a single multifunctional protein controlling cell-to-cell movement (Atabekov and Taliansky, 1990; Scholthof *et al.*,1995). However, members of the carla-, furo-, hordei-, and potexviruses encode three polypeptides that together fulfil the functions of the TMV MP (Atabekov and Taliansky, 1990; Mushegian and Koonin, 1993). The ORFs encoding these three proteins partially overlap and are called the triple gene block (TGB). The first ORF is usually the largest and encodes the TGB1 protein (Mushegian and Koonin, 1993). The other two ORFs encode the TGB2 and TGB3 proteins. The gene overlaps may allow for co-ordinated translation of the three TGB proteins (Morozov *et al.*,1991). Members of the carmo- and necroviruses encode two overlapping ORFs controlling virus movement, which may represent a truncated TGB (Mushegian and Koonin, 1993). Location of MP gene and TGB in genome of plus-sense RNA plant viruses is shown in various figures of the chapter 3 of this book. The bipartite geminiviruses also harbour two MP genes, BL1 and BR1 (Sanderfoot and Lazarowitz, 1996). Therefore, plant viruses usually encode one or more genes mediating cell-to-cell movement. Regardless of their differences, the MPs encoded by the majority of plant viruses share many features: they are non-structural proteins, are involved in movement but not replication of plant viruses, have homologous sequences, are often transiently expressed, are nucleic acid-binding proteins, are generally not virus-specific, are usually localised to PD, modify PD SEL, move through PD themselves, and facilitate movement of nucleic acids through PD.

Being non-structural proteins, MPs are generally not part of, nor contained within, the viral capsid (Atabekov and Taliansky, 1990) particularly of viruses that do not require coat protein (CP) for cell-to-cell spread (Giesman-Cookmeyer *et al.*, 1995). Many viruses require CPs to move from cell-to-cell (Rao, 1997). Even in these cases, MP is distinct from the virus particle. Potyviruses are a possible exception to this principle because their coat proteins possess some of the properties of an MP (Rojas *et al.*, 1997).

In addition to being independent of the virus particle, MPs are generally not required for virus replication (Meshi *Et al.,* 1987). For example, a mutation in cauliflower mosaic virus (CaMV) gene I (encoding the MP) had no effect on virus replication in protoplasts but the virus was unable to spread in infected plants (Thomas *et al.,* 1993). This CaMV mutant could replicate at the edges of leaf disks but did not spread to their interior. Thus, the mutant was impaired in virus cell-to-cell movement but not replication.

Plant viral MPs have homologous sequences. After the discovery of the TMV MP, a number of putative MPs of several other plant viruses were identified by sequence homology (Atabekov and Taliansky, 1990). The N-terminal two-thirds of may putative tobamoviruses MPs contains two highly conserved blocks (I and II), while the C-terminal one-third harbours three short conserved regions (A, B, and C, Fig. 2) (Saito *et al.,* 1988). A computer-assisted comparison of many MP sequences

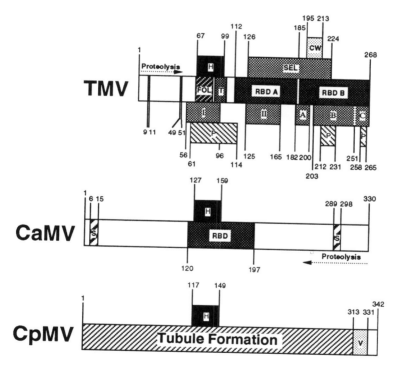

Figure 2. Structure of movement proteins of TMV, CaMV, and CPMV. Numbers indicate amino acid positions in the MP. H, regions of homology to other MPs (Mushegian and Koonin, 1993); RBD, RNA-binding domain (Citovsky *et al.,* 1992; Thomas and Maule, 1995a); dotted arrows indicate extent of proteolysis (Hughes *et al.,* 1995; Thomas and Maule, 1995b). For TMV: Roman numerals and letters A-C, regions of tobamovirus MP homology (Saito *et al.,* 1988); P, regions of MP phosphorylation (Citovsky *et al.,* 1993; Haley *et al.,* 1995); CW, targeting to cell wall (Berna, 1995); SEL, needed for changing PD SEL (Waigmann *et al.,* 1994); FOL, region from amino acids 65086 controlling protein folding (Citovsky *et al.,* 1992); regions 9-11, 49-51, and T (amino acids 88-101), targeting to plasmodesmata and/or cytoskeleton (Kahn *et al.,* 1998). For CaMV: S, found on the surface of tubules (Thomas and Maule, 1995b). For CPMV: V, incorporation of virions into tubules (Lekkerkerker *et al.,* 1996).

allowed a phylogenetic tree to be developed containing three clusters: the TMV cluster (containing the tobamo- and tobraviruses), the caulimovirus cluster (containing alfamo-, ilar-, bromo-, and cucumoviruses), and the caulimoviruses cluster (containing different members of the caulimovirus family) (Melcher, 1990). Subsequent work, discovered that MPs fell into 2 large families (Koonin *et al.,* 1991). Family I contains members of the tobamo-, tobra-, caulimo-, como-, and trichoviruses. These MPs share 2 motifs, one centered about a conserved glycine and the second surrounding a conserved aspartate residue. A subset of family I MPs also share sequence homology with cellular heat shock protein 90 (HSP90). Family II contains members of alfamo-, ilar -, bromo-, and cucumo-, and dianthoviruses. Later analysis showed that most virus MPs Belong to a large superfamily of proteins that share a common sequence motif, which may be a hydrophobic interaction domain (Mushegian and Koonin, 1993). An analysis of the predicted structure showed that MPs were divided into three regions: an N-terminal region of variable length, a C-terminal region lacking a defined secondary structure, and a conserved central region of shared sequence homology with the HIV-1 proteinase (Melcher, 1993). Using proteinase as a model, the structures of MPs were predicted to function as homodimers held together through the central conserved region.

In addition to their homology to HSP90, many MPs also were similar to other cellular proteins. MPs share sequence homology with ATP-binding domains of protein kinases (Maule, 1991). While kinase activity for viral MPs has yet to be demonstrated, a potexvirus MP possesses ATPase activity (Rouleau *et al.,* 1994). However, the TMV MP binds GTP rather than ATP, suggesting that different cell signalling pathways may be used by various MPs for PD transport (Li and Palukaitis, 1996). Interestingly, a furovirus MP contains a consensus sequence, found in ATP/GTP-dependent helicases, that is essential for MP activity (Bleykasten *et al.,* 1996).

Viral MPs are often transiently expressed. They are often expressed early after inoculation and then the levels decline later in infection (Watanabe *et al.,* 1984). The promoter and/or leader sequence located upstream of the TMV MP gene regulates the transient pattern of expression (Lehto *et al.,* 1990). Delaying synthesis by replacing the TMV MP promoter/leader with that from the coat protein had adverse effects on virus movement. Therefore, it is the timing rather than the amount of TMV MP that is critical for movement. In fact, only about 30% of the amount of MP normally produced by the virus is necessary for the establishment of multicellular infection sites and only about 2% was needed for subsequent spread of TMV through the inoculated leaf (Arce-Johnson *et al.,* 1995). In contrast to TMV, other viral MPs are expressed throughout the infection cycle. (Ritzenthaler *et al.,* 1995a). Thus, the duration of MP expression can vary with the virus.

Plant viral MPs are nucleic acid-binding proteins. For MPs to fulfil their function, they must transport viral nucleic acids through PD. While some MPs facilitate virus particle movement through PD, others aid in the transport of nucleic acid in a non-virion form (Deom *et al.,* 1992). In the latter situation, MPs must be nucleic acid binding proteins. Many MPs preferentially bind to single-stranded nucleic acids and their association with double-stranded (DNA or RNA) is usually of lower affinity (Citovsky *et al.,* 1990; Fujita *et al.,* 1998). Most MPs bind to single-stranded DNA and RNA with equal affinity (Citovsky *et al.,* 1990), while others preferentially associate with RNA (Citovsky *et al.,* 1991). Geminivirus MPs have been shown to bind to either double- or single-stranded DNA (Noueiry *et al.,* 1994; Sanderfoot and Lazarowitz, 1996).

Binding of MPs to nucleic acids is usually co-operative and sequence non-specific, with each protein molecule binding to approximately 3-7 nucleotides (Citosvsky *et al.,* 1990; Li and Palukaitis, 1996; Osman *et al.,* 1992). Even though MPs may bid co-operatively, they may not completely coat the viral RNA (Citovsky *et al.,* 1990; Karpova *et al.,* 1997). The co-operative binding of the TMV MP to single-stranded nucleic acids is thought to remove most of their secondary structure, elongate the RNA, and generate a ribonucleoprotein (RNP) complex with a cross sectional diameter of approximately 1.5 to 2 nm (Citovsky *et el.,* 1990; Citovsky *et al.,* 1992). Removal of secondary structure of viruses other than TMV may be mediated by MP helicase activity as suggested by sequence homology (Bleykasten *et al.,* 1996). In some cases, however, co-operative binding and the ability to remove RNA secondary structure are dispensable for MP function (Giesman-Cookmeyer and Lommel, 1993; Fujiwara *et al.,* 1993).

MP RNA binding is generally mediated by one or more RNA-interaction domains. The TMV MP harbours two RNA binding domains, termed A and B (Fig. 2) (Citovsky *et al.,* 1992). While domain A requires the entire MP to function, the B region can bind to RNA independent of the rest of the protein. The MPs of other viruses, such as CaMV, contain a single RNA-binding domain (Thomas and Maule, 1995a). RNA-binding domains are usually located in hydrophilic regions of MPs with high surface probability (Citovsky *et al.,* 1992; Fujita *et al.,* 1998; Schoumacher *et al.,* 1994; Thomas and Maule, 1995a).

Once MPs are associated with viral RNAs, translation is inhibited (Karpova *et al.,* 1997). Therefore, MPs were proposed to sequester a specific population of RNA molecules in a cell into a movement form that is independent of other viral genomes.

Plant viral MPs are generally not virus-specific so that, in many cases, they appear to be non-virus specific (reviewed in Atabekov and Taliansky, 1990). This is perhaps not surprising since many MPs are sequence non-specific nucleic acid-binding proteins. It has been shown that a virus incapable of spreading from cell-to-cell can move if the plant is inoculated with a second ("Helper") virus capable of spreading. For example, many tobamoviruses are able to assist the movement of each other in a host non-permissive to one of the virus strains. Coinoculation with a tobamovirus also allows certain movement-deficient mutants to spread or phloem-limited viruses to exit vascular tissues. One must be somewhat cautious about interpreting co-infection data since the helper virus may be inhibiting host defence responses as well as assisting cell-to-cell spread (Atabekov and Dorokhov, 1984). Transgenic plants expressing MPs have been used to examine specificity in the absence of the rest of the virus (Rao *et al.,* 1998). This work demonstrated the functional equivalence of several viral MPs (Giesman Cookmeyer *et al.,* 1995). Studies performed with chimeric genomes, in which the MP gene of a virus was replaced with that from another, also indicated that viral MPs may use similar mechanisms for transporting viral genomes from cell-to-cell (De Jong and Ahlquist, 1992). Thus, many MPs are relatively non-virus specific and they may interact with related host factors for potentiating viral cell-to-cell movement (Atabekov and

Taliansky, 1990). Since the creation of chimeras also changed the plant species these artificial viruses could infect, this shows that MPs can be powerful host-range determinants (Fenczik *et al.*, 1995).

Although many viruses may move from cell-to-cell by similar means, there are some that spread via different mechanisms. For example the TMV MP cannot facilitate the cell-to-cell movement of all viruses (Rao *et al.*, 1998). The form of movement that a virus uses also determines the specificity of complementation. While viruses that spread from cell-to-cell in a non-virion form can often assist each other, these viruses generally are not aided by those that form tubules (Lauber *et al.*, 1998a). Interestingly, the TGB MPs act as a functional unit whose organisation cannot be disrupted without detriment to virus movement. Thus, viruses that use the TGB MPs require specific interactions amongst the three proteins for cell-to-cell spread.

Plant viral MPs are usually localised to PD. Due to their role in cell-to-cell spread, it is no surprise that MPs colocalise with the cell wall fraction (Berna, 1995; Deom *et al.*, 1990) While some viral MPs have not been found in this cellular compartment (Davies *et al.*, 1993), the cell wall association of most of these proteins was explained by localisation to the PD (Fujita *et al.*, 1998; Tomenius *et al.*, 1987). PD localisation is critical for TMV MP function, since mutations disrupting this process also interfere with cell-to-cell spread (Berna, 1995).

Plant viral MPs modify PD SEL. For viruses to spread between cells, they must move through PD (Esau, 1948; Hull, 1989), yet the SEL of these structures is smaller than either virus nucleic acids or virions (Citovsky *et al.*, 1990). Therefore, MPs must increase PD SEL to permit cell-to-cell movement of viral nucleic acids. Indeed, the PD SEL is increased during virus infection and in transgenic plants expressing the MP (Oparka *et al.*, 1996b; Wolf *et al.*, 1989). MC PD in transgenic tobacco expressing the TMV MP gene exhibited a ten fold increase in SEL, permitting molecules with a stokes radius of 2.4 nanometers to pass through these channels (Waigmann *et al.*, 1994; Wolf *et al.*, 1989). The enlarged PD SEL is critical for some viruses since MP mutants unable to increase SEL did not spread from cell-to-cell (Fujiwara *et al.*, 1993). The domain responsible for increasing PD SEL was mapped to the C-terminal part of the TMV MP (Fig. 2) (Waigmann *et al.*, 1994). The increased SEL may be mediated by interaction of the C-terminal region with actin filaments located in the PD (Carrington *et al.*, 1996; Ding, 1997).

Increases in PD SEL are transient and occur at the leading edge of infection sites (Oparka *et al.*, 1997). TMV MP was able to cause changes in SEL only in mature leaves, showing that PD modification is dependent on leaf age (Deom *et al.*, 1990). One possible explanation could be that the TMV MP accumulates only within secondary PD, which are more common in mature leaves (Ding *et al.*, 1992; Lucas and Gilbertson, 1994). Since TMV spreads within young leaves (Deom *et al.*, 1990), either increased PD SEL is not important for short-distance movement (McLean *et al.*, 1997), or other viral factors play a role in cell-to-cell transport through these organs. It is also possible that the transient nature of SEL increase may be more extreme in young than in old leaves.

Plant viral MPs move through PD. Microinjection experiments performed with MPs indicated that increased PD SEL occurred several cells distant from the infection site (Waigmann *et al.*, 1994). These data suggested that either the TMV

MP induced the formation of a diffusible signal that trafficked between cells causing an increased PD SEL or that the protein itself may spread from cell-to-cell. A great deal of evidence now shows that virus MPs can traffic through PD (Ding *et al.*, 1995; Fujiwara *et al.*, 1993; Noueiry *et al.*, 1994). Movement of microinjected MPs from the infection site to nearby cells is very rapid and requires no other viral genes. However, how far from the infected cell MPs actually move during a natural infection is still unclear (Padgett *et al.*, 1996).

Plant viral MPs facilitate movement of nucleic acids through PD. Due to their ability to bind viral nucleic acids and move through PD, it would be reasonable for MPs to facilitate movement of viral DNA and RNA from cell-to-cell. This is indeed the case. Coinjection of fluorescent-labelled nucleic acids with MP, or injection of labelled nucleic acids into transgenic plants expressing the MP, resulted in rapid movement of nucleic acids from cell-to-cell (Ding *et al.*, 1995; Fujiwara *et al.*, 1993; Noueiry *et al.*, 994). Movement was also selective. For example, a dianthovirus MP transported RNA but not DNA between cells, while a geminivirus BL1 protein facilitated the movement of double-stranded DNA molecules but not their single-stranded counterparts (Fujiwara *et al.*, 1993; Noueiry *et al.*, 1994).

TUBULES

In addition to the characteristics of MPs described above, another property should be noted. The MPs of members of alfamo-, bromo-, caulimo-, como-, nepo-, gemini-, and tospovirus groups induce formation of tubules in protoplasts and/or infected plants, for transport of virus nucleic acids (Perbal *et al.*, 1993; Wellink *et al.*, 1993; Wieczorek and Sanfacon, 1993; Ritzenthaler *et al.*, 1995b; Storms *et al.*, 1995; Kasteel *et al.*, 1997a; Ward *et al.*, 1997). Nucleic acids are transported through tubules either in the form of virus particles or as nucleoprotein complexes. The formation of tubules is critical for movement of comoviruses since mutations that inhibit tubule formation also prevent virus spread (Lekkerkerker et *al.*, 1996). In infected plants, the tubular structures appear to pass through modified PD that lack a DT and have a pore diameter of approximately 40-80 nanometers (van Lent *et al.*, 1991). The tubules themselves are unbranched, with a wall 3-4 nanometres thick, and are sheathed by the plasma membrane. The tubule itself is immediately beneath the plasma membrane and it appears to be made up exclusively of the MP (Kasteel *et al.*, 1997b). Within this inner proteinaceous tubule, the virus particles, arranged in a single row, are located (van Lent *et al.*, 1991). However, virus particles are not required for tubule formation (Wellink *et al.*, 1993). Cell type and development specificity for tubule formation has been reported in infected plants (Kim and Lee, 1992; Ward *et al.*, 1997). Tubule formation can occur in protoplasts from non-host plants and insect cells (Kastel *et al.*, 1996; Wellink *et al.*, 1993).

MODELS OF CELL-TO-CELL MOVEMENT

Certain viruses, such as cowpea mosaic virus (CPMV) require coat protein and form tubules through which virions spread from cell-to-cell (van Lent *et al.*, 1990). Other viruses, such as TMV and red clover necrotic mosaic virus (RCNMV) neither require CP nor form tubules for short-distance movement (Giesman-Cookmeyer *et al.*, 1995). Therefore, two major mechanisms of virus cell-to-cell spread were proposed: movement as a nucleoprotein complex independent of the coat protein and as a virion through tubules (Deom *et al.*, 1992). It is likely that other viruses move by more than these two mechanisms. For example, certain geminiviruses form tubules but do not move as virions (Ward *et al.*, 1997) and potexviruses require CP but move as RNP complexes (Lough *et al.*, 1998)

The movement of tobamoviruses has been the paradigm for understanding virus cell-to-cell spread (Lucas and Gilbertson, 1994). TMV appears to move from cell-to-cell as a RNP complex (Atabekov and Dorokhov, 1984; Citovsky *et al.*, 1990). The TMV RNP complex must be transported from the sites of replication to the transport machinery, then to and through the PD, to enter adjacent cells (Carrington *et al.*, 1996; Ghoshroy *et al.*, 1997). Colocalization with the replicase proteins to cortical ER membranes shows that the TMV MP probably binds to newly synthesised viral RNA genomes as they are made (Heinlien *et al.*, 1998). Binding of the MP to the newly synthesised TMV genomes inhibits translation (Karpova *et al.*, 1997) and removes most of the RNA secondary structure, giving the RNP complex a narrow cross-sectional diameter (Citovsky *et al.*, 1992). The newly formed RNP complexes also interact with microtubules (Ghoshroy *et al.*, 1997; Heinlein *et al.*, 1998; McLean *et al.*, 1997). The association with microtubules may allow the RNP complexes to be transported to the PD. The transport may be accomplished through moving ER-derived membranes, with their associated RNPs to PD. Alternatively, the RNP complexes may dissociate from the ER membrane once bound to microtubules and move independent of membranes. The PD targeting domain was localised to the C-terminal portion of the TMV MP, in a region containing an amphipathic α-helix (Fig. 2). This region may interact with either membranes or cellular proteins (Berna, 1995). Regions in the N-terminal portion also influence PD- and microtubule-association (Kahn *et al.*, 1998).

The RNP complexes may then move along microtubules to the PD, using motor proteins such as kinesins or dynesins (Carrington *et al.*, 1996). The RNP complexes may be transported along specific microtubules directed at the PD possibly due to the PD targeting activity of the MP (Ghoshroy *et al.*, 1997; McLean *et al.*, 1997). As they approach the PD, the RNP complexes likely associate with the filamentous actin cytoskeleton. Association with actin probably provides the motive force for delivery of the RNP complexes to the PD. The rapidity with which MPs are targeted to PD suggests that they may utilise basic host pathways for cell-to-cell movement. Microtubules and ER play a role in mRNA localization in animal cells (Deshler *et al.*, 1997). Perhaps plants posses a similar system that viruses exploit for PD localization.

Once the RNP complexes have reached the PD, they dock with specific proteins, probably through interactions with the basic residues in the C-terminal part of the TMV MP (Ghoshroy *et al.*, 1997; Kragler *et al.*, 1998). TMV MP, either in its free form or as part of the RNP complex, then alters the PD SEL through its association with host proteins, possibly actin (Ghoshroy *et al.*, 1997). Interestingly, the cross-sectional diameter of the RNP complexes is small enough to fit through the cytoplasmic sleeve of PD with an enlarged SEL (Citovsky *et al.*, 1992). It is likely that the transit of the RNP complex through the cytoplasmic sleeve is not merely a passive process (Carrington *et al.*, 1996). The actin filaments and myosin present within the PD may be used to transport the viral RNP complexes through the cytoplasmic sleeve into an adjacent cell (Carrington *et al.*, 1996; Ghoshroy *et al.*, 1997). Since the MP is an ER-associated protein (Heinlein *et al.*, 1998), it is possible that RNP complexes transit through the PD as part of the DT. The GTP binding activities of the TMV MP (Li and Palukaitis, 1996) may play a regulatory role in this process or provide the energy for driving RNP movement through the cytoplasmic sleeve. In all, it is likely that the movement of RNP complexes through PD may be similar to transport of nucleic acids through nuclear pores (Ghoshroy *et al.*, 1997).

The TMV MP may contain a "Change Conformation" signal permitting the protein to alter its configuration during its transit through PD (Kragler *et al.*, 1998; McLean *et al.*, 1997). This may explain why some plant virus MPs share sequence homology with chaperones (Koonin *et al.*, 1991). The putative chaperone sequences may assist in the unfolding of viral MPs for transit through PD.

After entry into a new cell through PD, the RNP complex must be disassembled to permit translation of viral nucleic acids (Karpova *et al.*, 1997). Removal from the RNA is probably mediated through MP phosphorylation by cell wall-associated kinases. Phosphorylation of the TMV MP destabilises the RNP complex, permitting translation (Dr. J. Atabekov, personal communication). After phosphorylation, the MP accumulates in secondary PD (Ghoshroy *et al.*, 1997), or it is targeted for degradation by association with microtubules (Heinlein *et al.*, 1998).

Even though it is not needed for TMV short-distance spread, other viruses do require coat protein for cell-to-cell movement. Coat protein may provide a function important for cell-to-cell movement, or it may be needed because these viruses spread as virions. Viruses employing the TGB strategy appear to require coat protein even though they move from cell-to-cell as RNP complexes (Angell *et al.*, 1996; Lough *et al.*, 1998). Potexvirus TGB1 proteins can increase the PD SEL and move from cell-to-cell on its own. However, TGB2 and TGB3 are required to transport potexvirus RNA from cell-to-cell (Lough *et al.*, 1998). The TGB2 and TGB3 proteins are relatively small, are hydrophobic, and may play a role in modifying PD SEL (Angell *et al.*, 1996). All three of the TGB proteins are still insufficient for movement of RNA from cell-to-cell; the coat protein is also required (Lough *et al.*, 1998). Interestingly potexvirus coat protein is targeted to PD in infected plants but it does not increase the SEL (Oparka *et al.*, 1996b). It is likely that an RNP complex consisting of TGB1, coat protein, and potexviral RNA is trafficked to the PD orifice via the cytoskeleton (Lough *et al.*, 1998). TGB2 and TGB3 may be associated with internal cellular membranes that are trafficked to the PD, and may facilitate movement of the RNP complex. Alternatively, the TGB2 and TGB3 proteins may mediate interaction of the RNP complex with a putative docking protein located in the PD. Once at the PD orifice, the TGB1 protein increases the SEL, permitting movement of the RNP complex to an adjacent cell. Other viruses using the TGB probably spread in a similar manner.

Certain members of the bromovirus group also require coat protein for cell-to-cell movement, even though they are thought to move as RNP complexes (Kaplan *et al.*, 1998; Rao, 1997). For these viruses, the coat protein appears to aid in the transport of viral nucleic acids through PD. Brome mosaic virus (BMV) requires CP for cell-to-cell movement for a different reason. Since the BMV MP has been shown to form tubules in protoplasts, it is possible that the CP is required to allow this virus to move from cell-to-cell in the form of virions (Kasteel *et al.,* 1997a). Therefore, viruses within the same taxonomic group may use different mechanisms for cell-to-cell movement.

Viruses of the como-, nepo-, and tospovirus groups induce tubule formation in infected cells and require the CP for cell-to-cell movement (Ritzenthaler *et al.,* 1995b; Storms *et al.,* 1995; Wellink *et al.,* 1993; Wieczorek and Sanfacon, 1993). These viruses probably move in a manner similar to CPMV. The CPMV MP is transported to the periphery of the cell and anchored there, possibly by interactions with the cytoskeleton (van Lent *et al.,* 1990, 1991). Interestingly, the CPMV MP shares amino acid sequence homology with HSP90, a cellular protein known to interact with the cytoskeleton (Czar *et al.,* 1996; Koonin *et al.,* 1991). Once anchored, a tubule begins to form by polymerisation of MPs. In the cytoplasm near the tail end of these tubules, small electron-dense regions occur that contain the CPMV MP and may be the sites for addition of new MPs (van Lent *et al.,* 1991, 1990). The tubules have a definite polarity; they originate near the plasma membrane and then project through the cell wall into the cytoplasm of nearby cells. As the tubule MP extends, it is enveloped by plasma membrane. This extending MP tubule then needs to penetrate the cell wall to enter adjacent cells. This occurs by either inducing PD formation or by using those already present.

If the tubules are targeted to pre-existing PD, they must induce removal of the DT; an activity probably mediated by the MP (Lucas and Gilbertson, 1994). The tubule could then extend through the modified PD and enter the adjacent cell. Since PD are not required for tubule formation, it is also possible that the MP may induce the formation of new PD-like structures (van Lent *et al.,* 1991). The growing tubule may activate cell wall loosening or digesting enzymes as well as inducing the activity of plasma membrane fusion protein, permitting the elongating MP tubule to pass into the adjacent cell. CPMV particles are probably inserted into the tubules by the C-terminal domain of the MP (Fig. 2) (Lekkerkerker *et al.,* 1996). Whether virus particles are loaded as the MP tubules are forming or after their formation is not clear. In addition, the mechanisms driving the actual movement of virions through the tubule remain to be answered. The region of the MP sharing homology to nucleotide-binding proteins (Koonin *et al.,* 1991) may possess an ATPase activity providing the energy needed for virion movement through tubules. Alternatively, virus particles may be transported through MP tubules by a process similar to "treadmilling" shown by actin filaments (Dr. J. Wellink, personal communication).

Geminiviruses appear to use tubules for cell-to-cell movement, but do not require CP (Ward *et al.,* 1997). Because they replicate within the nucleus, cell-to-cell movement of bipartite geminiviruses requires two MPs. The BR1 MP carries the viral single-stranded DNA out of the nucleus and into the cytoplasm, presumably by its effects on the nuclear pore complexes (Sanderfoot and Lazarowitz, 1996). The BL1 MP transports the nucleic acid from cell-to-cell through PD. BL1 is associated with the cell wall and plasma membrane (Noueiry *et al.,* 1994). This protein also appears to induce the formation of ER-derived tubules (Ward *et al.,* 1997). Once in the cytoplasm, the DNA-BR1 DNP (deoxyribonucleoprotein) complex then interacts with BL1 which targets it to the periphery of the cell (Sanderfoot and Lazarowitz, 1996), by interactions with ER but not the cytoskeleton (Ward *et al.*, 1997). Upon reaching the cell periphery, the BL1-DNP complex presumably moves through BL1-induced tubules to the adjacent cell. The tubules made by geminiviruses project into cytoplasm of leaf cells in both directions indicating that they are probably made in a manner different from CPMV.

After entering a neighbouring cell, BL1 dissociates from the DNP complex allowing BR1 to then carry the viral DNA into the nucleus (Sanderfoot and Lazarowitz, 1996). Alternatively, once BR1 has carried the viral DNA into the cytoplasm, it may also be displaced by BL1, which then carries the nucleic acid through the PD (Carrington *et al.*, 1996). In support of this latter hypothesis, BL1 was found to transport double-stranded DNA through PD (Noueiry *et al.*, 1994).

Some viruses may employ more than one mechanism for short-distance movement. CaMV has been proposed to move from cell-to-cell in two different ways, either as an RNP or as a virion through tubules (Citovsky *et al.*, 1991). The CaMV MP can assist the cell-to-cell spread of tobamoviruses (Dr. U. Melcher, personal communication), indicating that these virus proteins are functionally equivalent. The CaMV MP also binds RNA but not DNA, so this virus was proposed to move from cell-to-cell as RNP complex in a manner similar to TMV (Citovsky *et al.*, 1991). However, caulimoviruses form tubules containing virions, both in protoplasts and virus infected plants (Perbal *et al.*, 1993). These structures probably form in a manner similar to those of CPMV. In infected cells, however, the 75 C-terminal amino acids are often removed from the CaMV MP, which disrupts tubule formation but does not interfere with RNA binding (Thomas and Maule, 1995b). Perhaps CaMV exploits proteolytic cleavage to shift from one type of movement to another.

The CaMV MP is not the only one that can bind to RNA and induce tubule formation; those of alfamo- and bromoviruses also perform these functions (Fujita *et al.*, 1998; Kasteel *et al.*, 1997a; Schoumacher *et al.*, 1994). In addition, certain geminiviruses may utilise both a tubule and a non-tubule transport pathway at different times in infection (Kim and Lee, 1992). TMV also may use more than one mechanism for cell-to-cell transport (Heinlein *et al.*, 1998). This virus forms tubules in a limited number of infected protoplasts. The tubules either may not be formed in infected leaves, or they may be needed to infect other hosts.

Other viruses, such as the potyviruses, appear to have complex gene requirements for cell-to-cell movement. Potyviral helper component-proteinase (HC-Pro) and capsid proteins are similar to MPs in that they can modify SEL, move through PD, and assist the cell-to-cell movement of RNAs (Rojas *et al.*, 1997). The HC-Pro also appears to aid CP for cell-to-cell spread. The potyviral CI protein also plays a role in movement. The CI pinwheel inclusion bodies are centrally located over PD apertures in cells at the leading edge of a spreading infection, where they may assemble the cell-to-cell movement form of the virus (Roberts *et al.*, 1998). This transport form presumably contains CP and HC-Pro (Rojas *et al.*, 1997), but it is not clear whether virions or RNP complexes move from cell-to-cell. The transport form only occurs during the active phase of virus replication and may be positioned by the inclusion bodies (Roberts *et al.*, 1998). After transport, the inclusion bodies dissociate and the CI proteins accumulate in the cytosol.

MOVEMENT PROTEIN-HOST INTERACTIONS

Viral MPs, due to their ability to influence PD SEL, can exert powerful effects on host physiology. Transgenic plants constitutively expressing viral MPs show increases in sugar and starch levels in mature leaves as well as other alterations in photoassimilate transport (Lucas *et al.*, 1993b; Balachandran *et al.*, 1997). These changes are not due to a change in PD SEL *per se*. Rather, it is thought that a physiological signal responsible for controlling photoassimilate is being disrupted in these plants.

Since they disrupt plant physiological processes, it would behove plants to control the activities of viral MPs. This regulation may be achieved by phosphorylation, proteolytic cleavage, or by affecting MP levels. The TMV MP is phosphorylated at three separate regions (Citovsky *et al.*, 1993; Haley *et al.*, 1995) (Fig. 2). The region closest to the C-terminus is phosphorylated by a cell wall-associated protein kinase that increases in activity as leaf development progresses and that may be regulated by cyclic AMP (Atabekov and Taliansky, 1990; Citovsky *et al.*, 1993). Kinases phosphorylating other sites on the TMV MP as well as other viral MPs have also been reported (Haley *et al.*, 1995; Sokolova *et al.*, 1997). Phosphorylation has been proposed to be a way of inactivating the TMV MP by sequestering it in secondary PD (Citovsky *et al.*, 1993). In *Arabidopsis thaliana*, the TMV MP is not phosphorylated but it is proteolytically cleaved (Hughes *et al.*, 1995). Cleavage occurs at the N-terminus and the processed form of MP is non-functional. In addition, plants can also influence MP levels. Actinomycin D treatment of infected plants results in elevated MP levels, presumably by inhibiting a host factor that reduces MP expression (Blum *et al.*, 1989).

Because they interact with PD proteins, it would not be surprising for expression of dysfunctional MPs in plants to interfere with virus movement. In fact, transgenic plants expressing dysfunctional MPs inhibit virus spread, probably by competing with their functional counterparts for PD sites (Baulcombe, 1996). This type of resistance appeared to be active against several different viruses, possibly reflecting the non-specific nature of these proteins. Therefore, the understanding of viral MPs has led to new strategies for interfering with virus infection.

VIRAL GENES INFLUENCING VASCULAR MOVEMENT

While many of the viral genes involved in cell-to-cell spread have been defined, those controlling virus long-distance movement are less clear. One possible reason for this may be due to the number of different types of cellular interfaces, through which a virus must pass to enter, spread through, and exit the vasculature. Indeed, the PD at various cellular interfaces respond differently to MPs (Ding *et al.*, 1992; McLean *et al.*, 1997). A large number of viral genes play a role in long-distance virus movement (Nelson and van Bel, 1998; Seron and Haenni, 1996).

In addition to their role in cell-to-cell movement, MPs take part in long-distance spread as well. For example, certain RCNMV MP mutants were unaffected in cell-to-cell movement but unable to invade the SE-CCs (Wang *et al.*, 1998). However, the role MP plays in TMV vascular transport is less clear, since one set of experiments showed that concomitant synthesis of the MP was not needed for long-distance movement while a second study indicated that it was required for spread of virus through the phloem in plant stems (Arce-Johnson *et al.*, 1997). It is possible that MP may not be needed for viruses to spread through phloem and invade the upper portions of the plant, but it may be required for a systemic infection to be sustained.

A genetic determinant controlling host-specific long-distance movement is nested within the tomato bushy stunt virus MP gene (Scholthof *et al.*, 1995). The p19 gene encodes a soluble protein required for long-distance spread. This protein may mediate these effects by inducing cellular leakage, allowing the virus to overcome barriers to long-distance movement, or it may inhibit host defenses.

CP plays important role in virus long-distance movement. Mutations in the TMV CP gene or in the origin of assembly impair long-distance spread, probably by blocking entry into the SE-CCs or beyond (reviewed in Nelson and van Bel, 1998). The CP could affect virus long-distance movement in at least two ways. First, the CP may be needed to protect the viral nucleic acid in its transport through the phloem sap (Leisner and Turgeon, 1993). Second, the CP may play an active role in phloem invasion, transit, or exit (Nelson and van Bel, 1998). For example, certain bromovirus CP mutants incapable of forming virus particles can still spread through the vascular system (Schneider *et al.*, 1997). The CP requirement for vascular transport is not absolute and may vary with the host (Dalmay *et al.*, 1992).

Replicase proteins also influence virus long-distance movement (Nelson and van Bel, 1998). The masked strain of TMV harbors a mutation in the gene encoding the replicase proteins and shows reduced long-distance movement; even though it appears to replicate normally in inoculated leaves. This mutation regulates TMV accumulation in VPCs and CCs. The mechanisms by which the replicase proteins regulate long-distance movement remain to be determined; perhaps they act as MPs to facilitate vascular transport. The TMV replicase proteins are found in cytoplasmic inclusions called X-bodies (Heinlein *et al.*, 1998). Interestingly, the major inclusion body protein of CaMV, gene VI product, has also been proposed to control virus long-distance movement (Schoelz *et al.*, 1991). Perhaps the formation of cytoplasmic inclusion bodies is a pre-requisite for vascular transport and these structures play a role similar to that of the potyviral CI protein in short-distance movement as discussed earlier.

Tobamoviruses are not the only viruses that appear to have more than one long-distance movement determinants. At least three different potyvirus genes, those encoding the virus genome-linked protein (VPg), CP, and HC-Pro, play a role in vascular transport. The VPg protein probably interacts with host components to potentiate potyvirus long-distance movement by an unknown mechanism (Schaad *et al.*, 1997). The role of the CP and HC-Pro in long-distance movement is also not clear. Mutations in the CP gene indicated that the variable N- and C-terminal regions are essential for virus long-distance movement (Dolja *et al.*, 1995). Perhaps the CP exploits its MP-like characteristics to facilitate vascular transport (Rojas *et al.*, 1997). However, the CP functions in cell-to-cell movement are probably different from those involved in long-distance spread (Dolja *et al.*, 1995). The HC-Pro gene also controls vascular transport (Cronin *et al.*, 1995). HC-Pro mutants were impaired in either invasion or exit from SEs and it is possible that HC-Pro proteins may be required for both activities (Cronin *et al.*, 1995; Kasschau *et al.*, 1997). The mechanisms by which HC-Pro mediates long-distance movement is complicated by the fact that this protein is multifunctional. While the HC-Pro can act like an MP (Rojas *et al.*, 1997), it also inhibits host defences (Pruss *et al.*, 1997). Thus, the HC-Pro could also possibly potentiate long-distance movement by either mechanism.

Viral *cis*-acting sequences also play a role in vascular transport (Lauber *et al.*, 1998b). RNA 3 is required for furovirus long-distance movement. While mutations blocking translation of each of the open reading frames encoded by this RNA did not interfere with long-distance movement, an internal deletion abolished vascular spread. Thus, vascular movement of this virus is dependent on a feature of the sequence of RNA 3 rather than the proteins that it encodes.

The experiments with TMV illustrate an important point: a process as complex as vascular transport requires the orchestration of several viral gene products rather than the activity of a single protein. Practically every gene in the TMV genome plays a role in long-distance movement. Because complex functions such as vascular transport requires the coordination of many gene activities, it may be appropriate to view viruses as complex adaptive systems in which the whole is more than the sum of the parts (Eigen, 1993). Viruses are probably gene networks whose products interact and whose interdependence generates a stable genetic system. Systems approaches have been taken to describe different biological phenomena (Ioannidi *et al.*, 1998; McAdams and Arkin, 1998). The application of such approaches may be useful for understanding some of the complex processes that viruses meditate.

CONCLUSIONS

Virus movement within a plant host is a complex process requiring transport of viral nucleic acids from one cell to another within an infected organ and long-distance spread through the vascular system. Movement from cell-to-cell requires the activities of one or more virus encoded gene products called MPs. MPs are among the most important genes needed by the viruses for their survival and spread. Nearly all plant viruses contain them in their genomes (Chapters 1 and 3). They are often multifunctional proteins that share many characteristics such as binding to nucleic acids, targeting to and modification of the PD SEL, and transport of the viral genomes through the intercellular connections. Viruses move through PD by at least two different mechanisms: as nucleoprotein complexes or as virions. Transport of plant viruses to other portions of the plant requires that the virus gains access to the vascular system. The process is different from cell-to-cell movement and requires the participation of many viral proteins. With the development of new technologies for manipulating viral genomes, for following spread within hosts, and for studying specific interactions between virus and host proteins, the field of movement is about to enter a golden age of discovery.

ACKNOWLEDGEMENTS

The author thanks Drs. R. Beachy, U. Melcher, J. Wellink, and J. Atabekov for providing unpublished data and Dr. L. Shemshedini for critical reading of the manuscript. This work was supported by USDA grant number 96-3503-3284.

REFERENCES

Angell, S. M., Davies, C., and Baulcombe, D. C. (1996). Cell-to-cell movement of potato virus X is associated with a change in the size-exclusion limit of plasmodesmata in trichome cells of *Nicotiana clevelandii. Virology* 216, 197-201.

Arce-Johnson, P., Kahn, T. W., Reimann-Philipp, U., Rivera-Bustamante, R., and Beachy, R. N. (1995). The amount of movement protein produced in transgenic plants influences the establishment, local movement, and systemic spread of infection by movement protein-deficient tobacco mosaic virus. *Mol. Plant-Microbe Interact.* 8, 415-423.

Arce-Johnson, P., Reimann-Phillipp, U., Padgett, H. S., Rivers-Bustamante, R., and Beachy R. N. (1997). Requirement of the movement protein for long distance spread of tobacco mosaic virus in grafted plants. *Mol. Plant-Microbe Interact.* 10, 691-699.

Atabekov, J. G., and Dorokhov, Y. L. (1984). Plant virus-specific transport function and resistance of plants to viruses. *Adv. Virus Res.* 29, 313-364.

Atabekov, J. G., and Taliansky, M. E. (1990). Expression of a plant virus-encoded transport function by different viral genomes. *Adv. Virus Res.* 38, 201-248.

Balachandran, S., Hull, R. J., Martins, R. A., Vaadia, Y., and Lucas, W. J. (1997). Influence of environmental stress on biomass partitioning in transgenic tobacco plants expressing the movement protein of tobacco mosaic virus. *Plant Physiol.,* 114, 475-481.

Baulcombe, D. C. (1996). Mechanisms of pathogen-derived resistance to viruses in transgenic plants. *Plant Cell* 8, 1833-1844.

Berna, A. (1995). Involvement of residues within putative *a*-helix motifs in the behaviour of the alfalfa and tobacco mosaic virus movement proteins. *Phytopathology* 85, 1441-1448.

Bleykasten, C., Gilmer, D., Guilley, H., Richards, K. E., and Jonard, G. (1996). Beet necrotic yellow vein virus 42 kDa triple gene block protein binds nucleic acid *in vitro. J. Gen. Virol.* 77, 889-897.

Blum, H., Gross, H. J., and Beier, H. (1989). The expression of the TMV-specific 30 kDa protein in tobacco protoplasts is strongly and selectively enhanced by actinomycin. *Virology* 169, 51-61.

Carrington, J. C., Kasschau, K. D., Mahajan, S. K., and Schaad, M. C. (1996). Cell-to-cell and long-distance transport of viruses in plants. *Plant Cell* 8, 1669-1681.

Citovsky, V. (1993). Probing plasmodesmata transport with plant viruses. *Plant Physiol.* 102, 1071-1076.

Citovsky, V., and Zambruski, P. (1993). Transport of nucleic acids through membrane channels: Snaking through small holes. *Annu. Rev. Microbiol.* 47, 167-197.

Citovsky, V., Knorr, D., Schuster, G., and Zambryski, P. (1990). The P30 movement protein of tobacco mosaic virus is a single-strand nucleic acid binding protein. *Cell* 60, 637-647.

Citovsky, V., Knorr, D., and Zambryski, P. (1991). Gene I, a potential cell-to-cell movement locus of cauliflower mosaic virus, encodes an RNA-binding protein. *Proc. Natl. Acad. Sci. USA* 88, 2476-2480.

Citovsky, V., Wong, M. L. Shaw, A. L., Prasad, B. V. V., and Zambryski, P. (1992). Visualization and characterization of tobacco mosaic virus movement protein binding to single-stranded nucleic acids. *Plant Cell* 4, 397-411.

Citovsky, V., McLean, B. G., Zupan, J. R., and Zambryski, P. (1993). Phosphorylation of tobacco mosaic virus cell-to-cell movement protein by a developmentally regulated plant cell wall-associated protein kinase. *Genes Develop.* 7, 904-910.

Cohen, M., Siegel, A., Zaitlin, M., Hudson, W. R., and Wildman, S. G. (1957). A study of tobacco mosaic virus strain predominance and an hypothesis for the origin of systemic virus infection. *Phytopathology* 47, 694-702.

Cronin, S., Verchot, J., Haldeman-Cahill, R., Schaad, M. C., and Carrington, J. C. (1995). Long-distance movement factor: A transport function of the potyvirus helper component-protienase. *Plant Cell* 7, 549-559.

Czar, M. J., Welsh, M. J., and Pratt, W. B. (1996). Immunofluorescence localization of the 90-kDa heat-shock protein to cytoskelton. *Eur. J. Cell Biol.* 70, 322-330.

Dalmay, T., Rubino, L., Burgyan, J., and Russo, M. (1992). Replication and movement of a coat protein mutant of cymbidium ringspot tomobusvirus. *Mol. Plant-Microbe Interact.* 5, 379-383.

Daubert, S. (1992). Molecular determinants of plant-virus interaction. *In* "Molecular Signals in Plant-Microbe Communications" (D. P. S. Verma, Ed.). CRC Press, Boca Raton.

Davies, C., Hills, G., and Baulcombe, D. C. (1993). Sub-cellular localization of the 25 kDa protein encoded in the triple gene block of potato virus X. *Virology* 197, 166-175.

De Jong, W., and Ahlquist, P. (1992). A hybrid plant RNA virus made by transferring the noncapsid movement protein from a rod-shaped to an icosahedral virus competent for systemic infection. *Proc. Natl. Acad. Sci. USA.* 89, 6806-6812.

Deom, C. M., Lapidot, M., and Brachy, R. N. (1992). Plant virus movement proteins. *Cell* 69, 221-224.

Deom, C. M., Oliver, M. J., and Beachy, R. N. (1987). The 30-kilodation gene product of tobacco mosaic virus potentiates virus movement. *Science* 237, 389-393.

Deom, C. M., Schubert, K. L., Wolf, S., Holt, C. A., Lucas, W. J., and Beachy, R. N. (1990). Molecular characterization and biological function of the movement protein of tobacco mosaic virus in transgenic plants. *Proc. Natl. Acad. Sci. USA* 87, 3284-3288.

Deshler, J. O., Highett, M. I., and Schnapp, B. J. (1997). Localization of *Xenopus* Vgl mRNA by Vera protein and the endoplasmic reticulum. *Science* 276, 1128-1131.

Ding, B. (1997). Cell-to-cell transport of macromolecules through plasmodesmata: A novel signalling pathway in plants. *Trends Cell Biol.* 7, 5-9.

Ding, B., Haudenshield, J. S., Hull, R. J., Wolf, S., Beachy, R. N., and Lucas, W. J. (1992). Secondary plasmodesmata are specific sites of localization of the tobacco mosaic virus movement protein in transgenic tobacco plants. *Cell* 4, 915-928.

Ding, B., Li, Q., Nguyen, L. Palukaitis, P., and Lucas, W. J. (1995). Cucumber mosaic virus 3a protein potentiates cell-to-cell trafficking of CMV RNA in tobacco plants. *Virology* 207, 345-353.

Dolja, V. V., Haldeman-Cahill, R., Montgomery, A. E., Vabdenbosch, K. A., and Carrington, J. C. (1995). Capsid protein determinants involved in cell-to-cell and long-distance movement of tobacco etch virus. *Virology* 206, 1007-1016.

Eigen, M. (1993). The origin of genetic information: Viruses as models. *Gene* 135, 37-47.

Esau, K. (1948). Some anatomical aspects of plant virus disease problems. II. *Bot. Rev.* 14, 413 - 449.

Esau, K. (1964). "Plant Anatomy." Sixth ed. John Wiley & Sons Inc., New York.

Fenczik, C. A., Padgett, H. S., Holt, C. A., Casper, S. J., and Beachy, R. N. (1995). Mutational analysis of the movement protein of Odonotoglossum ringspot virus to identify a host-range determinant. *Mol. Plant-Microbe Interact.* 8, 666-673.

Fujita, M., Mise, K., Kajiura, Y., Dohi, K., and Furusawa, I. (1998). Nucleic acid-binding properties and subcellular localization of the 3a protein of brome mosaic bromovirus. *J. Gen. Virol.* 79, 1273 - 1280.

Fujiwara, T., Giesman-Cookmeyer, D., Ding, B., Lommel, S. A., and Lucas, W. J. (1993). Cell-to-cell trafficking of macromolecules through plasmodesmata potentiated by the red clover necrotic mosaic virus movement protein. *Plant Cell* 5, 1783-1794.

Ghoshroy, S., Lartey, R., Sheng, J., and Citovsky, V. (1997). Transport of proteins and nucleic acids through plasmodesmata. *Annu. Rev. Plant Physiol. Plant Mol. Biol.* 48, 27-50.

Giesman-Cookmeyer, D., and Lommel, S. A. (1993). Alanine scanning mutagenesis of a plant virus movement protein identifies three functional domains. *Plant Cell* 5, 973-982.

Giesman-Cookmeyer, D., Silver, S., Vaewhongs, A. A., Lommel, S. A., and Deom, C. M. (1995). Tobamovirus and dianthovirus movement proteins are functionally homologous. Virology 213, 38-45.

Gilbertson, R. L., and Lucas, W. J. (1996). How do viruses traffic on the "Vascular Highway"? *Trends Plant Sci.* 1, 260-268.

Haley, A., Hunter, T., Kiberstis, P., and Zimmern, D. (1995). Multiple serine phosohorylation on the 30 kDa Cell-to-cel l movement protein synthesized in tobacco protoplasts. *Plant J.* 8, 715-724.

Heinlein, M., Padgett, H. S., Gens, J. S., Pickard, B. G., Casper, S. J., Epel, B. L., and Beachy, R. N. (1998). Changing patterns of localization of the tobacco mosaic virus movement protein and replicase to the endoplasmic reticulum and microtubules during infection. *Plant Cell* 10, 1107 - 1120.

Hughes, R. K., Perbal, M. C. Maule, A. J., and Hull, R. (1995). Evidence for proteolytic processing of tobacco mosaic virus movement protein in *Arabidopsis thaliana. Mol. Plant Microbe Interact.* 8, 658-665.

Hull, R. (1989). The movement of viruses in plants. *Annu. Rev. Phytopathol.* 27, 213-240.

Hull, R. (1991). The movement of viruses within plants. *Semin. Virol.* 2, 89-95.

Ioannidis, J. P. A., McQueen, P. G., Goedert, J. J. and Kaslow, R. A. (1998). Use of neural networks to model complex immunogenetic associations of disease: Human leukocyte antigen impact on the progression of human immunogdeficiency virus infection. *Am. J. Epidemiol.* 147, 464-471.

Kahn, T. W., Lapidot, M., Heinlein, M., Reichel, C., Cooper, C., Gafney, R., and Beachy, R. N. (1998). Domains of the TMV movement protein involved in subcellular localization. *Plant J.* 15, 15 - 25.

Kaplan, I. B., Zhang, L., and Palukaitis, P. (1998). Characterization of cucumber mosaic virus V. Cell-to-cell movement requires capsid protein but not virions. *Virology* 246, 221-231.

Karpova, O. V., Ivan ov, K. I., Rodionova, N. P., Dorokhov, Y. L., and Atabekov, J. G. (1997). Nontranslatability and dissimilar behaviour in plants and protoplasts of viral RNA and movement protein complexes formed *in vitro*. *Virology* 230, 11-21.

Kasschau, K. D., Cronin, S., and Carrington, J. C. (1997). Genome amplification and long-distance movement functions associated with the central domain of tobacco etch potyvirus helper component-proteinase. *Virology* 228, 251-262.

Kasteel, D, T. J., Perbal, M. C., Boyer, J. C., Wellink, J., Goldbach, R. W., Maule, A. J., and van Lent, J. W. M. (1996). The movement proteins of cowpea mosaic virus and cauliflower mosaic virus induce tubular structures in plant and insect cells. *J. Gen. Virol.* 77, 2857-2864.

Kasteel, D. T. J., Van Der Wel, N. N., Jansen, K. A. J., Goldbach, R. W., and Van Lent, J. W. M. (1997a). Tubule forming capacity of the movement proteins of alfalfa mosaic virus and brome mosaic virus. *J. Gen. Virol.* 78, 2089-2093.

Kasteel, D. T. J., Wellink, J., Goldbach, R. W., and van Lent, J. W. M. (1997b). Isolation and charactersization of tubular structures of cowpea mosaic virus. *J. Gen. Virol.*, 78, 3167 - 3170.

Kim, K. S., and Lee, K. W. (1992). Geminivirus-induced macrotubules and their suggested role in cell-to-cell movement. *Phytopathology* 82, 664-669.

Koonin, E. V., Mushegian, A. R., Ryabov, E. V., and Dolja, V. V. (1991). Diverse groups of plant RNA and DNA viruses share related movement proteins that may possess chaperone-like activity. *J. Gen. Virol.* 72, 2985-2903.

Kormelink, R., Stroms, M., Van Lent, J., Peters, D., and Goldbach, R. (1994). Expression and subcellular location of the NSm protein of tomato spotted wilt virus (TSMV), a putative viral movement protein. *Virology* 200, 56-65.

Kragler, F., Monzer, J., Shash, K., Xoconostle-Cazares, B., and Lucas, W. J. (1998). Cell-to-cell transport of proteins: Requirements for unfolding and characterization of binding to a putative plasmodesmatal receptor. *Plant J.* 15, 367-381.

Lauber, E., Bleykasten-Grosshans, C., Erhardt, M., Bouzouba, S., Jonard, G., Richards, K. E., and Guilley, H. (1998a). Cell-to-cell movement of beet necrotic yellow vein virus: I. Heterologus complementation experiments provide evidence for specific interactions among the triple gene block proteins. *Mol. Plant-Microbe Interact.* 11, 618-625.

Lauber, E., Guilley, H., Tamada, T., Richards, K. E., and Jonard, G. (1998b). Vascular movement of beet necrotic yellow vein virus in *Beta macrocarpa* is probably dependent on an RNA 3 sequence domain rather than a gene product. *J. Gen. Virol.* 79, 385 -393.

Lehto, K., Grantham, G. L., and Dawson, W. O. (1990). Insertion of sequences containing the coat protein subgenomic RNA promter and leader in front of the tobacco mosaic virus 30K ORF delays its expression and causes defective cell-to-cell movement. *Virology* 174, 145 -157.

Leisner, S. M., and Turgeon, R. (1993). Movement of virus and photoassimilate in the phloem: Comparative analysis. *BioEssays* 15, 741-748.

Lekkerkerker, A., Wellink, J., Yuan, P., van Lent, J., Goldbach, R., and van Kammen, A. (1996). Distinct functional domains in the cowpea mosaic virus movement protein. *J. Virol.* 70, 5658 -5661.

Li, Q., and Palukatis, P. (1996). Comparison of the nucleic acid- and NTP-binding properties of the movement protein of cucumber mosaic cucumovirus and tobacco mosaic tobamovirus. *Virology* 216, 71-79.

Lough, T. J., Shash, K., Xoconostle-Cazares, B., Hofstra, K. R., Beck, D. L. Balmori, E., Forster, R. L. S., and Lucas, W. J. (1998). Molecular dissection of the mechanisms by which potexvirus triple gene block proteins mediate cell-to-celltransport of infectious RNA. *Mol. Plant-Microbe Interact.* 11, 801-814.

Lucas, W. J. (1995). Plasmodesmata : Intercellular channels for macromolecular transport in plants. *Curr. Opin. Cell Biol.* 7, 673-680.

Lucas, W. J., and Gilbertson, R. L. (1994). Plasmodesmata in relation to viral movement within leaf tissues. *Annu. Rev. Phytopathol.* 32, 387- 411.

Lucas, W. J., Ding, B., and Van Der Schoot, C. (1993a). Plasmodesmata and the supracellular nature of plants. *New Phytol.* 125, 435-476.

Lucas, W. J., Olesinski, A., Hull, R. J., Hauldenshield, J. S., Deom, C. M., Beachy, R. N., and Wolf, S. (1993b). Influence of the tobacco mosaic virus 30-kDa movement protein on carbon metabolism and photosynthate partitioning in transgenic tobacco plants. *Planta* 190, 88-96.

Maule, A. J. (1991). Virus movement in infected plants. *CRC Crit. Rev. Plant Sci.* 9, 457-473.

McAdams, H. H, and Arkin, A. (1998). Simulation of prokaryotic genetic circuits. *Annu. Rev. Biophys. Biomol. Struct.* 27, 199-224.

McLean, B. G., Hempel, F. D., and Zambryski, P. C. (1997). Plant intercellular communication via plasmodesmata. *Plant Cell* 9, 1043-1054.

Melcher, U. (1990). Similarties between putative transport proteins of plant viruses. *J. Gen. Virol.* 71, 1009-1018.

Melcher, U. (1993). HIV-I proteinase as a structural model of intercellular transport proteins of plant viruses. *J. Theor. Biol.* 162, 61-74.

Meshi, T., Watanabe, Y., Saito, T., Sugimoto, A., Maeda, T., and Okada, Y. (1987). Function of the 30kd protein of tobacco mosaic virus: involvement in cell-to-cell movement and dispensability for replication. *EMBO J.* 6, 2557-2563.

Mezitt, L. A., and Lucas, W. J. (1996). Plasmodesmatal cell-to-cell transport of proteins and nucleic acids. *Plant Mol. Biol.* 32, 251-273.

Morozov, S. Y., Microhnichenko, N. A., Solovyev, A. G., Fedorkin, O. N., Zelenina, D. A., Lukasheva, L. I. Karasev, A. V. Dolja, V. V., and Atabekov, J. G. (1991). Expression strategy of the potato virus X triple gene block. *J. Gen. Virol.* 72, 2039-2042.

Mushegian, A. R., and Koonin, E. V. (1993). Cell- to-cell movement of plant viruses. Insights from amino acid sequence comparisons of movement proteins and from homologies with cellular transport systems. *Arch. Virol.* 133, 239-257.

Nelson, R. S., and van Bel, A. J. E. (1998). The mystery of virus trafficking into, through and out of vascular tissue. *Prog. Botany* 59, 476-533.

Noueiry, A.O., Lucas, W. J., and Gilbertson, R. L. (1994). Two proteins of a plant DNA virus co-ordinate nuclear and plasmodesmatal transport. *Cell* 76, 925-932.

Opalka, N., Brugidou, C., Bonneau, C., Nicole, M., Beachy, R. N., Yeager, M., and Fauquet, C. (1998). Movement of rice yellow mottle virus between xylem cells through pit membranes. *Proc. Natl. Acad. Sci. USA* 95, 3323 -3328.

Oparka, K. J., Boveink, P., and Santa Cruz, S. (1996a). Studying the movement of plant viruses using the green fluorescent protein. *Trends Plant Sci.* 1, 412-417.

Oparka, K. J., Roberts, A. G., Roberts, I. M., Prior, D. A. M., and Santa Cruz, S. (1996b). Viral coat protein is targeted to, but does not gate plasmodesmata during cell-to-cell movement of potato virus X. *Plant J.* 10, 805-813.

Oparka, K. J., Prior, D. A. M., Santa Cruz, S., Padgett, H. S., and Beachy, R. N. (1997). Gating of epidermal plasmodesmata is restricted to the leading edge of expanding infection sites of tobacco mosaic virus (TMV). *Plant J.* 12, 781-789.

Osman, T. A. M., Hayes, R. J., and Buck, K. W. (1992). Co-operative binding of the red clover necrotic mosaic virus movement protein to single-stranded nucleic acids. *J. Gen. Virol.* 73, 223 - 227.

Overall, R. L., and Blackman, L. M. (1996). A model of the macromolecular structure of plasmodesmata. *Trends Plant Sci.* 1, 307-311.

Padgett, H. S., Watanabe, Y., and Beachy, R. N. (1997). Identification of the TMV replicase sequence that activates the *N* gene-mediated hypersensitive response. *Mol. Plant-Microbe Interact.* 10, 709 - 715.

Perbal, M. C., Thomas, C. L., and Maule, A. J. (1993). Cauliflower mosaic virus gene I product (PI) forms tubular structures which extend from the surface of infected protoplasts. *Virology* 195, 281-285.

Pruss, G., Ge, X., Ming Shi, X., Carrington, J. C., and Bowman Van ce, V. (1997). Plant viral synergism: The potyviral genome encodes a broad-range pathogenicity enhancer that transactivates replication of heterologous viruses. *Plant Cell* 9, 859-868.

Rao, A. L. N. (1997). Molecular studies on bromovirus capsid protein. III. Analysis of cell-to-cell competence of coat protein defective variants of cowpea chlorotic mottle virus. *Virology* 232, 385-395

Rao, A. L. N., Cooper, B., and Deom. C. M. (1998). Defective movement of viruses in the family *Bromoviridae is* differentially complemented in *Nicotiana benthamiana* expressing tobamovirus or dianthovirus movement proteins. *Phytopathology* 88, 666-672.

Ritzenthaler, C., Pinck, M., and Pinck, L. (1995a). Grapevine fanleaf nepovirus P38 putative movement protein is not tramsiently expressed and is a stable final maturation product *in vivo. J. Gen. Virol* 76, 907-915.

Ritzenthaler, C., Schmidt, A., Michler,, P., Stussi-Garaud, C., and Pinck, L. (1995b). Grapevine fanleaf nepovirus P38 putative movement protein is located on tubules *in vivo. Mol. Plant-Microbe Interact.* 8, 379-387.

Roberts, I. M., Wang, D., Findlay, K., and Maule, A. J. (1998). Ultrastructural and temporal observations of the potyvirus cylindrical inclusions (Cis) show that the CI protein acts transiently in aiding virus movement. *Virology* 245, 173-181.

Rojas, M. R., Zerbini, F. M., Allison, R. F., Gilbertson, R. L., and Lucas, W. J. (1997). Capsid protein and helper component-proteinase function as potyvirus cell-to-cell movement proteins. *Virology* 237, 283-295.

Rouleau, M., Smith, R. J., Bancroft, J. B., and Mackie, G. A. (1994). Purification, properties, and subcellular localization of foxtail mosaic potexvirus 26 kDa protein. *Virology* 204, 254 - 265.

Saito, T., Imai, Y., Meshi, T., and Okada, Y. (1998). Interviral homolgies of the 30K proteins of tobamoviruses. *Virology* 167, 653-656.

Samuel, G. (1934). The movement of tobacco mosaic virus within the plant. *Ann. Appl. Biol.* 21, 90 - 111.

Sanderfoot, A. A., and Lazarowitz, S. G. (1996). Getting it together in plant virus movement: Co-operative interactions between bipartite geminivirus movement proteins. *Trends Cell Biol.* 6, 353 - 358.

Schaad, M. C., Lellis, A. D., and Carrington, J. C. (1997). VPg of tobacco etch potyvirus is a host genetype-specific determinant for long-distance movement. *J. Virol .* 71, 8624-8631.

Schneider, W. L., Greene, A. E., and Allison, R. F. (1997). The carboxy-terminal two-thirds of the cowpea chlorotic mottle bromovirus capsid protein is incapable of virion formation yet supports systemic movement. *J. Virol.* 71, 4862-4865.

Schoelz, J. E., Goldberg, K. B., and Kiernan, J. (1991). Expression of cauliflower mosaic virus (CaMV) gene VI in transgenic *Nicotiana biglovii* complements a strain of CaMV defective in long-distance movement in non-transformed *N. biglovii. Mol. Plant-Microbe Interact.* 4, 350-355.

Scholthof, H. B., Scholthof, K. B. G., Kikkert, M., and Jackson, A. O. (1995). Tomato bushy stunt virus spread is regulated by two nested genes that function in cell-to-cell movement and host-dependent systemic invasion. *Virology* 213, 425-438.

Schoumacher, F., Giovan e, C., Maira, M., Poirson, A., Godefroy-Cloburn, T., and Berna, A. (1994). Mapping of the RNA-binding domain of the alfalfa mosaic virus movement protein. *J. Gen. Virol.* 75, 3199 -3202.

Seron, K., and Haenni, A. L. (1996). Vascular movement of plant viruses. *Mol. Plant-Microbe Interact.* 9, 435-442.

Shalla, T. A., Petersen,, L. J., and Zaitlin, M. (1982). Restricted movement of a temperature-sensitive virus in tobacco leaves is associated with a reduction in numbers of plasmodesmata. *J. Gen. Virol.* 60, 355-358.

Sokolova, M., Prufer, D., Tacke, E., and Rohde, W. (1997). The potato leafroll virus 17K movement protein is phosphorylated by a membrane-associated protein kinase from potato with the biochemical features of protein kinase C. *FEBS Lett.* 400, 201-205.

Storms, M. M. H., Kormelink, R., Peters, D., v an Lent, J. W. M., and Goldbach, R. W. (1995). The non-structural NSm protein of tomato spotted wilt virus induces tubular structures in plant and insect cells. *Virology* 314, 485-493.

Thomas, C. L., and Maule, A. J. (1995a). Identification of the cauliflower mosaic virus movement protein RNA-binding domain. *Virology* 206, 1145 -1149.

Thomas, T. J., and Maule, A. J. (1995b). Identification of structural domains within the cauliflower mosaic virus movement protein by scanning deletion mutagenesis and epitope tagging. *Plant Cell* 7, 561-572.

Thomas, C. L., Perbal, C., and Maule, A. J. (1993). A mutation of cauliflower mosaic virus gene I interferes with virus movement but not virus replication. *Virology* 192, 415-421.

Tomenius, K., Clapham, D., and Meshi, T. (1987). Localization by immunogold cytochemistry of the virus-coded 30K protein in plasmodesmata of leaves infected with tobacco mosaic virus. *Virology* 160, 363-371.

van Lent, J., Wellink, J., and Goldbach, R. (1990). Evidence for the involvement of the 58K and 48K proteins in the intercellular movement of cowpea mosaic virus. *J. Gen. Virol.* 71, 219-223.

van Lent, J., Storms, M., vgan Der Meer, F., Wellink, J., and Goldbach, R. (1991). Tubular structures involved in movement of cowpea mosaic virus are also formed in infected cowpea protoplasts. *J. Gen. Virol.* 72, 2615 -2623.

Waigmann, E., Lucas, W. J., Citvosky, V., and Zambryski, P. (1994). Direct functional assay for tobacco mosaic virus cell-to-cell movement protein and identification of a domain involved in increasing plasmodesmatal permeability. *Proc. Natl. Acad. Sci. USA* 91, 1433 -1437.

Wang, H. L., Wang, Y., Giesman-Cookmeyer, D., Lommel, S. A., and Lucas, W. J. (1998). Mutations in viral movement protein alter systemic infection and identify an intercellular barrier to entry into the phloem long-distance transport system. *Virology* 245, 75 - 89.

Ward, B. M., Medville, R., Lazarwitz, S. G., and Turgeon, R. (1997). The geminivirus BL1 movement protein is associated with endoplasmic reticulum-derived tubules in developing phloem cells. *J. Virol.* 71, 3726-3733.

Watanabe, Y., Emori, Y., Ooshika, I., Meshi, T., Ohno, T., and Okada, Y. (1984). Synthesis of TMV-specific RNAs and proteins at the early stage of infection in tobacco protoplasts: Transient expression of the 30K proteins its mRNA. *Virology* 133, 18 -24.

Wellink, J., Van Lent, J. W. M., Vermer, J., Sijen, T., Goldbach, R. W., and Van Kammen, A. (1993). The cowpea mosaic virus M RNA-encoded 48-kilodalton protein is responsible for induction of tubular structures in protoplasts. *J. Virol.* 67, 3660 -3664.

Wieczorek,A., and Sanfacon, H. (1993). Characterization and subcellular localization of tomato ringspot nepovirus putative movement protein. *Virology* 194, 734 -742.

Wobbe, K. K., Akgoz, M. Dempsey, D. M. A., and Klessig, D. F. (1998). A single amino acid change in the turnip crinkle virus movement protein p8 affects RNA binding and virulence on *Arabidopsis thaliana*. *J. Virol.* 72, 6247 - 6250.

Wolf, S., Deom, C. M., Beachy, R. N., and Lucas, W. J. (1989). Movement protein of tobacco mosaic virus modifies plasmodesmatal size exclusion limit. *Science* 246, 337 -339.

8 MOLECULAR BASIS OF VIRUS TRANSMISSION

Johannes F.J.M. van den Heuvel,
Alexander W.E. Franz, and Frank van der Wilk

INTRODUCTION

Knowledge of the molecular mechanisms underlying virus transmission is a prerequisite to the creation of new approaches to modulate vector competence and to reduce the multiplication of initial infection sites in seed and pollen. In this chapter the recent advances made in identifying the virus, vector and host determinants involved in transmission processes are described.

TRANSMISSION BY ARTHROPOD VECTORS

Arthropod vectors transmit more than 70% of the plant viruses. They do so in circulative and noncirculative manner. The circulative viruses may be propagative and non-propagative depending on whether or not a virus replicates in its vector. Viruses which are transmitted in a circulative manner not only have to cross epithelial cell linings of the gut and salivary glands but they also have to resist the potentially hostile environment of the vector. Some of the viruses replicate in their vector during this cyclic passage, i.e. are propagative. The majority of the plant viruses, however, is transmitted in a noncirculative manner. Noncirculative viruses do not replicate in their vectors as they are only associated with the cuticular linings of the mouthparts and the anterior part of the alimentary tract. Generally such viruses are mechanically transmissible as well.

Phloem-feeding homopteran insects play a prominent role in the dissemination of insect-borne viruses. Over the past decades, a detailed descriptive framework concerning the fate of a virus in an invertebrate vector and the transmission barriers it encounters has been built, providing the groundwork for recent studies on the molecular bases of the various events in virus transmission. The high degree of vector specificity and differences in tissue tropism and retention sites, commonly observed among insect-borne viruses, suggests an intimate association in which both virus-encoded proteins and vector components are involved. Identification of the vector determinants involved in the different phases of the transmission process represent a nascent, but rapidly developing research area. Despite the vast body of literature dealing with host receptors for mammalian viruses, there is only limited information on vector proteins involved in recognition, attachment and retention of plant viruses, and only recently, have insect proteins with a capacity to bind plant viruses been identified.

Viral determinants governing transmissibility of viruses are primarily located in the structural proteins (Hull, 1994; Ammar, 1994). The domains involved are usually identified by comparative studies of the encoding genes of transmissible and nontransmissible forms. The latter often arise naturally when vector pressure is lifted. A functional role of the transmission domains in virus transmission is then confirmed by mutational analysis of the viral genome or of infectious viral cDNA clones. In this way viral determinants involved in vector transmission have been identified for noncirculative (e.g. Pirone and Blanc, 1996; Perry *et al.*, 1998) and circulative insect-borne viruses (e.g. Briddon *et al.*, 1990; Brault *et al.*, 1995; Chay *et al.*, 1996; Demler *et al.*, 1997).

Noncirculative Virus Transmission by Insects

The transmission of some noncirculative viruses not only depends on the viral coat protein (CP; Perry *et al.*, 1998) but also on the presence of active virus-encoded helper factors which must be present before the virus or simultaneously with it (e.g. Pirone and Blanc, 1996). Helper factors of the potyviruses and caulimoviruses are among the best-characterized viral products. The helper component (HC) of a potyvirus is believed to act as a 'bridge' with affinity for sites on both the viral capsid and the surface of the stylet food canal of an aphid thus allowing retention of the virus (Pirone and Blanc, 1996; Wang *et al.*, 1996). Following this hypothesis, binding of HC to virus particle and/or stylet must be reversible, so that the virus can subsequently be released and inoculated. The highly conserved 'DAG' motif near the N-terminus of the CP is involved in potyvirus transmission by aphids. Amino acid substitutions in and around this motif reduced or abolished the aphid transmissibility of mutants of e.g. tobacco vein mottling virus (TVMV; Atreya *et al.*, 1990) and zucchini yellow mosaic virus (ZYMV; Gal-On *et al.*, 1992). DAG and neighboring amino acid residues mediate specific binding between the TVMV CP and HC *in vitro* as well, and, interestingly, the degree of HC/CP binding correlated well with the frequency of aphid transmission (Blanc *et al.*, 1997). Moreover, transmission was always associated with retention of virus particles in the food canal of the aphid (Wang *et al.*, 1996). Taken together these data provide strong support for the bridge hypothesis although a direct interaction between HC and the aphid stylets at the site where the virions are retained is still to be demonstrated.

Wang *et al.* (1998) provided direct evidence that the specificity of virus transmission can be regulated by HC. *Lipaphis erysimi*, which transmits turnip mosaic virus (TuMV) but not tobacco etch virus (TEV) from infected plants, became an efficient vector of purified TEV when assisted by the TuMV HC. Also, transmission correlated well with retention of the virus in the stylets, thus indicating that components of (or in) the aphid food canal, e.g. attachment sites of the stylet surface or constituents of saliva are involved (Martín *et al.*, 1997; Wang *et al.*, 1998). As yet, the nature of these components is not known.

Comparing the amino acid sequences of HCs from transmissible and nontransmissible potyviruses revealed conserved amino acid motifs in different regions of HC: a 'PTK' motif, which is located in the central region of the HC, is involved in virus-binding of ZYMV (Peng *et al.*, 1998), and a highly conserved 'KITC' motif and neighboring sequences located near the N-terminus of HC, which are also associated with HC activity (Atreya *et al.*, 1992; Atreya and Pirone, 1993; Huet *et al.*, 1994). However, the role of these amino acid residues in binding and productive folding of HC is yet to be resolved.

The aphid transmission factor (ATF) of cauliflower mosaic virus (CaMV) is presumed to act in a similar fashion as the potyvirus HC (Pirone and Blanc, 1996). However, recent work indicates that a second transmission factor, a 15 kDa protein (P15) encoded by gene III of CaMV, may also be involved in caulimovirus transmission. Adding recombinant P15 to mixtures of purified virus and ATF significantly increased the efficiency of transmission (Leh *et al.*, 1998). Moreover, *in vitro* binding assays show that the C-terminal region of ATF interacts with the N-terminus of P15 (Leh *et al.*, 1998), which may indicate that the CaMV ATF is not directly interacting with CP but with P15. P15 is associated with purified CaMV particles, and by using antibodies to a synthetic peptide covering the 19 N-terminal amino acids of P15 it has previously been shown that this part of the protein is exposed on the surface of a virion (Giband *et al.*, 1986). Single amino acid substitutions in the C-terminal region of ATF have shown that the amino acid residues at position 157 and 159 determine virion binding (Schmidt *et al.*, 1994).

Helper factor dependency has been reported for *Sequiviridae* as well (Hunt *et al.*, 1988; Murphy *et al.*, 1995) and has been suggested to be important for several other viruses including the circulatively transmitted *Nanovirus* spp. (Chu and Helms, 1988) which loose their ability to be transmitted by their vectors after purification from plant tissue. Thus far, virus-encoded helper factors of circulative viruses have not been identified.

Circulative Virus Transmission by Insects

Significant advances have been made in identifying viral determinants and vector components involved in the transmission of virus species of the families *Bunyaviridae*, *Geminiviridae* and *Luteoviridae*, and the genus *Phytoreovirus*, which are carried systemically by their vectors.

Long term vegetative propagation of rice infected with rice dwarf virus (RDV; *Phytoreovirus* sp.) results in the development of a transmission-defective isolate that lacks the outer CP, P2, owing to a point mutation in the genome segment encoding this protein (Tomaru *et al.*, 1997). Recent work has shown that P2 is involved in the adsorption of the virus to monolayer-cultured cells of its vector *Nephotettix cincticeps* (Omura *et al.*, 1998). Insect transmissibility of African cassava mosaic virus (ACMV; *Begomovirus* sp.; *Geminiviridae*), and tomato spotted wilt virus (TSWV; *Tospovirus* sp.; *Bunyaviridae*) is also lost after several rounds of mechanical inoculation of these viruses because of defects in CP (Liu *et al.,* 1997) or lack of the lipid envelope (Goldbach and Peters, 1996). The CP of a geminivirus is involved in the molecular recognition of the virus by its vector. It is the basic determinant of virus acquisition although other gene products can influence the accumulation of the virus in the whitefly (Liu *et al.,* 1997; Höfer *et al.,* 1997). Exchanging the CP gene of ACMV with that of beet curly top virus (*Curtovirus* sp.) altered the insect specificity of the former virus from whiteflies to leafhopper (Briddon *et al.,* 1990). Thrips transmission of the envelope-minus TSWV isolate, NL-04, which is highly infectious upon mechanical inoculation, has so far failed, suggesting that structural components in envelope such as glycoproteins G1 and G2 may be involved in recognition of a receptor in the thrips midgut.

Putative Receptors

Binding between a virion and a cell surface receptor provides the initial physical association required for virus entry into vector cells. By analogy to virus entry into animal cells, viruses may enter insect cells either by receptor-mediated endocytosis, generally through clathrin-coated vesicles, as has been proposed for luteoviruses (Gildow, 1987), or by direct fusion of the viral envelope with the cell membrane (Marsh and Helenius, 1989). Based on observations made by electron microscopy, the latter mechanism has been proposed for bunyaviruses. Glycoproteins present in the membrane of the enveloped bunyaviruses have been suggested to mediate attachment (Ludwig *et al.,* 1991; Bandla *et al.,* 1998; Kikkert *et al.,* 1998). Anti-idiotypic antibodies mimicking the glycoproteins of TSWV specifically labeled a 50 kDa band in thrips homogenates on Western blots, and on the plasmamembrane of the larval thrips midgut (Bandla *et al.,* 1998). As purified virus displays an affinity for a similar-sized protein in an overlay assay of thrips proteins, the 50 kDa protein may be a cellular receptor (Bandla *et al.,* 1998). Simultaneously, a second thrips protein of about 94 kDa with a TSWV binding capacity has been identified by Kikkert and coworkers (1998). This protein firmly binds the G2 glycoprotein in virus overlay assays. G2 contains the highly conserved amino acid sequence RGD near the N-terminus, which is an important determinant for cellular attachment of a number of mammalian viruses and pathogens like foot-and-mouth disease virus (Berinstein *et al.,* 1995), human coxsackievirus A9 (CAV-9; Roivainen *et al.,* 1996), and the causal agent of Lyme disease, the spirochete *Borrelia burgdorferi* (Coburn *et al.,* 1998). Owing to the protease sensitivity of TSWV G2 and the high concentration of proteolytic enzymes that probably occur in the gut lumen, G2 is not expected to be involved in cellular attachment in the midgut of a thrips (Kikkert *et al.,* 1998). However, studies on CAV-9 have shown that even when the RGD-containing capsid domain is cleaved off, the virus remains infectious because it can bypass the RGD-recognizing vitronectin receptor (Roivainen *et al.,* 1996). Likewise, TSWV might employ alternative routes to enter thrips midgut cells depending on its phenotype, which in turn may be regulated by thrips midgut proteases. Further characterization and localization of the 94 kDa protein is required to elucidate its functional role in the cyclic passage of TSWV in thrips.

Several aphid proteins with relative molecular masses ranging from 31 to 85 kDa immobilize purified luteoviruses *in vitro* (Van den Heuvel *et al.*, 1994; Wang and Zhou, 1998). A Chinese barley yellow dwarf virus-MAV isolate (BYDV-MAV-Ch; *Luteovirus* sp.) displayed a strong affinity for two proteins of 31 (P31) and 44 (P44) kDa from the vector aphids *Sitobion avenae* and *Schizaphis graminum*, but not from *Rhopalosiphum padi*. The latter is unable to transmit BYDV-MAV-Ch. Antisera to P31 and P44 reacted specifically with extracts of the accessory salivary glands (ASG) of vector aphids, suggesting that these proteins might be involved in luteovirus recognition at the ASG. Several lines of research indicate that vector specificity of luteoviruses is determined at the level of the ASG (Gildow, 1987). Upon contacting the basal lamina of the ASG virus particles may be transported through this gland (Peiffer *et al.*, 1997), eventually arriving in the salivary duct from which they are excreted with the saliva when the aphid feeds (Gildow and Gray, 1993).

Receptors accounting for the high degree of vector specificity, which is commonly observed among invertebrates transmitting viruses, still remain to be characterized. It will be of great importance to reveal the true nature of these determinants because they offer new opportunities for disease control, including genetic manipulation of the vector and transmission neutralization by means of recombinant proteins expressed in transgenic plants.

Persistence in Hemolymph of Vectors

Species of the family *Luteoviridae* are solely transmitted by aphids in a circulative non-propagative manner. The hemolymph acts as a reservoir in which acquired virus particles are retained in an infective form for the aphid's lifespan, without replication. Species from all three genera of the family *Luteoviridae* display a strong affinity for *Buchnera* GroEL, which is produced by the primary endosymbiotic bacterium (*Buchnera* sp.) of aphids, and readily detected in the hemolymph of an aphid (Van den Heuvel *et al.*, 1994, 1997; Filichkin *et al.* 1997). The Gram-negative *Buchnera* spp. are present in specialized polyploid host cells, called mycetocytes, in the hemocoel (Buchner, 1953). *Buchnera* GroEL homologues are immunologically closely related, and share >80% sequence identity with the *Escherichia coli* heat shock protein GroEL, a member of the Hsp-60 family of molecular chaperons (Ohtaka *et al.*, 1992; Van den Heuvel *et al.*, 1994; Filichkin *et al.*, 1997; Hogenhout *et al.*, 1998). Structural and functional characteristics of *Buchnera* GroEL are highly similar to those of GroEL of *E. coli* (Ohtaka *et al.*, 1992).

Both GroEL from vector and non-vector aphid species, and that of *E. coli* have been shown to bind luteoviruses (Van den Heuvel *et al.*, 1994, 1997; Filichkin *et al.*, 1997). Mutational analysis of the gene coding for *Buchnera* GroEL of *Myzus persicae*, the major vector of potato leafroll virus (PLRV; *Polerovirus* sp.), shows that the virus-binding site was located in the equatorial domain of the subunit (Hogenhout *et al.*, 1998). The minor capsid protein of a luteovirus, the readthrough domain (RTD), determines the interaction with GroEL (Filichkin *et al.*, 1997; Van den Heuvel *et al.*, 1997). The RTD is exposed on the surface of a luteovirus particle and contains determinants necessary for virus transmission by aphids (Brault *et al.,* 1995; Jolly and Mayo, 1994; Chay *et al.,* 1996; Bruyère *et al.,* 1997). Sequence comparison of the RTDs of luteoviruses (including pea enation mosaic virus) has identified highly conserved amino acid residues in the N-terminal region of the RTD which are potentially important in the interaction with *Buchnera* GroEL (Van den Heuvel *et al.,* 1997).

Direct injection of beet western yellows virus (BWYV; *Polerovirus* sp.) mutants into *M. persicae* has shown that virions devoid of RTD, which are unable to interact with *Buchnera* GroEL, are significantly less persistent in the aphid hemolymph than wild-type virions (Van den Heuvel *et al.,* 1997). Moreover, a treatment of *M. persicae* larvae with antibiotics that interfere with prokaryotic protein synthesis significantly reduces *Buchnera* GroEL levels in the hemolymph, inhibits the transmission, and results in the loss of capsid integrity in the hemolymph (Van den Heuvel *et al.,* 1994). Collectively, these observations indicate that the luteovirus-GroEL interaction retards proteolytic breakdown and is essential for virus retention in the hemolymph of an aphid (Van den Heuvel *et al.,* 1994, 1997; Filichkin *et al.,* 1997). It is reasonable to assume that GroEL-mediated retention also applies to viruses of the genus *Umbravirus*, because umbravirus RNA can only be transmitted by aphids if it is packaged in the capsid of a helper virus, commonly a luteovirus (Murphy *et al.,* 1995).

The Israeli strain of tomato yellow leafcurl virus (TYLCV; *Begomovirus* sp.) interacts with an endosymbiotic GroEL homologue from its whitefly vector *Bemisia tabaci* (Morin *et al.,*1998). Feeding whiteflies with anti-*Buchnera* GroEL antiserum prior to virus acquisition reduces the transmission of TYLCV by more than 80% relative to control insects which fed on normal serum. Active antibodies have been recovered from the insects hemolymph, providing near direct evidence that the antibody interfered with the interaction between TYLCV and the GroEL homologue *in vivo.*

Endosymbiotic bacteria from leafhoppers have also been implicated in vertical transmission of RDV (Nasu, 1965).

TRANSMISSION BY SOIL-INHABITING NEMATODES

Tobravirus and *Nepovirus* spp. are transmitted by soil-inhabiting trichodorid and longidorid nematodes, respectively. Once acquired, virus particles attach to the surface of the food canal of a vector, which may be lined by carbohydrate-containing material that may be involved in the retention and release process although esophageal gland secretions during salivation have been proposed to be involved as well (Robertson and Henry, 1986). Viruses can be retained from several weeks up to nearly a year by non-feeding nematodes, however, they are lost upon molting. Vector specificity is highly developed among nematode-transmitted viruses and recent reviews on nematode-virus interactions have covered this issue in great detail (Brown *et al.,* 1995, 1996; Taylor and Brown, 1997).

The production of pseudorecombinants of nepo- or tobraviruses, by mixing RNA-1 from one isolate and RNA-2 from another, closely related isolate with different transmission features has been instrumental in unraveling the transmission determinants of these viruses. Vector specificity and transmissibility, as well as serological characteristics are determined by RNA-2, thus suggesting a prominent role for the CP in the transmission process by nematodes (Harrison, 1964; Harrison *et al.,* 1974; Harrison and Murant, 1977; Ploeg *et al.,* 1993). The construction of full-length infectious cDNA clones of RNA-1 and RNA-2 of the tobraviruses {tobacco rattle virus (TRV) and pea early browning virus (PEBV)} has further advanced our knowledge on viral transmission determinants. Mutagenesis studies of a nematode-transmissible isolate of PEBV revealed that deletion of the distal 15 amino acids of the 29-amino-acid 'arm-like' C-terminal CP domain did not disrupt virion formation but abolished nematode transmission (MacFarlane *et al.,* 1996). This flexible domain is, based on epitope mapping and NMR analysis of TRV CP, exposed on the surface of a virus particle (Legorburu *et al.,* 1995; Mayo *et al.,* 1994).

Vector transmission of tobraviruses, however, is not determined exclusively by CP. In addition to the CP, RNA-2 of the nematode-transmissible TRV isolate, PpK20, codes for two proteins of 29.4 and 32.8 kDa (Hernández *et al.,* 1995), whereas the RNA-2 of isolate PLB, which is not transmitted by nematodes, only encodes an additional 16 kDa protein (Angenent *et al.,* 1989). Mutational analysis using infectious cDNA clones of RNA-2 of PpK20 has shown that affecting the open reading frame (ORF) coding for the 29.4 kDa protein abolished nematode transmission, whereas a large deletion in the ORF coding for the 32.8 kDa protein had no effect on nematode transmission (Hernández *et al.,* 1997). The 29.4 kDa protein has been suggested to act in a similar fashion as the potyvirus HC: bridging the TRV particles to potential receptor sites in the food canal of the vector (Hernández *et al.,* 1997).

In contrast to TRV, all proteins encoded by RNA-2 of PEBV seem to play a role in nematode transmission (MacFarlane *et al.,* 1995, 1996). Mutations introduced into the non-structural proteins generally reduce or completely inhibit the transmission by nematodes. The most clear-cut results have been obtained for a 29.6 kDa protein, which, when affected, abolishes PEBV transmission, suggesting that it may act as a nematode transmission factor. Comparison of the derived amino acid sequences of the 29.6 kDa gene products of a nematode-transmissible and a non-transmissible isolate of PEBV revealed two changes that probably cause the difference in transmissibility (MacFarlane and Brown, 1995). Mutagenesis studies of the 23 kDa of PEBV are more difficult to interpret. Results indicate that this protein is involved by maintaining the frequency of transmission, but is not essential (MacFarlane *et al.,* 1996; Schmitt *et al.,* 1998). The 23 kDa protein may act in concert with esophageal gland secretions to release virus particles from their retention sites in the vector's food canal, while the 29.6 kDa protein facilitates binding. An alternative role for the 23 kDa protein may be in enhancing the virus titer and/or optimizing the distribution of the virus particles in the roots to increase availability of the virus to nematodes at their feeding sites (Schmitt *et al.,* 1998).

Comparison of the deduced amino acid sequences of RNA-2 of four nepoviruses revealed substantial similarities in the portion of the polyprotein N-terminal from the CP. Since amino acid similarity was only evident between viruses which are transmitted by the same vector these polypeptides may influence nepovirus transmission (Blok *et al.,* 1992). Non-structural 'transmission factors' of nepoviruses are still to be identified.

TRANSMISSION BY ZOOSPORIC FUNGI

Several economically important plant viruses are transmitted by soil-inhabiting zoosporic fungi, classified as species of the Plasmodiophoromycetes (*Polymyxa graminis, P. betae* and *Spongospora subterranea*) and the Chytridiomycetes (*Olpidium brassicae* and *O. bornovanus*). Although unrelated, both Chytridiomycetes and Plasmodiophoromycetes share some general characteristics. They are obligate parasites, live entirely within the host cells, have similar developmental stages including a plasmodial stage, survive in soil by formation of resting spores or cysts, and produce motile zoospores to infect roots of the host plants (Alexopoulos *et al.,* 1996). Two types of fungal transmission are distinguished based on either the mode of acquisition or the site of virus retention (Teakle, 1983). In non-persistent transmission the virus particles are adsorbed to the surface of the zoospores and are not present in the resting spores, whereas in persistent transmission the virus is carried internally into the resting spore and zoospores. In addition to the terminology persistent and non-persistent transmission, the terms *in vivo* and *in vitro* transmission are also still in use. Both the sets of terms have their flaws (Campbell, 1993, 1996), but the terms persistent/non-persistent appear to be the currently most accepted and will therefore be used in the following section.

Persistent Transmission

Plasmodiophoraceous vectors carry virus particles internally to the zoospores and the resting spores, and the virus is acquired by the vector when infecting a virus-infected host. Virus particles are retained in an infective form by the resting spores for a long period of time without replication. The mechanisms of virus acquisition, retention and transmission are not elaborated. Viruses of the genera *Bymovirus* (*Potyviridae*), *Furovirus, Pecluvirus,* and *Pomovirus* (all rod-shaped genera with bipartite genome) and *Benyvirus* (rod-shaped genus with a genome of four to five RNA molecules), are transmitted by plasmodiophoraceous fungi. Bymoviruses, like the other viruses mentioned, loose their ability to be transmitted by the fungus after being repeatedly mechanically transmitted (Adams *et al.,* 1988).

The capsid of beet necrotic yellow vein virus (BNYVV; *Benyvirus* sp.), potato mop-top virus (PMTV; *Pomovirus* sp.) and soil-borne wheat mosaic virus (SBWMV; *Furovirus* sp.) is composed of two components, a major CP and a minor readthrough protein (RT). Comparison of wild-type BNYVV with non-transmissible BNYVV strains revealed that the non-transmissible strains contained truncated RT proteins (Tamada and Kusume, 1991). Mutational analysis indicated that while the N-terminal part of the RT was involved in virus assembly, the C-terminal part was dispensable for assembly but essential for fungus transmission (Tamada *et al.*, 1996). Alanine-scanning mutagenesis showed that a peptide motif (KTER) in the RT protein is involved in fungal transmission (Tamada *et al.*, 1996). A similar motif is also present in the RT of SBWMV, but is absent in PMTV. Peanut clump virus (PCV; *Pecluvirus* sp.) lacks a RT. Therefore, the molecular mechanism underlying the transmission of pecluviruses most likely differs from that of the other viruses, and possibly resembles the strategy of the bymoviruses.

Comparing the nucleotide sequences of the RNA2 of a transmissible and a non-transmissible strain of barley mild mosaic virus (BaMMV; *Bymovirus* sp.; viral capsid of bymoviruses is composed of a single protein of 28 to 33 kDa) revealed a deletion of approximately 1000 nucleotides in the C-terminal P2 gene (Dessens *et al.*, 1995; Peerenboom *et al.*, 1997), whereas no significant differences were found in the RNA1 molecules of these strains. The P2 protein contains an N-terminal 'pseudo' capsid protein domain, but is not a component of the viral capsid and its function is unknown (Dessens *et al.*, 1995). Comparison of the bymovirus P2 proteins and the RTs of the persistently transmitted viruses revealed the presence of conserved domains, and an evolutionary relationship has been suggested (Dessens and Meyer, 1996). A conserved C-terminal hydrophobic domain is putatively involved in fungal transmission of these viruses. In view of the hydrophobic character of several domains of the RTs and P2, they may mediate attachment to or passage through the fungal membrane.

Olpidium spp. persistently transmit a group of only partly characterized viruses (*Varicosavirus* sp.) of which lettuce big-vein virus (LBVV) and tobacco stunt virus are the most prominent members. Only for LBVV have Koch's postulates been confirmed (Huijberts *et al.*, 1990), but little is known about their nature.

Non-persistent Transmission

The chytrid fungi *O. brassicae* and *O. bornovanus* are the major vectors of isometric viruses from the *Tombusviridae*. Most of the research has been done on tobacco necrosis virus (TNV; *Necrovirus* sp.) and cucumber necrosis virus (CNV; *Tombusvirus* sp.). Their transmission by *Olpidium* spp. clearly differs from virus transmission by plasmodiophoreceous vectors since the virus is acquired by adsorption to the zoospore membrane, and not internalized. Adsorption takes place in soil water or *in vitro* since virus particles and zoospores are released independently from infected roots. To initiate a fungal infection, zoospores retract their flagella, encyst on the surface of the host cells and inject their protoplast in the host cell. The fungal protoplast is not surrounded by a host membrane and lives entirely intracellularly in the host cell.

The mechanism of entry of the adsorbed virus into the host cell has not been established unequivocally. Temmink (1971) stated that upon entry of the fungal protoplast the zoospore plasmalemma remains in the cyst and a new membrane is *de novo* synthesized. This rules out the most simple explanation of release of the adsorbed virus particles from the zoospore membrane in the root cell cytoplasm. Therefore, it was suggested that the adsorbed virus particles enter the zoospore protoplast on withdrawal of flagellum and are transmitted to the root cell during or after the discharge of virus-containing protoplasm from the encysted spore. This was supported by the observation that virus particles adsorb both to the plasmalemma of the zoospore body and to the axonemal sheath, which surrounds the flagellum (Temmink *et al.,* 1970; Stobbs *et al.,* 1982). The axonemal sheath is taken in during the flagellar retraction, resulting in the appearance of a 'whorl of membranes' in the cytoplasm of the protoplast (Temmink and Campbell, 1969; Stobbs *et al.,* 1982). The presence of virus particles adhered to this structure has been confirmed (Stobbs *et al.,* 1982).

However, the observation of *de novo* synthesis of the protoplast plasmalemma is based upon electron microscopical studies and no further evidence has been presented. It is difficult if not impossible to determine from these studies whether the protoplast plasmalemma is indeed completely newly synthesized. It can not be excluded that the plasmalemma of the penetrating zooplast is not entirely new and virus particles are present on the membrane or that transport of virus particles takes place along the outside of the plasmalemma to the new part of the membrane (Adams, 1991). The presence of a specific relationship between the virus and its fungal vector has been well established. Different strains or isolates of *O. brassicae* differ in their efficiency to transmit TNV, ranging from good, poor to non-vectors (Temmink *et al.,* 1970). The observed differences in efficiency of transmission were shown to be linked to the ability of the virus particles to adhere to the zoospores, and were independent of the host range of the fungus. Moreover, adsorption of virus particles was restricted to fungus-transmitted viruses (Temmink *et al.,* 1970). Isolates of *O. bornovanus* and *O. brassicae* differing in their ability to vector different carmoviruses (*Tombusviridae*) have been identified (Campbell *et al.,* 1995).

The presence of the viral CP is a prerequisite for fungal transmission, since purified genomic CNV RNA was not transmitted by zoospores to cucumber (Stobbs *et al.,* 1982). McLean and coworkers (1994) have investigated the role of CP in virus transmission by reciprocal exchange of the CP genes of tomato bushy stunt virus (TBSV; *Tombusvirus* sp.) and CNV in infectious cDNA clones. CNV is transmitted by *O. bornovanus*, while TBSV is not fungus-transmissible. However, the recombinant TBSV genome encapsidated by CNV CP was efficiently transmitted by *O. bornovanus*, whereas the CNV genome in TBSV CP was not. Although in the recent years considerable progress has been made in characterizing the viral determinants involved in fungal transmission, data on the fungal determinants involved in virus recognition, virus binding, and transport through the zoospore membrane is lacking.

TRANSMISSION BY SEED AND POLLEN

Seed transmission has been reported for more than 100 plant viruses, and is highly dependent on virus isolate, host plant, environmental conditions and time of inoculation relative to flowering and age of the host (Mandahar, 1981; Mink, 1993). All seed-transmitted viruses, except two, are carried within the embryo. The presence of tobacco mosaic virus (TMV; *Tobamovirus* sp.) in the seed coat or endosperm tissues leads to virus-infected progeny (Taylor *et al.,* 1961). Seed-borne melon necrotic spot virus (MNSV; *Carmovirus* sp.) requires the presence of its vector, the soil-borne fungus, *O. bornovanus*, to infect the seedling (vector-assisted seed transmission; Campbell *et al.,* 1996). All other viruses must be able to infect the embryo and to be able to survive in the embryo during seed maturation and storage (Bowers and Goodman, 1979).

There are two possible pathways for virus infection of embryonic tissues. Either the gametes are infected and fusion results in an infected embryo (indirect invasion) or the virus invades the embryo after fertilization (direct invasion). The principles controlling the two pathways are different and it has been suggested that different host genes are involved (Maule and Wang, 1996).

Direct invasion of the embryo is complicated by the fact that embryos are physically separated from mother plant tissues, thus preventing invasion of the embryo by cell-to-cell movement of the virus. Therefore, virus invasion has to occur early in the development of the embryo before the symplastic connections between the maternal and progeny tissues are disrupted. Pea seed-borne mosaic virus (PSbMV; *Potyvirus* sp.; *Potyviridae*) does not infect the gametes and is dependent on direct invasion of the embryo for seed transmission. Electron microscopical studies have shown that PSbMV invades the embryo through the embryonic suspensor before it disintegrates during maturation of the embryo (Wang and Maule, 1994, 1997). Failure of the virus to reach the suspensor in time or to invade the suspensor prevents seed transmission. However, there remains the question how the virus invades the suspensor. There is no symplastic connection between the embryogenic suspensor and the maternal cells and transport of nutrients is believed to occur apoplastically (Raghavan, 1986).

Indirect seed transmission occurs when the virus invades the reproductive tissues prior to embryogenesis. Infection of both maternal and paternal reproductive organs can lead to seed transmission. Tobacco ringspot virus (*Nepovirus* sp.; *Comoviridae*) is predominantly transmitted through the megagametophyte, while pollen transmission for bean common mosaic virus (*Potyvirus* sp.) is more effective (Morales and Castano, 1987). Cucumber mosaic virus (*Cucumovirus* sp.; *Bromoviridae*) is both efficiently seed-transmitted through infected pollen or the infected megagametophyte (Yang *et al.,* 1997). Indirect seed transmission of barley stripe mosaic virus (BSMV; *Hordeivirus* sp.) and lettuce mosaic virus (*Potyvirus* sp.) depends on early invasion of the floral meristem and the ability of the virus to infect the megaspore mother cell or pollen mother cells (Caroll and Mayhew, 1976; Hunter and Bowyer, 1997). Most likely seed transmission by indirect invasion of the embryo is governed by the ability of the virus to infect meristematic tissues.

Viral Genetic Determinants

The viral genetic determinants of seed transmission have been identified for three viruses belonging to different taxonomic groups and employing different strategies for embryo invasion. Although, and not surprisingly, genes involved in replication and movement of viruses affect seed transmission to a certain extent in all these cases. The bipartite PEBV is seed-transmitted by indirect invasion of the embryo. Mutational analysis of its genome, by making deletions in the different genes in infectious clones of both RNAs, revealed a crucial role for the RNA2-encoded 12 kDa protein in seed transmission (Wang *et al.,* 1997). A mutant with a deleted 12 kDa gene caused more severe symptoms on leaves. However, it could not be detected in the pollen and ovules and only poorly accumulated in anther tissues and carpels while the wild-type virus accumulated at high levels in all tissues, and incidence of its seed transmission was below 1%. Presumably the 12 kDa protein is required for invasion of the gametes or is necessary to sustain its replication in the meristematic reproductive tissues of the host.

BSMV is primarily dependent on seed transmission in barley for its survival in nature, which is believed to occur primarily by indirect invasion of the embryo. Its tripartite genome is identified with RNAα, RNAβ and RNAγ. The RNAα encodes replicase, the RNAβ encodes CP and the triple gene block involved in viral movement, and the RNAγ codes for a polymerase (γa protein) and the γb protein of 17 kDa. The role of all three RNAs and their gene products in seed transmission was investigated by constructing chimeric viruses using infectious cDNA clones of a transmissible and a poorly-transmissible virus isolate (Edwards, 1995). It was shown that major determinants of seed transmission were located on the RNAγ since mutations in the 5' untranslated region (5'UTR), a repeat in the γa gene and the γb protein had pronounced effects on seed transmission. Mutations in the 5'UTR, and consequently changes in its secondary structure presumably affect replication and translation. Presence of a repetitive sequence in the γa protein blocked seed transmission completely, and it was suggested that absence of the repeat sequence enhanced replication and movement. Mutations in the γb gene affected both seed transmission efficiency and symptom expression. The γb gene affects virulence and regulates viral gene expression thus influencing virus movement. However, its precise role in the infection cycle has not been elucidated (Donald and Jackson, 1994). The protein contains two cysteine-rich 'zinc finger-like' domains and has been shown to have RNA-binding capacity (Donald and Jackson, 1996).

The construction of chimeric viruses between a highly seed-transmitted and a rarely seed-transmitted isolate of PSbMV revealed that the helper-component protease (HC-Pro) was a major determinant of seed transmission (Johansen *et al.,* 1996). It has been suggested that HC-Pro affects replication and/or long distance movement in reproductive tissues and thus the infection rate of embryos (Kasschau *et al.,* 1997). Like the two other viral proteins identified as determinants for seed transmission, HC-Pro contains a cysteine-rich domain.

The significance of the presence of these 'zinc-finger-like' domains remains to be determined. Zinc-finger domains are generally involved in binding of nucleic acids, suggesting two possible functions for the three viral proteins: virus movement or regulation of viral gene expression. Indeed, both the potyviral HC-Pro and the BSMV γb protein have RNA-binding activity (Donald and Jackson, 1996; Maia and Bernardi, 1996). The γb protein is believed to regulate expression of BSMV genes by binding to genomic RNA, while potyviral HC-Pro is both involved in genome amplification and long distance movement, the latter by facilitating transport of the virus through the plasmodesmata between companion cells and sieve elements. It remains to be investigated which precise function of the proteins controls seed transmissibility.

Host Genetic Determinants

Resistance to seed transmission can be caused by inability of the virus to directly or indirectly infect the embryo or its inability to survive maturation of the seed. For soybean mosaic virus (SMV; *Potyvirus* sp.) it was shown that the virus was inactivated in seeds of soybean cultivars showing resistance to seed transmission (Bowers and Goodman, 1979) The host genetic determinants involved in seed transmission of BSMV and PSbMV have been partly characterized (Caroll *et al.,* 1979; Wang and Maule, 1994). Cross-pollination experiments between pea cultivars susceptible to seed transmission of PSbMV and resistant to seed transmission showed that the genotype of the progeny did not influence seed transmission efficiency and the genes controlling resistance segregated as quantitative trait loci, possibly involving only a few genes (Wang and Maule, 1994). This is in contrast with the findings of Caroll and coworkers (1979) who identified the host genetic determinant in barley controlling resistance to BSMV seed transmission as a single recessive gene. However, PSbMV is transmitted by direct embryo invasion and BSMV is predominantly transmitted by indirect embryo invasion. Presumably, different host genes are involved in both pathways of seed transmission.

REFERENCES

Adams, M.J. (1991). Transmission of plant viruses by fungi. *Ann. Appl. Biol.* 118, 479-492.

Adams, M.J., Swaby, A.G., and McFarlane, I. (1998). The susceptibility of barley cultivars to barley yellow mosaic virus (BaYMV) and its fungal vector *Polymyxa graminis. Ann. Appl. Biol.* 109,561-572.

Alexopoulos, C.J., Mims, C.W., and Blackwell, M. (1996). Introductory Mycology 4th ed. John Wiley & Sons, New York.

Angenent, G.C., Posthumus, E., Brederode, F.T., and Bol, J.F. (1989). Genome structure of tobacco transmissibility and helper-component activity correlates with non-retention of virions in aphid stylets rattle virus strain PLB: further evidence on the occurrence of RNA recombination among tobraviruses. *Virology* 171, 271-274.

Ammar, E.D. (1994). Propagative transmission of plant and animal viruses by insects: factors affecting vector specificity and competence. *Adv. Dis. Vector Res.* 10, 289-331.

Atreya, C.D., and Pirone, T.P. (1993). Mutational analysis of the helper component-proteinase gene of a potyvirus: effects of amino acid substitutions, deletions, and gene replacement on virulence and aphid transmissibility. *Proc. Natl. Acad. Sci. U.S.A.* 90, 11919-11923.

Atreya, C.D., Raccah, B., and Pirone, T.P. (1990). A point mutation in the coat protein abolishes aphid transmissibility of a potyvirus. *Virology* 178, 161-165.

Atreya, C.D., Atreya, P.L., Thornbury, D.W., and Pirone, T. (1992). Site directed mutagenesis in the potyvirus HC-Pro gene affect helper component activity, virus accumulation and symptom expression in infected tobacco plants. *Virology* 191, 106-111.

Bandla, M.D., Campbell, L.R., Ullman, D.E., and Sherwood, J.L. (1998). Interaction of tomato spotted wilt tospovirus (TSWV) glycoproteins with a thrips midgut protein, a potential cellular receptor for TSWV. *Phytopathology* 88, 98-104.

Berinstein, A., Roivainen, M., Hovi, T., Mason, P., and Baxt, B. (1995). Antibodies to the vitronectin receptor (integrin avB3) inhibit binding and infection of foot-and-mouth disease virus to cultured cells. *J. Virol.* 69, 2664-2666.

Blanc, S., Lopez-Moya, J.-J., Wang, R., García-Lampasona, S., Thornbury, D., and Pirone, T.P. (1997). A specific interaction between coat protein and helper component correlates with aphid transmission of a potyvirus. *Virology* 231, 141-147.

Blok, V.C., Wardell, J., Jolly, C.A., Manoukian, A., Robinson, D.J., Edwards, M.L., and Mayo, M.A. (1992). The nucleotide sequence of RNA-2 of raspberry ringspot nepovirus. *J. Gen. Virol.* 73, 2189-2194.

Bowers, G.R., and Goodman, R.M. (1979). Soybean mosaic virus: Infection of soybean seed parts and seed transmission. *Phytopathology* 69, 569-572.

Brault, V., Van den Heuvel, J.F.J.M., Verbeek, M., Ziegler-Graff, V., Reutenauer, A., Herrbach, E., Garaud, J.-C., Guilley, H., Richards, K., and Jonard, G. (1995). Aphid transmission of beet western yellows virus requires the minor capsid read-through protein P74. *EMBO J.* 114, 650-659.

Briddon, R.W., Pinner, M.S., Stanley, J., and Markham, P.G. (1990). Geminivirus coat protein gene replacement alters insect specificity. *Virology* 177, 85-94.

Brown, D.J.F., Robertson, W.M., and Trudgill, D.L. (1995). Transmission of viruses by plant nematodes. *Ann. Rev. Phytopath.* 33, 223-249.

Brown, D.J.F., Trudgill, D.L., and Robertson, W.M. (1996). "Nepoviruses: transmission by nematodes". In *The Plant Viruses Vol. V. Polyhedral Virions and Bipartite RNA Genomes* (B. D. Harrison and A. F. Murant, Eds.), pp. 187-209. Plenum, New York.

Bruyère, A., Brault, V., Ziegler-Graff, V., Simonis, M.-T., Van den Heuvel, J.F.J.M., Richards, K., Guilley, H., Jonard, G., and Herrbach, E. (1997). Effects of mutations in the beet western yellows virus readthrough protein on its expression and packaging, and on virus accumulation, symptoms, and aphid transmission. *Virology* 230, 323-334.

Buchner, P. (1953). Endosymbiose der Tiere mit pflanzlichen Mikroorganismen. Birkhaeuser Verlag.

Campbell, R.N. (1993). Persistence: a vector relationship not applicable to fungal vectors. *Phytopathology* 83, 363-364.

Campbell, R.N. (1996). Fungal transmission of plant viruses. *Annu. Rev. Phytopathol.* 34, 87-108.

Campbell, R.N., Sim, S.T., and LeCoq, H. (1995). Virus transmission by host-specific strains of *Olpidium bornovanus* and *Olpidium brassicae*. *Eur. J. Plant Pathol.* 101, 273-282.

Campbell, R.N., Wipf-Scheibel, C., and Lecoq, H. (1996). Vector-assisted seed transmission of melon necrotic spot virus in melon. *Phytopathology* 86, 1294-1298.

Carroll, T.W., Gossel, P.L., and Hockett, E.A. (1979). Inheritance of resistance to seed transmission of barley stripe mosaic virus in barley. *Phytopathology* 69, 431-433.

Caroll, T.W., and Mayhew, D.E. (1976). Anther and pollen infection in relation to the seed transmissibility of two strains of barley stripe mosaic virus in barley. *Can. J. Bot.* 54, 1604-1621.

Chay, C.A., Gunasinge, U.B., Dinesh-Kumar, S.P., Miller, W.A., and Gray, S.M. (1996). Aphid transmission and systemic plant infection determinants of barley yellow dwarf luteovirus-PAV are contained in the coat protein readthrough domain and 17-kDa protein, respectively. *Virology* 219, 57-65.

Chu, P.W.G. and Helms, K. (1988). Novel virus-like particles containing circular single-stranded DNAs associated with subterranean clover stunt disease. *Virology* 167, 38-49.

Coburn, J., Magoun, L., Bodary, S.C., and Leong, J.M. (1998). Integrins alpha(v)beta3 and alpha5beta1 mediate attachment of Lyme disease spirochetes to human cells. *Infect Immun.* 66, 1946-1952.

Demler, S.A., Rucker-Feeney, D.G., Skaf, J.S., and De Zoeten, G.A. (1997). Expression and suppression of circulative aphid transmission in pea enation mosaic virus. *J. Gen. Virol.* 78, 511-523.

Dessens, J.T., and Meyer, M. (1996). Identification of structural similarities between putative transmission proteins of *Polymyxa* and *Spongospora* transmitted bymoviruses and furoviruses. *Virus Genes* 12, 95-99.

Dessens, J.T., Nguyen, M., and Meyer, M. (1995). Primary structure and sequence analysis of RNA2 of a mechanically transmitted barley mild mosaic virus isolate: An evolutionary relationship between bymo- and furoviruses. *Arch. Virol.* 140, 325-333.

Donald, R.K.G., and Jackson, A.O. (1994). The barley stripe mosaic virus yb gene encodes a multifunctional cysteine-rich protein that affects pathogenesis. *Plant Cell* 6, 1593-1606.

Donald, R.K.G., and Jackson, A.O. (1996). RNA-binding activities of barley stripe mosaic virus yb fusion proteins. *J. Gen. Virol.* 77, 879-888.

Edwards, M.C. (1995). Mapping of the seed transmission determinants of barley stripe mosaic virus. *Mol. Plant-Microbe Interact.* 6, 906 -915.

Filichkin, S.A., Brumfield, S., Filichkin, T.P., and Young, M.J. (1997). In vitro interactions of the aphid endosymbiotic SymL chaperonin with barley yellow dwarf virus. *J. Virol.* 71, 569-577.

Gal-On, A., Antignus, Y., Rosner, A., and Raccah, B. (1992). A zucchini yellow mosaic virus coat protein gene mutation restores aphid transmissibility but has no effect on multiplication. *J. Gen. Virol.* 73, 2183-2187.

Giband, M., Mesnard, J.M., and Lebeurier, G. (1986). The gene III product (P15) of cauliflower mosaic virus is a DNA-binding protein while an immunologically related P11 polypeptide is associated with virions. *EMBO J.* 5, 2433-2438.

Gildow, F.E. (1987). Virus-membrane interactions involved in circulative transmission of luteoviruses by aphids. *Curr. Top. Vector Res.* 4, 93-120.

Gildow, F., and Gray, S.M. (1993). The aphid salivary gland basal lamina as a selective barrier associated with vector-specific transmission of barley yellow dwarf luteovirus. *Phytopathology* 83, 1293-1302.

Goldbach, R., and Peters, D. (1996). "Molecular and biological aspects of tospoviruses." In *The Bunyaviridae* (R.M. Elliot, Ed.), pp. 127-159. Plenum, New York.

Harrison, B.D. (1964). Specific nematode vectors for serologically distinctive forms of raspberry ringspot and tomato black ring viruses. *Virology* 22, 544-550.

Harrison, B.D. and Murant, A.F. (1977). Nematode transmissibility of pseudo-recombinant isolates of tomato black ring virus. *Ann. Appl. Biol.* 86, 209 -212.

Harrison, B.D., Murant, A.F., Mayo, M.A., and Roberts, I.M. (1974). Distribution of determinants for symptom production, host range and nematode transmissibility between the two RNA components of raspberry ringspot virus. *J. Gen. Virol.* 22, 233-247.

Hartl, F.U. (1996). Molecular chaperones in cellular protein folding. *Nature* 381, 571-580.

Hernández, C., Mathis, A., Brown, D.J.F., and Bol, J.F. (1995). Sequence of RNA 2 of a nematode-transmissible isolate of tobacco rattle virus. *J. Gen. Virol.* 76, 2847-2851.

Hernández, C., Visser, P.B., Brown, D.J.F., and Bol, J.F. (1997). Transmission of tobacco rattle virus isolate PpK20 by its nematode vector requires one of the two non-structural genes in the viral RNA 2. *J. Gen. Virol.* 78, 465-467.

Höfer, P., Bedford, I.D., Markham, P.G., Jeske, H., and Frischmuth, T. (1997). Coat protein gene replacement results in whitefly transmission of an insect nontransmissible geminivirus isolate. *Virology* 236, 288-295.

Hogenhout, S.A., Van der Wilk, F., Verbeek, M., Goldbach, R.W., and Van den Heuvel, J.F.J.M. (1998). Potato leafroll virus binds to the equatorial domain of the aphid endosymbiotic GroEL homolog. *J. Virol.* 72, 358-365.

Huet, H., Gal-On, A., Meir, E., Lecoq, H., and Raccah, B. (1994). Mutations in the helper component protease gene of zucchini yellow mosaic virus affect its ability to mediate aphid transmissibility. *J. Gen.. Virol..* 75, 1407-1414.

Huijberts, N., Blystad, D.R., and Bos, L. (1990). Lettuce big-vein virus: Mechanical transmission and relationships to tobacco stunt virus. *Ann. appl. Biol.* 116, 463-475.

Hull, R. (1994). Molecular biology of plant virus-vector interactions. *Adv. Dis. Vector Res.* 10, 361-386.

Hunt, R.E., Nault, L.R., and Gingery, R.E. (1988). Evidence for infectivity of maize chlorotic dwarf virus and for a helper component in its leafhopper transmission. *Phytopathology* 78, 499-504.

Hunter, D.G., and Bowyer, J.W. (1997). Cytopathology of developing anthers and pollen mother cells from lettuce plants infected by lettuce mosaic potyvirus. *J. Phytopathology* 145, 521-524.

Johansen, I.E., Dougherty, W.G., Keller, K.E., Wang, D., and Hampton, R.O. (1996). Multiple viral determinants affect seed transmission of pea seedborne mosaic virus in *Pisum sativum. J. Gen. Virol.* 77, 3149-3154.

Jolly, C.A., and Mayo, M.A. (1994). Changes in the amino acid sequence of the coat protein readthrough domain of potato leafroll luteovirus affect the formation of an epitope and aphid transmission. *Virology* 201, 182-185.

Kasschau, K.D., Cronin, S., and Carrington, J.C. (1997). Genome amplification and long distance movement functions associated with the central domain of tobacco etch potyvirus helper component proteinase. *Virology* 228, 251-262.

Kikkert, M., Meurs, C., Van de Wetering, F., Dorfmüller, S., Peters, D., Kormelink, R., and Goldbach, R. (1998). Binding of tomato spotted wilt virus to a 94-kDa thrips protein. *Phytopathology* 88, 63-69.

Legorburu, F.J., Robinson, D.J., Torrance, L., and Duncan, G.H. (1995). Antigenic analysis of nematode-transmissible and non-transmissible isolates of tobacco rattle tobravirus using monoclonal antibodies. *J. Gen. Virol.* 76, 1497-1501.

Leh, V., Jacquot, E., Geldreich, A., Leclerc, D., Cerutti, M., Yot, P., Keller, M., and Blanc, S. (1998). A second helper protein is required for the aphid transmission of cauliflower mosaic virus. Abstracts, Volume 1, 7th International Congress of Plant Pathology; August 9 – 16; Edinburgh.

Liu, S., Bedford, I.D., Briddon, R.W., and Markham, P.G. (1997). Efficient whitefly transmission of African cassava mosaic geminivirus requires sequences from both genomic components. *J. Gen. Virol.* 78, 1791-1794.

Ludwig, G.V., Israel, B.A., Christensen, B.M., Yuill, T.M., and Schultz, K.T. (1991). Role of La Crosse virus glycoproteins in attachment of virus to host cells. *Virology* 181, 564-571.

MacFarlane, S.A., and Brown, D.J.F. (1995). Sequence comparison of RNA2 of nematode-transmissible and nematode-non-transmissible isolates of pea early-browning virus suggests that the gene encoding the 29 kDa protein may be involved in nematode transmission. *J. Gen. Virol.* 76, 1299-1304.

MacFarlane, S.A., Brown, D.J.F., and Bol, J.F. (1995). The transmission by nematodes of tobraviruses is not determined exclusively by the virus coat protein. *Eur. J. Plant Pathol.* 101, 535-539.

MacFarlane, S.A., Wallis, C.V., and Brown, D.J.F. (1996). Multiple virus genes involved in the nematode transmission of pea early browning virus. *Virology* 219, 417-422.

Maia, I.G., and Bernardi, F. (1996). Nucleic acid binding properties of a bacterially expressed potato virus Y helper component protease. *J. Gen. Virol.* 77, 869-877.

Mandahar, C. L. (1981). Virus transmission through seed and pollen. *In* "Plant Diseases and Vectors: Ecology and Epidemiology" (K. Maramorosch and K. F. Harris, Eds.). pp. 241-292. Academic Press, New York.

Marsh, M., and Helenius, A. (1989). Virus entry into animal cells. *Adv. Virus Res.* 36, 107-151.

Martín, B., Collar, J.L., Tjallingii, W.F., and Fereres, A. (1997). Intracellular ingestion and salivation by aphids may cause the acquisition and inoculation of non-persistently transmitted plant viruses. *J. Gen.. Virol.* 78, 2701-2705.

Maule, A.J., and Wang, D. (1996). Seed transmission of plant viruses: A lesson in biological complexity. *Trends Microbiol.* 4, 153-158.

Mayo, M.A., Robertson, W.M., Legorburu, J., and Brierly, K.M. (1994). "Molecular approaches to an understanding of the transmission of plant viruses by nematodes." In *Advance in Molecular Plant Nematology* (F. Lamberti, C. de Giorgi, and D.McK. Bird, Eds.), pp. 277-293. Plenum, New York.

McLean, M.A., Campbell, R.N., Hamilton, R.I., and Rochon, D.M. (1994). Involvement of the cucumber necrosis virus coat protein in the specificity of fungus transmission by *Olpidium bornovanus*. *Virology* 204, 840-842.

Mink, G.I. (1993). Pollen-transmitted and seed-transmitted viruses and viroids. *Annu. Rev. Phytopathol.* 31, 375-402.

Morales F.J., and Castano M. (1987). Seed transmission characteristics of selected bean common mosaic virus strains in differential bean cultivars. *Plant Disease* 71:51-53.

Morin, S., Ghanim, M., Zeidan, M., Czosnek, H., Verbeek, M., and van den Heuvel, J.F.J.M. (1998). A GroEL homologue from endosymbiotic bacteria of *Bemisia tabaci* is implicated in the circulative transmission of *Tomato yellow leaf curl virus* (TYLCV-Is). Submitted.

Murphy, F.A., Fauquet, C.M., Bishop, D.H.L., Ghabrial, S.A., Jarvis, A.W., Martelli, G.P., Mayo, M.A., and Summers, M.D. (1995). *Virus Taxonomy*, Springer-Verlag, Wien.

Nasu, S. (1965). Electron microscopic studies on transovarial passage of rice dwarf virus. *Jpn. J. Appl. Entomol. Zool.* 9, 225-237.

Ohtaka, C., Nakamura, H., and Ishikawa, H. (1992). Structures of chaperonins from an intracellular symbiont and their functional expression in *Escherichia coli groE* mutants. *J. Bacteriol.* 174, 1869-1874.

Omura, T., Yan, J., Zhong, B., Wada, M., Zhu, Y., Tomaru, M., Maruyama, W., Kikuchi, A., Watanabe, Y., Kimura, I., and Hibino, H. (1998). The P2 protein of rice dwarf phytoreovirus is required for adsorption of the virus to cells of the insect vector. *J. Virol.* 72, 9370-9373.

Peerenboom, E. Cartwright, E.J., Foulds, I., Adams, M.J., Stratford, R., Rosner, A., Steinbiss, H.-H., and Antoniw, J.F. (1997). Complete RNA1 sequences of two UK isolates of barley mild mosaic virus: a wild-type fungus-transmitted isolate and a non-fungus-transmissible derivative. *Virus Res.* 50, 175-183.

Peiffer, M.L., Gildow, F.E., and Gray, S.M. (1997). Two distinct mechanisms regulate luteovirus transmission efficiency and specificity at the aphid salivary gland. *J. Gen. Virol.* 7, 495-503.

Peng, Y.-H., Kadoury, D., Gal-On, A., Huet, H., Wang, Y., and Raccah, B. (1998). Mutations in the HC-Pro gene of zucchini yellow mosaic potyvirus: effects on aphid transmission and binding to purified virions. *J. Gen. Virol.* 79, 897-904.

Perry, K.L., Zhang, L., and Palukaitis, P. (1998). Amino acid changes in the coat protein of cucumber mosaic virus differentially affect transmission by the aphids *Myzus persicae* and *Aphid gossypii*. *Virology* 242, 204-210.

Pirone, T.P., and Blanc, S. (1996). Helper-dependent vector transmission of plant viruses. *Annu. Rev. Phytopathol.* 34, 227-247.

Ploeg, A. T., Robinson, D. J., and Brown, D. J. F. (1993). RNA-2 of tobacco rattle virus encodes the determinants of transmissibility by trichodorid vector nematodes. *J. Gen. Virol.* 74, 1463-1466.

Raghavan, V. (1986) Embryogenesis in angiosperms. Cambridge University Press.

Robertson, W.M., and Henry, C.E. (1986). An association of carbohydrates with particles of arabis mosaic virus retained with *Xiphinema diversicaudatum*. *Ann. Appl. Biol.* 109, 299-305.

Roivainen, M., Piirainen, L., and Hovi, T. (1996). Efficient RGD-independent entry process of coxsackievirus A9. *Arch. Virol.* 141, 1909-1919.

Schmidt, I., Blanc, S., Esperandieu, P., Kuhl, G., Devauchelle, G., Louis, C., and Cerutti, M. (1994). Interaction between the aphid transmission factor and virus particles is a part of the molecular mechanism of cauliflower mosaic virus aphid transmission. *Proc. Natl. Acad. Sci. U.S.A.* 91, 8885-8889.

Schmitt, C., Mueller, A.-M., Mooney, A., Brown, D., and MacFarlane, S. (1998). Immunological detection and mutational analysis of the RNA2-encoded nematode transmission proteins of pea early browning virus. *J. Gen. Virol.* 79, 1281-1288.

Stobbs, L.W., Cross, G.W., Manocha, M.S. (1982). Specificity and methods of transmission of cucumber necrosis virus by *Olpidium radicale* zoospores. *Can. J. Plant Pathol.* 4, 134-142.

Tamada, T., and Kusume, T. (1991). Evidence that the 75K readthrough protein of beet necrotic yellow vein virus RNA-2 is essential for transmission by the fungus *Polymyxa betae*. *J. Gen. Virol.* 72, 1497-1504.

Tamada, T., Schmitt, C., Saito, M., Guilley, H., Richards, K., and Jonard, G. (1996). High resolution analysis of the readthrough domain of beet necrotic yellow vein virus readthrough protein: A KTER motif is important for efficient transmission of the virus by *Polymyxa betae*. *J. Gen. Virol.* 77, 1359-1367.

Teakle, D.S. (1983). "Zoosporic fungi and viruses, double trouble." In *Zoosporic Plant Pathogens, a modern perspective* (S.T. Buczacki, Ed.), pp. 233-248. Academic Press, London.

Taylor, C.E. and Brown, D.J.F. (1997). Nematode Vectors of Plant Viruses. CAB International, London. 286 pp.

Taylor, R.H., Grogan, R.G., and Kimble, K.A. (1961). Transmission of tobacco mosaic virus in tomato seed. *Phytopathology* 51, 837-842.

Temmink, J.H.M. (1971). An ultrastructural study of *Olpidium brassicae* and its transmission of tobacco necrosis virus. *Meded. Landbouwhogeschool Wageningen* 71-6.

Temmink, J.H.M., and Campbell, R.N. (1969). The ultrastructure of *Olpidium brassicae*. II. Zoospores. *Can. J. Bot.* 47, 421-424.

Temmink, J.H.M., Campbell, R.N., and Smith, P.R. (1970). Specificity and site of in vitro acquisition of tobacco necrosis virus by zoospores of *Olpidium brassicae*. *J. Gen. Virol.* 9, 201-203.

Tomaru, M., Maruyama, W., Kikuchi, A., Yan, J., Zhu, Y., Suzuki, N., Isogal, M., Oguma, Y., Kimura, I., and Omura, T. (1997). The loss of outer capsid protein P2 results in nontransmissibility by the insect vector of rice dwarf phytoreovirus. *J. Virol.* 71, 8019-8023.

Van den Heuvel, J.F.J.M., Bruyère, A., Hogenhout, S.A., Ziegler-Graff, V., Brault, V., Verbeek, M., Van der Wilk, F., and Richards, K. (1997). The N-terminal region of the luteovirus readthrough domain determines virus binding to *Buchnera* GroEL and is essential for virus persistence in the aphid. *J. Virol.* 71, 7258-7265.

Van den Heuvel, J.F.J.M., Verbeek, M., and Van der Wilk, F. (1994). Endosymbiotic bacteria associated with circulative transmission of potato leafroll virus by *Myzus persicae*. *J. Gen. Virol.* 75, 2559-2565.

Wang, D., MacFarlane, S.A., and Maule, A.J. (1997). Viral determinants of pea early browning virus seed transmission in pea. *Virology* 234, 112-117.

Wang, D., and Maule, A.J. (1994). A model for seed transmission of a plant virus: Genetic and structural analysis of pea embryo invasion by pea seed-borne mosaic virus. *Plant Cell* 5, 777-787.

Wang, D., and Maule, A.J. (1997). Contrasting patterns in the spread of two seed-borne viruses in pea embryos. *Plant J.* 11, 1333-1340.

Wang, R.Y., Ammar, E.D., Thornbury, D.W., Lopez-Moya, J.J., and Pirone, T.P. (1996). Loss of potyvirus. *J. Gen. Virol.* 77, 861-867.

Wang, R.Y., Powell, G., Hardie, J., and Pirone, T.P. (1998). Role of the helper component in vector-specific transmission of potyviruses. *J. Gen. Virol.* 79, 1519-1524.

Wang, X., and Zhou, G. (1998). Identification of the proteins associated with circulative transmission of barley yellow dwarf luteoviruses from *Sitobion avenae* and *Schizaphis graminum*. Abstracts, Volume 2, 7th International Congress of Plant Pathology; 1998 August 9 – 16; Edinburgh.

Yang, Y., Kim, K.S., and Anderson, E.J. (1997). Seed transmission of cucumber mosaic virus in spinach. *Phytopathology* 87, 924-931.

9 MOLECULAR BASIS OF SYMPTOMATOLOGY

A. L. N. Rao

INTRODUCTION

Viral pathogenesis is the mechanism by which viruses enter host plants, establish infection, and cause disease. It encompasses several events such as entry into the host by being deposited in a cell by viral vectors such as insects, nematodes etc. that feed on host plants or by mechanical damage to cell wall and plasma membrane. It is followed by replication and assembly of the daughter virus particles at the specific site(s), spread from the site of infection to neighboring healthy cells (cell-to-cell movement) followed by invasion of distal parts of the plant (long-distance movement) (Chapter 7), and disease induction during which specific symptoms are produced.

Symptoms are usually described as local or systemic. Local symptoms appear at the site of inoculation. Local lesions can be chlorotic or necrotic and also range in size from small pinpoint areas to large irregular spreading patches. Chlorotic lesions result from loss of chlorophyll and other pigments (Culver et al., 1991). In most cases chlorotic local lesions are not very distinct in appearance since they are of a slightly paler shade of green or yellow. They also can expand with time, coalesce to yield blotches and hence are difficult to quantitate. In contrast, necrotic local lesions produced in some virus-host combinations are often very distinct, easy to visualize, and are not considered to be economically significant but are important for biological assays. Local lesion hosts offer several advantages to plant virologists. They are useful in separating viruses from mixed infections. Since the number of local lesions is directly proportional to infectivity of a given inoculum, the relative infectivity of various mutants can be compared to that of wild type. Most importantly, since each lesion initiates in one or a few cells and typically results from a single infection event, the local lesion assay is analogous to a bacterial plaque assay and therefore useful in isolating specific mutants, revertants or recombinants (Rao et al., 1990).

GENETIC BASIS OF SYMPTOM EXPRESSION

Symptom expression is a complex process and can be influenced by a variety of interactions between gene products of the virus and the host. One or more viral gene products are involved in the induction of symptoms. A list of selected symptom phenotypes controlled by a specific viral gene product for several plant RNA viruses is given in Table 1. Early evidence that specific viral genes are involved in symptom expression came from genetic studies involving pseudorecombinants of

Table 1. Symptom phenotypes specified by viral genes

Virus[a]	Host	Phenotype[b]	Viral gene[c]	References
Monopartite Viruses				
TMV	*N. tabacum*	Yl; Nec	CP	Dawson *et al.*, 1988
	N. tabacum	Chl	CP	Banerjee *et al.*, 1995
	N. sylvestris	HR	CP	Culver and Dawson, 1992
TCV	*N. benthamiana*	Att	CP	Heaton *et al.*, 1991
	A. thaliana	Att	CP	Kong *et al.*, 1997
TYMV	Chinese cabbage	SM	MP	Tsai and Dreher, 1993
Bipartite Viruses				
RRSV	*C. quinoa*	LL	CP	Harrison *et al.*, 1974
RCNMV	*C. quinoa*	LL	CP	Rao and Hiruki, 1987
Tripartite Viruses				
BMV	*C. hybridum*	LL	CP	Bancroft and Lane, 1973
	C. quinoa	LL	CP	Rao and Grantham, 1995b
	N. benthamiana	VChl	MP	Rao and Grantham, 1995a
AlMV	*N. tabacum*	Nec	CP	van der Vossen *et al.*, 1994
CMV	*N .tabacum*	Chl	CP	Shintaku *et al.*, 1992;
				Suzuki *et al.*, 1995
	G. globosa	LL	CP	Rao and Francki, 1982
	N. edwardsonii	Nec	Rep	Rao and Francki, 1982
	Cucurbita pepo	Ra	Rep	Roossinck and Palukaitis, 1990
	Pea and cowpea	NLL	Rep	Edwards *et al.*, 1983

[a] AlMV=alfalfa mosaic virus; BMV= brome mosaic virus; CMV= cucumber mosaic virus; RCNMV=red clover necrotic mosaic virus; RRSV=raspberry ring spot virus; TCV= turnip crinkle virus; TMV= tobacco mosaic virus; TYMV= turnip yellow mosaic virus
[b] Att, attenuation; Chl, chlorosis; HR, hypersensitive response; LL, local lesion; Nec, necrosis; NLL, necrotic local lesions; Ra, rapid appearance of symptoms; SM, severe mosaic; VChl, vein chlorosis; Yl, yellowing.
[c] CP= Coat Protein; MP= movement protein; Rep= replicase protein

multicomponent viruses. Such studies helped to characterize specific functions dictated by each RNA component either individually or collectively in virus-host interactions. Bancroft (1972) constructed the first pseudorecombinant virus by mixing RNAs of two bromoviruses - RNAs 1 and 2 of brome mosaic virus (BMV) and RNA3 of cowpea chlorotic mottle virus (CCMV). This pseudorecombinant virus exhibited the antigenic property of CCMV, indicating that coat protein gene is located on RNA3 while its infectivity assays on *Chenopodium quinoa* revealed that lesion morphology is also similar to that of CCMV. By contrast, inoculation of the same pseudorecombinant to *C. hybridum* resulted in the induction of small necrotic local lesions distinct from the parents. Additional pseudorecombinants constructed between BMV and CCMV mapped the symptom phenotype in *C. hybridum* to the CP gene (Table 1; Bancroft and Lane, 1973).

Following the pseudorecombinant experiments of Bancroft and Lane (1973), a number of workers constructed pseudorecombinants either between different strains of cucumber mosaic cucumovirus (CMV) or between two distinct cucumoviruses, CMV and tomato aspermy viruses (TAV). A pseudorecombinant constructed from RNA1 and 2 of CMV strain Q and RNA3 of TAV strain V was serologically indistinguishable from VTAV and produced symptoms on cucumber and several *Nicotiana* spp. characteristic of VTAV (Habili and Francki, 1974; Mossop and Francki, 1977). Thus it was suggested that, RNA1and/or RNA2 carried genes determining host reactions. Mossop and Francki (1977) observed that the symptoms produced on a range of plant hosts by pseudorecombinants constructed between CMV strains Q and M were due to the genetic information located on RNA1 and/or RNA2, as well as RNA3.

By using highly purified genomic RNAs of CMV, Rao and Francki (1982) constructed a total of eighteen pseudorecombinants in all possible combinations between pairs of three CMV strains that differ significantly in host range and symptom expression. It was observed that symptom expression of CMV in plant is a complex interaction of the genetic material of the virus with that of the host genome. Some host reactions are controlled by a single RNA species. For example, the systemic necrosis in *Nicotiana edwardsonii*, ability to infect maize systemically and leaf blistering and distortion in several host species are determined by RNA2 alone. Similarly the production of chlorotic local lesions in *Gomphrena globosa* and brown necrotic local lesions in *Vicia faba* are determined by RNA3 alone. Some host reactions were determined by a combination of RNA2 and RNA3 while in some instances the symptom expression appears to involve interaction between genetic information on any two or perhaps even all three RNA segments of the virus (Rao and Francki, 1982).

MOLECULAR BIOLOGY OF SYMPTOM EXPRESSION

Role of Viral Coat Protein

As discussed above, pseudorecombinant genetic studies were only possible with viruses containing divided genomes and therefore the functions of viral gene products encoded by the monopartite viruses in various virus-host interactions remained obscure. However, the advent of recombinant DNA technology has provided many new approaches and now it is relatively simple to manipulate and analyze RNA genomes of any size to determine the functions of their gene products in virus-host interactions. By applying these techniques in several viral systems such as tobamo-, bromo-, cucumo-, and carmoviruses, viral CPs were shown to be intimately associated with a wide spectrum of disease symptoms in plants (Culver *et al.*, 1991; Dawson,1992; Heaton *et al.*,1991; Kong *et al.*, 1997; Rao and Grantham, 1995b, 1996; Schmitz and Rao, 1998; Shintaku *et al.*, 1992). Inoculation of several mutant transcripts of TMV with defined mutations or deletions in the CP gene exhibited varying degrees of symptom phenotypes in tobacco (Dawson *et al.*, 1988; Banerjee *et al.*, 1995).

By contrast, an incompatible interaction between the virus and the host is manifested by the development of necrotic local lesions at the site of infection. This reaction is often referred to as a hypersensitive response (HR). It is generally accepted that in an HR the plant recognizes some product of the pathogen as an "elicitor" and turns on a cascade of defense reactions (Keen, 1990). The HR conferred by the *N'* gene of *Nicotiana sylvestris* and directed against most tobamoviruses is one of the better studied HR systems (Culver *et al.*, 1991). In TMV, specific amino acid substitutions within the virus CP were responsible for host recognition and HR elicitation (Knorr and Dawson, 1988; Culver and Dawson, 1989). Furthermore, the integration and expression of HR-eliciting CP open reading frames into *N. sylvestris* resulted in transgenic plants that displayed HR phenotypes (Culver and Dawson, 1991). It was also found that structural alterations that affect the stability of the CP quaternary structure but not tertiary structure lead to host cell recognition and HR elicitation (Culver *et al.*, 1994).

Mutational analysis of CP genes of cucumo- and bromoviruses specifically identified regions of CP that affect symptom expression. For example, using a chimeric RNA3 assembled from parts of M-CMV RNA3 (which induces chlorosis on tobacco; Mossop *et al.*, 1976) and Fny-CMV RNA3 (which induces green mosaic on tobacco), the chlorosis inducing phenotype was mapped to the CP gene of M-CMV (Shintaku and Palukaitis, 1990; Shintaku, 1991; Shintaku *et al.*, 1992). Additional experiments involving site-directed mutagenesis of the CP gene of CMV further delineated that single amino acid changes are sufficient to alter symptom phenotype and host range (Shintaku, 1991; Shintaku *et al.*, 1992). Comparison of nucleotide and amino acid sequences of four chlorosis inducing CMV strains with those of four green mosaic inducing CMV strains, showed that the later strains contain a proline at CP amino acid 129, whereas the former strains contain either a leucine or serine at this position (Palukaitis *et al.*, 1992). It was subsequently demonstrated that chlorosis is induced by the CP and not the nucleic acid sequence of the CP gene.

The CP ORF of BMV, in addition to the normal initiating codon (IAUG), also contains a second AUG codon (IIAUG) separated by seven amino acids (Fig.1). In a normal wild type BMV infection, most of the wild type mature CP (CP1) is translated from the normal initiating AUG codon (IAUG) of the subgenomic RNA4. In addition, detectable amount of a truncated CP component (CP2), accumulating to approximately 4% of total CP (Sacher and Ahlquist, 1989), is also translated because ribosomes can bypass the first AUG of RNA4 and initiate at the second AUG codon (Rao and Grantham, 1995b; Sacher and Ahlquist, 1989). Deletion of amino terminal residues 1 to 7 from the mature BMV CP has no deleterious effect on encapsidation, long distance movement and symptom phenotype in barley plants (Flasinski, *et al.*, 1995; Rao and Grantham, 1995b; Sacher and Ahlquist,1989). To unravel the biological significance of this region of the BMV, Rao and Grantham (1995b) engineered changes into this gene such that the subgenomic RNA4 resulting from each RNA3 variant infection would produce either CP1 (i.e. CP translated from the first AUG codon only) or CP2 (i.e. CP lacking the first 7 amino terminal residues due to initiation of translation at the second AUG codon only) but not both. Thus, mutant B3IIAUA was constructed to yield only CP1 and mutants B3IAUA and B3ΔCP7 were constructed to yield only CP2. In mutant B3IIAUA, the second AUG codon was modified (^{1275}AUG^{1277} → ^{1275}AUA^{1277}) in order to prevent

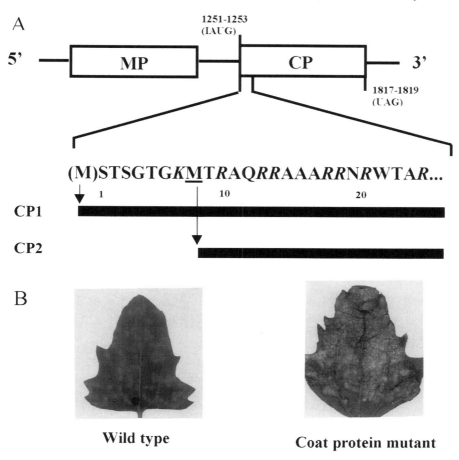

Figure 1. (A) The structure of BMV RNA3 is shown, with noncoding sequences represented as single lines and movement protein (MP) and coat protein (CP) as open boxes. The positions of initiating methionine (IAUG), 2nd methionine (IIAUG) and the CP termination codon (UAG) are shown. The first 25 basic N-terminal amino acids of BMV CP are shown. Positively charged side chains that are also trypsin cleavage sites are italicized. In bromoviruses, the initiating methionine (enclosed in parentheses) is removed and the resultant N-terminal serine is acetylated in the mature CP. The second methionine located at position 8 is underlined. In BMV wild type infections, 96% of the wild type CP (CP1) is initiated from the first methionine whereas 4% of a truncated CP (CP2) is initiated at second methionine due to leaky translational event (Rao and Grantham, 1995a; Sacher and Ahlquist, 1989). (B) Symptom phenotypes induced in *C. quinoa* by wild-type BMV and a variant harboring a deletion in the N-terminal basic amino acid region of the CP.

translation of CP2 from the second AUG codon of subgenomic RNA4. In B3IAUA, the first initiation AUG codon was altered from $^{1251}AUG^{1253} \rightarrow {}^{1251}AUA^{1253}$. In mutant B3ΔCP7, the 24 base sequence encompassing the first AUG codon and the first 7 basic amino terminal amino acids was deleted, fusing the second AUG codon to the 5' untranslated leader sequence. Barley plants inoculated with B3IAUA, B3Δ CP7 and B3IIAUA displayed characteristic mosaic symptoms on uninoculated young leaves on a time scale similar to the wild type. Inoculation of *C. hybridum* (purple) and *C. amaranticolor* with all three RNA3 variants induced characteristic necrotic lesions indistinguishable from those resulting from inoculation with wild type B3 (Rao and Grantham, 1995b). In marked contrast to the chlorotic expanding local lesions induced by wt BMV RNA3 in *C. quinoa*, variants expressing CP2 only (i.e. B3IAUA and B3ΔCP7) changed the phenotype to necrotic local lesions (Fig. 1). Additional mutational analysis of BMV CP N-terminus revealed that deletion of 11, 14 and 18 N-terminal amino acids also resulted in the induction of necrotic local lesions in *C. quinoa* (Rao and Grantham, 1996). Interestingly none of these mutants were able to infect barley. Based on these observations it was concluded BMV CP plays an important role in virus-host interactions and contributes differently to the virulence phenotype in different host plants.

Role of Movement Protein

Many plant viruses, if not all, encode non-structural movement proteins (MPs). The major function of this protein is to mediate cell-to-cell movement of the virus since inactivation of this protein results in subliminal infections with no spread beyond initially infected cells (Deom *et al.*, 1992; Schmitz and Rao, 1996; Rao, 1997). However, recent experimental evidence indicates that the MP can also modulate symptom expression (Table 1). For example, BMV causes symptomless infections in *N. benthamiana*. Rao and Grantham (1995a) characterized a variant of BMV, spontaneously generated under greenhouse conditions, capable of inducing mild vein chlorosis. Pseudorecombinants constructed by exchanging RNAs1 and 2 and RNA3 components between symptomless and vein chlorosis inducing strains of BMV indicated that the genetic determinant for vein chlorosis symptom phenotype in *N. benthamiana* is located on RNA3. Sequence analysis of progeny RNA3 recovered from symptomatic plants revealed that induction of vein chlorosis is due to a single nucleotide transition, which changes the codon Val-266 to Ile-266. Likewise, Fujita *et al.*, (1996) also observed that a mutation in the MP gene of a bromovirus hybrid RNA3 not only confers compatibility with a new host but also exacerbates symptom expression. Similarly, a single mutation in the MP gene of TYMV also resulted in increased symptom severity (Tsai and Dreher, 1993).

Role of Viral Replicases

Although the major function of the viral replicases is to amplify the genomic RNAs, several secondary characteristics have been associated with these proteins. For example, a mutant of TMV that produced symptomless infection in tobacco was found to have a single nucleotide change resulting in an amino acid substitution in the 126/183 kDa proteins (Lewandowski and Dawson, 1993). At least one other function that is specified by viral replicases is its involvement in cell-to-cell movement. For example, selected BMV variants with mutations in the 2a replicase gene replicated to near wild type levels in protoplasts but failed to move efficiently in whole plants (Traynor *et al.*, 1991).

Other Factors

Apart from viral genes some other factors that affect symptom expression are the presence of satellite RNAs found to be associated with some plant viruses. For example some CMV strains, in addition to genomic and subgenomic RNAs, also package an additional RNA component, often referred to as satellite RNA (sat-RNA). sat-RNA can replicate only in the presence of its helper CMV and is packaged by CMV CP. However, the presence of sat-RNA can affect CMV replication and symptom expression. A detailed description of how CMV sat-RNA can affect symptom expression in various hosts can be found in a review article of Palukaitis *et al.,* (1991). Likewise, sat-RNAs associated with turnip crinkle carmovirus (Oh *et al.*, 1995) and TMV have also been shown to modulate symptom expression (Routh *et al.*, 1995).

CONCLUSIONS

Despite significant technical advances made in manipulating the genomes of viruses and their hosts, our knowledge in understanding how viruses interact with their hosts and cause disease is still in infancy. Majority of viral genomes, because of their small size, have been completely sequenced and the function of their gene products dissected but that of plants remain obscured. Plant virologists are optimistic in fulfilling this phase of research by using much simpler hosts such as *Arabidopsis* as a model system.

REFERENCES

Bancroft, J. B. (1972). A virus made from parts of the genomes of brome mosaic and cowpea chlorotic mottle viruses. *J. Gen. Virol.* 14, 223-228.

Bancroft, J. B., and Lane, L. (1973). Genetic analysis of cowpea chlorotic mottle and brome mosaic viruses.*J. Gen. Virol.* 19, 381-389.

Banerjee, N., Wang, J. Y., and Zaitlin, M. (1995). A single nucleotide change in the coat protein gene of tobacco mosaic virus is involved in the induction of severe chlorosis. *Virology* 207, 234-239.

Cronin, S., Verchot, J., Haldeman-Cahil, R., Schaad, M. C., and Carrington, J. C. (1995). Long distance movement factor: A transport function of the potyvirus helper component proteinase. *Plant J.* 7, 549-559.

Culver, J. N., and Dawson, W. O. (1989). Tobacco mosaic virus coat protein: An elicitor of the hypersensitive reaction but not required for the development of mosaic symptoms in *Nicotiana sylvestris*. *Virology* 173, 755-758.

Culver, J. N., and Dawson, W. O. (1991). Tobacco mosaic virus elicitor coat protein genes produce hypersensitive phenotype in transgenic *Nicotiana sylvestris*. *Mol. Plant-Microbe Interact.* 2. 209-213.

Culver, J. N., Lindbeck, A. G. C., and Dawson, W. O. (1991). Virus-host interactions: Identifiation of chlorotic and necrotic responses in plants by tobamoviruses. *Annu. Rev. Phytopathol..* 29, 193-217.

Culver, J. N., Stubbs, G., and Dawson, W. O. (1994). Structure-function relationship between tobacco mosaic virus coat protein and hypersensitivity in *Nicotiana sylvestris*. *J. Mol. Biol.* 242, 130-138.

Dawson, W. O. (1992). Tobamovirus-plant interactions. *Virology* 186, 359-367.

Dawson, W. O., Bubrick, P., and Grantham, G. (1988). Modifications of the tobacco mosaic virus coat protein gene affecting replication, movement and symptomatology. *Phytopathology* 78, 783-789.

Deom, C. M., Lapidot, M., Beachy, R. N. (1992). Plant virus movement proteins. *Cell* 69, 221- 224.

Edwards, M. C., Gonsalves, D., and Provvidenti, R. (1982). Genetic analysis of cucumber mosaic virus in relation to host resistance: Location of determinants for pathogenicity to certain legumes and *Lactuca saligna*. *Phytopathology* 73, 269-273.

Flasinski, S., Dzianott, A., Pratt, S., and Bujarski, J. (1995). Mutational analysis of the coat protein gene of brome mosaic virus: Effects on replication and movement in barley and in *Chenopodium hybridum*. *Mol. Plant-Microbe Interact.* 8, 23-31.

Fujita, Y., Mise, K., Okuno, T., Ahlquist, P., and Furusawa, I. (1996). A single codon change in a conserved motif of a bromovirus movement protein gene confers compatibility with a new host. *Virology* 223, 283-291.

Gal-On, A., Kaplan, I., Roossinck., M. J., and Palukaitis, P. (1994). The kinetics of infection of zucchini squash by cucumber mosaic virus indicates a function for RNA1 in virus movement. *Virology* 205, 280-289.

Habili, N., and Francki, R. I. B. (1974). Comparative studies on tomato aspermy and cucumber mosaic viruses. III. Further studies on relationship and construction of a virus from parts of the two viral genomes. *Virology* 61, 443-449.

Harrison, B. D., Murant, A. F., Mayo, M. A., and Roberts, I. M. (1974). Distribution of determinants symptom production, host range and nematode transmissibility between the two RNA components of raspberry ringspot virus. *J. Gen. Virol.* 22, 233-247.

Heaton, L. A., and Laakso, M. M. (1995). Several symptom modulating mutations in the coat protein turnip crinkle carmovirus result in the particle with aberrant conformational properties. *J. Gen. Virol.* 76, 225-230.

Heaton, L. A., Lee, T. C., Wei, N., and Morris, T. J. (1991). Point mutations in the turnip crinkle virus capsid protein affect the symptoms expressed by *Nicotiana benthamiana*. *Virology* 183, 143-150.

Kong, Q., Oh, J. W., Carpenter, C. D., and Simon, A. E. (1997). The coat protein of turnip crinkle virus is involved in subviral RNA-mediated symptom modulation and accumulation. *Virology* 238, 478-485.

Kasteel, D. T. J., Van der Wel N., Jansen, K. A. J., Goldbach, R. W., and Van Lent, J. W. M. (1997). Tubule-forming capacity of the movement proteins of alfalfa mosaic virus and brome mosaic virus. *J.Gen. Virol.* 78, 2089-2093.

Keen, N. T. (1990). Gene-for-gene complementary in plant pathogens and symbionts. *Annu. Rev. Genet.* 24, 447-463.

Knorr, D. A. and Dawson, W. O. (1988). A point mutation in the tobacco mosaic virus capsid protein gene induces hypersensitivity in *Nicotiana sylvestris*. *Proc. Natl. Acda. Sci. USA* 85, 170-174.

Lewandowski, D. J and Dawson, W. O. (1993). A single amino acid change in tobacco mosaic virus replicase prevents symptom production. *Mol. Plant Microbe. Interact.* 6, 157-160.

Mise, K. and Ahlquist, P. (1995). Host specificity restriction by bromovirus cell-to-cell movement protein occurs after initial cell-to-cell spread of infection in nonhost plants. *Virology* 206, 276-286.

Mise, K., Allison, R. F., Janda, M., and Ahlquist, P. (1993). Bromovirus movement protein genes play a crucial role in host specificity. *J. Virol.* 67, 2815-2823.

Mossop, D. W., and Francki, R. I. B. (1977). Association of RNA3 with aphid transmission of cucumber mosaic virus. *Virology* 81, 177-181.

Mossop, D. W., Francki, R. I. B., and Grivell, C. J. (1976). Comparative studies on tomato aspermy and cucumber mosaic viruses. V. Purification and properties of a cucumber mosaic virus inducing severe chlorosis. *Virology* 74, 544-546.

Oh, J. W., Kong, Q., Song, S., Carpenter, C. D., Simon. A. E. (1995). Open reading frames of turnip crinkle virus involved in satellite symptom expression and incompatibility with *Arabidopsis thalianas* ecotype Dijon. *Mol. Plant-Microbe Interact.* 8, 979-987.

Osman, F., Grantham, G. L., and Rao, A. L. N. (1997). Molecular studies on bromovirus capsid protein. IV. Coat protein exchanges between brome mosaic and cowpea chlorotic mottle viruses exhibit neutral effects in heterologous hosts. *Virology* 238, 452-459.

Palukaitis, P., Roossnick, M. J., Shintaku, M. H., and Sleat, D. E. (1991). Mapping functional domains in cucumber mosaic virus and its satellite RNAs. *Can. J. Plant Pathol.* 13, 155-162.

Rao, A. L. N. (1997). Molecular studies on bromovirus capsid protein: III. Analysis of cell-to-cell movement competence of coat protein defective variants of cowpea chlorotic mottle virus. *Virology*, 385-395.

Rao, A. L. N. and Francki, R. I. B. (1981). Comparative studies on tomato aspermy and cucumber mosaic viruses. VI. Partial compatibility of genome segments from the two viruses. *Virology* 114, 573-575.

Rao, A. L. N. and Francki, R. I. B. (1982). Distribution of determinant for symptom production and host range on the three RNA components of cucumber mosaic virus. *J. Gen. Virol.* 61, 197-205.

Rao, A. L. N. and Hiruki, C. (1987). Unilateral compatibility of genome segments from two distinct strains of red clover necrotic mosaic virus. *J. Gen. Virol.* 68, 191-194.

Rao, A. L. N., and Grantham, G. L. (1995a). A spontaneous mutation in the movement protein gene of brome mosaic virus modulates symptom phenotype in *Nicotiana benthamiana. J. Virol.* 69, 2689-2691.

Rao, A. L. N., and Grantham, G. L. (1995b). Biological significance of the seven amino-terminal basic residues of brome mosaic virus coat protein. *Virology* 211, 42-52.

Rao, A. L. N., and Grantham, G. L. (1996). Molecular studies on bromovirus capsid protein: II. Functional analysis of the amino terminal arginine rich motif and its role in encapsidation, movement and pathology. *Virology* 226, 294-305.

Rao, A. L. N., Sullavan, B., and Hall, T. C (1990). Use of *Chenopodium hybridum* facilitates isolation of brome mosaic virus RNA recombinants. *J. Gen. Virol.* 71, 1403-1407.

Roossnick, M. J., and Palukaitis, P. (1990). Rapid induction and severity of symptoms in zucchini squash (*Cucurbita pepo*) map to RNA1 of cucumber mosaic virus. *Mol. Plant-Microbe Interact.* 3, 188-192.

Routh, G., Dodds, A. J., Fitzmaurice, L., and Mirkov, T. E. (1995). Characterization of deletion and frameshift mutants of satellite tobacco mosaic virus. *Virology* 212, 121-127.

Sacher, R., and Ahlquist, P. (1989). Effects of deletions in the N-terminal basic arm of brome mosaic virus coat protein on RNA packaging and systemic infection. *J. Virol.* 63, 4545-4552.

Schmitz, I., and Rao, A. L. N. (1996). Molecular studies on bromovirus capsid protein. I. Characterization of cell-to-cell movement-defective RNA3 variants of brome mosaic virus. *Virology* 226, 281-293.

Schmitz, I., and Rao, A. L. N. (1998). Deletions in the conserved amino-terminal basic arm of cucumber mosaic virus coat protein disrupt virion assembly but do not abolish infectivity and cell-to-cell movement. *Virology* 248, 323-331.

Scholthof, H. B., Scholthof, K.-B.G, Kikkert, M., and Jackson, A. O. (1995). Tomato bushy stunt virus spread is regulated by two nested genes that function in cell-to-cell movement and host-dependent systemic invasion. *Virology* 213, 425-438.

Shintaku, M. H. (1991). Coat protein gene sequence of two cucumber mosaic virus strains reveal a single amino acid change correlating with chlorosis induction. *J. Gen. Virol.* 72, 2587-2589.

Shintaku, M. H., and Palukaitis, P. (1990). Mapping determinants of pathogenicity and transmission of cucumber mosaic virus. *Phatopathology* 80. 1035.

Shintaku, M. H., Zhang, L., and Palukaitis, P. (1992). A single amino acid substitution in the coat protein of cucumber mosaic virus induces chlorosis in tobacco. *Plant Cell* 4, 751-757

Suzuki, M., Kuwata, S., Masuta, C., and Takanami, Y. (1995). Point mutations in the coat protein of cucumber mosaic virus affect symptom expression and virion accumulation in tobacco. *J. Gen. Virol.* 76, 1791-1799.

Traynor, P., Young, B. M., and Ahlquist, P. (1991). Deletion analysis of brome mosaic virus 2a protein: Effects on RNA replication and systemic spread. *J. Virol.* 65, 2807-2815.

Tsai, C.-H., and Dreher, T. W. (1993). Increased viral yield and symptom severity result from a single amino acid substitution in the turnip yellow mosaic virus movement protein. *Mol. Plant-Microbe Interact.* 6, 268-273.

van der Vossen, E. A. G., Neeleman, L., and Bol, J. F. (1994). Early and late functions of alfalfa mosaic virus coat protein can be mutated separately. *Virology* 202, 891-903.

10 GENE-FOR-GENE INTERACTIONS

Christopher D. Dardick and James N. Culver

INTRODUCTION

Traditionally, measures to prevent viral diseases have included breeding strategies to introduce natural forms of plant resistance from wild relatives into economically important crops. Thus, many crop plants have been selectively bred to incorporate specific resistance genes that target a wide variety of pathogens (Fraser, 1990). One shortcoming of this practice is the inability to transfer resistance across species barriers. Recent advances in biotechnology are overcoming this limitation as modern molecular techniques now provide the means to transfer resistance from one plant species to another (Whitham *et al.,*1996). However, as new resistance genes are introduced into plants, pathogens may evolve to overcome resistance. Fraser and Gerwitz (1987) examined over 50 virus-host interactions where resistance genes have been identified. They showed that fewer than 10% of these genes remained effective against long-term exposure to multiple virus strains. Understanding the molecular interactions between pathogens and the plant genes controlling resistance will allow for the development of new and better approaches to providing more effective long-term protection.

One of the most effective ways in which plants resist pathogen infection is through induction of the hypersensitive response (HR). The HR is an active defense mechanism that plants employ to prevent the spread of viral, bacterial, fungal, and nematode pathogens. An important feature of the HR is that it is a generalized response. Despite the different characteristics of the various types of pathogens, the same set of biochemical responses ensue: production of pathogenesis-related proteins, hydrolytic enzymes, callose and lignin precursors, oxidative bursts of H_2O_2, activation of systemic acquired resistance, etc. (Baker *et al.,* 1997). These responses act in concert to restrict pathogen infection and prevent systemic disease. The end result of HR induction is localized cell death and necrosis at site of pathogen infection (Fig. 1).

In contrast to the generalized defense responses, induction of this phenomenon occurs in a highly specific manner. The ability of a plant to respond to the presence of a particular pathogen and initiate defense mechanisms implies that there is a specific recognition event between plant and the invading pathogen. Observations by Flor (1971) that resistance segregated with single dominant loci in plants and single dominant loci in pathogens provided the framework for the current model termed the gene-for-gene hypothesis. This hypothesis suggests that plants carry specific resistance (*R*) genes, the products of which directly or indirectly interact with the products of pathogen encoded avirulence (*avr*) genes leading to induction of the HR (Keen, 1990).

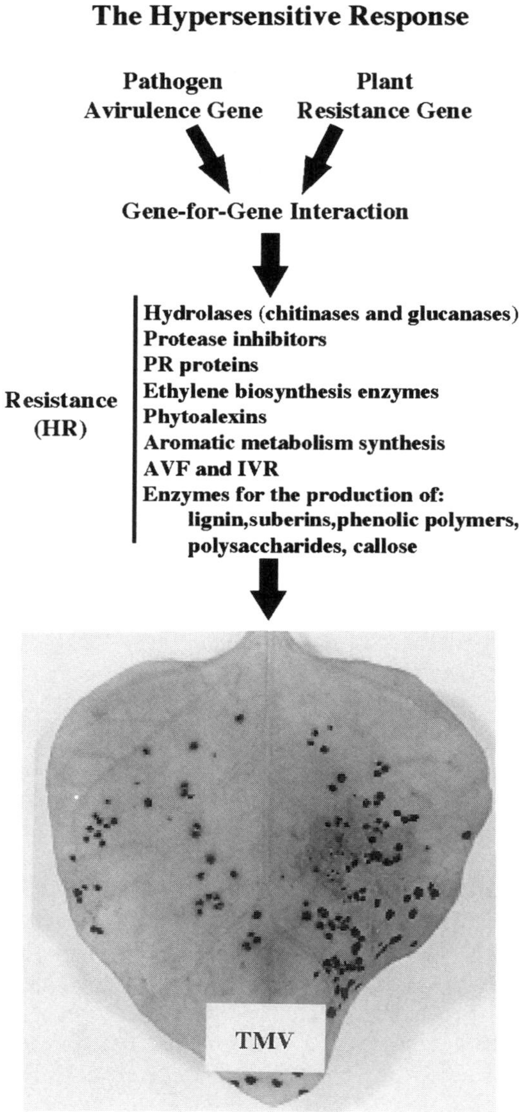

The Hypersensitive Response

Figure 1 Diagram outlining the steps leading to induction of the hypersensitive response and some of the biochemical responses involved. Leaf photograph shows an eggplant leaf that has been mechanically inoculated with TMV. Dark spots are local lesions resulting from HR induction.

In recent years, significant support for this model has come from the application of molecular techniques to the study of the plant-pathogen interactions. To date, a handful of *R* genes and their corresponding *avr* genes have been cloned and characterized, however, identification of genes involved in plant-virus interactions has been limited. In this chapter, we will summarize the current information regarding the host genes that function in the pathogen recognition and the virus factors involved in eliciting host responses.

HOST FACTORS

The development of transposon tagging as well as map based cloning strategies has allowed for the cloning and characterization of *R* genes from five different plant species that provide resistance against viral, bacterial, fungal, and nematode pathogens. Interestingly, many *R* genes encode proteins with similar features. Table 1 shows a partial list of identified *R* genes and their characteristic motifs.

Table 1. Identified *R* genes share similar features

R Gene	Motifs	Plant Host	Pathogen	Reference
N	LRR, NBS, Toll /IL-1R	tobacco	Viral	Whitham *et al.*, 1994
Rps2	LRR, NBS, leucine-zipper	*A. thaliana**	Bacterial	Mindrinos *et al.*, 1994
Rpm1	LRR, NBS, leucine-zipper	*A. thaliana*	Bacterial	Grant *et al.*, 1995
Rpp5	LRR, NBS, Toll/IL-1R	*A. thaliana*	Fungal	Parker *et al.*, 1994
L⁶	LRR, NBS, Toll/IL-1R	flax	Fungal	Lawrence *et al.*, 1995
Cf-9	LRR, mem-brane anchor	tomato	Fungal	Jones *et al.*, 1994
Xa21	LRR, trans-membrane domain, protein kinase	rice	Bacterial	Song *et al.*, 1995
Prf	LRR, NBS, leucine-zipper	tomato	Bacterial	Salmeron *et al.*, 1996
Pto	protein kinase	tomato	Bacterial	Martin *et al.*, 1993

**Arabidopsis thaliana*

The most common feature found in all *R* genes with one exception, described later, is a region of leucine rich repeats (LRR). LRR motifs are found in a vast array of proteins from both plants and animals and consist of tandem amino acid repeats, 20-29 residues in length, that have a high proportion of leucines. Although the role of LRR motifs is not entirely clear, nearly all LRR containing proteins are involved in protein-protein interactions and many of these proteins are known to function in signal transduction pathways (Kobe and Deisenhofer, 1995). Therefore, it is feasible that the LRRs are directly involved in pathogen recognition by *R* genes. However, there is currently no direct evidence to confirm this possibility. Other features that are common to *R* genes include nucleotide binding sites (NBS), domains similar to *Drosophila* Toll and Interleukin-1 Receptor (IL-1R) from mammals, leucine zippers, membrane anchoring domains, and protein kinases. The presence or absence of these features differentiates four distinct classes of *R* genes (Dangl and Holub, 1997).

Characterization of *R* Genes

The vast majority of identified *R* genes are designated as the NB-LRR (nucleotide binding site - leucine rich repeat) type. This group includes genes from rice, tobacco, flax, tomato, and *Arabidopsis thaliana* and provide resistance against viral, bacterial, fungal, and nematode pathogens. This group also includes the only viral *R* gene, the *N* gene from *Nicotiana glutinosa,* that has been cloned and characterized to date. The *N* gene confers resistance to tobacco mosaic tobamovirus (TMV) and contains a LRR domain, nucleotide binding site and a domain similar to *Drosophila* Toll and mammalian IL-1R (Whitham *et al.,*1994). Additional examples in the NB-LRR class include the bacterial resistance gene *RPS2* from *A. thaliana* (Mindrinos *et al.,* 1994) as well as the rust resistance gene *L6* from flax (Lawrence *et al.,* 1995). These genes contain a leucine zipper motif in place of the Toll/IL-1R motif that is found in the *N* gene. Currently, little is known about the functions of these different domains, however, it has been shown that mutations in the nucleotide binding site as well as the LRR can disrupt the resistance conferred by *N* (Baker *et al.,* 1997) Because *R* genes share similar features required for gene function, it is apparent that a common molecular mechanism exists by which plants recognize and respond to pathogens.

The *R* gene *Cf-9* from tomato conferring resistance to the fungal pathogen *Cladosporium fulvum* represents a second class of LRR containing *R* genes. *Cf-9* consists of an extracellular LRR domain along with a C-terminal membrane anchor, suggesting that *Cf-9* responds to an extracellular signal (Jones *et al.,* 1994). This is in contrast to *N* and *RPS2* which are predicted to encode cytoplasmic proteins (Mindrinos *et al.,*1994; Whitham *et al.,* 1994). Because *Cf-9* contains no apparent signaling domains, it is expected that *Cf-9* protein interacts with other proteins in order to initiate defense response (Jones *et al.,*1994).

A third class is represented by the rice gene *Xa21* that confers resistance against the bacterial pathogen *Xanthomonas oryzae.* The protein encoded by *Xa21* contains extracellular LRR motifs (like Cf-9), a membrane spanning domain, and a protein kinase domain (Song *et al.,* 1995). A model for *Xa21* mediated resistance has been proposed in which the extracellular LRR domain interacts with the product of the bacterial *avr* gene resulting in the activation of an intracellular kinase domain capable of initiating a signal transduction cascade. Although the structure of *Xa21* lends support to this model, the role of both the LRR and the kinase domain in establishing a resistance response has yet to be determined.

The *R* gene *Pto* and the Fenthion sensitivity gene *Fen* (application of the insecticide Fenthion on *Fen* gene-containing tomato plants results in an HR-like phenotype) are the only known *R* genes that do not encode LRR domains. Both *Pto* and *Fen* encode serine/threonine kinases that contain no apparent additional domains (Martin *et al.,* 1993, 1994). *Pto* is the only *R* gene in which physical interaction with the corresponding *avr* gene has been confirmed. This was shown by Scofield *et al.* (1996) and Tang *et al.* (1996) via the yeast two-hybrid protein-protein interaction trap. Within this system, Martin and colleagues have partially characterized the gene-for-gene pathway that leads to HR induction. The physical interaction between *Pto* and *avrPto* results in the activation of the *Pto* kinase domain. Once active, *Pto* is then capable of phosphorylating a second kinase designated *Pti1* (*Pto* interacting) as well as a set of transcription factors designated *Pti* 5 and 6 that are thought to specifically activate pathogenesis-related (PR) defense genes (Zhou *et al.,* 1997). Thus, in this system, it appears that a specific interaction between an *R* gene product and its corresponding pathogen elicitor leads to initiation of plant defense responses. However, this model is still incomplete in that the HR induced by both the *Pto* and *Fen* genes requires the presence of an additional gene, *Prf. Prf* falls into NB-LRR class of *R* genes and its role in *Pto-* and *Fen*-mediated resistance is not yet known (Salmeron *et al.,* 1996).

Although the structure of *R* genes provide compelling models for plant-pathogen recognition, further insights are needed to elucidate how these genes function in conferring resistance. The role of LRR motifs in pathogen recognition as well as how recognition leads to resistance are still not understood. However, because of the close similarities between known *R* genes, it seems likely that plants have common molecular mechanisms that provide resistance to a broad range of pathogens.

Similarities to Pathogen Resistance in Animals

The finding that the tobacco *N* gene and the flax L^6 gene encode domains similar to the cytoplasmic domains of *Drosophila* Toll and IL-1R receptors in animals has fueled speculation that plants and animals share related resistance mechanisms. IL-1R pathways have been characterized in a number a different animal systems including human, mouse, murine, and chicken (Sims *et al.,* 1989). Upon interaction with the cytokine IL-1, the IL-1R activates a transcription factor, NF-kB, that in turn stimulates the production of an assortment of defense-related proteins (Mittler and Lam, 1996). Likewise, the Toll protein has been implicated in a similar signal transduction cascade in *Drosophila* (Heguy *et al.,* 1992). Thus, plant *R* genes may function in a similar manner to initiate pathogen defense responses. Interactions with pathogen *avr* gene products may result in the stimulation of transcription factors that specifically activate defense-related genes (as described above for *Pto*). In addition, the physiological aspects of HR cell death show significant similarities to apoptosis in animals. Fragmentation of nuclear DNA, activation of nucleases, condensation of cytoplasm, and other physiological responses are characteristics shared by programmed cell death in both plants and animals (Mittler and Lam, 1996). Although HR cell death and apoptosis in animals have a number of distinguishing features, a growing body of evidence is revealing that these mechanisms may be related.

Organization of Pathogen Recognition Loci

The genetic dissection and direct sequencing of *R* gene loci has demonstrated that *R* genes are organized as clusters of both functional and non-functional alleles. This type of organization is thought to promote meiotic recombination resulting in duplications and deletions among different alleles. Consequently, the swapping or reorganization of *R* gene domains may allow for the evolution of new pathogen recognition specificities (Dangl and Holub, 1997). Evidence revealing such a mechanism comes from the fungal rust resistance loci *Rp1* in maize within which frequent recombination between alleles was shown to give rise to new resistance profiles (Bennetzen *et al.,* 1988). Furthermore, sequence analysis of a 1.9 megabase region of chromosome 4 from *A. thaliana* revealed the presence of a cluster of eight *R* gene homologues (Bevan *et al.,* 1998). Several of these genes are non-functional in that they either lack an ATG initiator codon, contain frameshifts, or are interrupted by the presence of transposons. The remaining genes were shown to be homologues of known *R* genes that confer resistance against the fungal pathogen *Pernospora parasitica.* Additional putative *R* genes were identified as single copy sequences at other locations including genes similar to the tobacco *N* and *Arabidopsis RPS2* genes previously discussed. In fact, homologues of cloned *R* genes are known to be present in diverse plant species, however, in most cases it is not clear whether or not they are functional. Further sequence analysis and characterization of the *Arabidopsis* genome will shed more light on the precise organization of *R* genes and the role of *R* gene clusters in generating new specificities.

VIRUS FACTORS

The recognition of pathogens by plants is dependent upon the production of pathogen-encoded molecules called elicitors (Keen, 1990). A wide variety of viral proteins are known to function as specific elicitors of the HR (Table 2). These include viral replicase proteins, movement proteins, coat proteins as well as a number of proteins with unknown functions. Thus, it is clear that almost any viral protein could potentially serve as a target for host recognition. A brief description of several specific virus encoded elicitors is given below.

Capsid Protein as a Virulence Factor

Tobamovirus Capsid Proteins as Elicitors of Hypersensitive Response

The best characterized HR elicitor is the CP of TMV. Studies have shown that the TMV CP elicits the HRs conferred by the *N'* gene from *Nicotiana sylvestris* (tobacco), the *L¹*, *L²*, and *L³* genes from *Capsicum* spp. (pepper), as well as a genetically uncharacterized HR in *Solanum melongena* (eggplant) (Table 2) (Culver and Dawson, 1989; Berzal-Herranz *et al.,* 1995; de la Cruz *et al.,* 1997; Dardick *et al.,* 1997). TMV mutants missing a CP initiator codon do not elicit the HR in tobacco, pepper, and eggplant. In contrast, expression of the CP as a transgene results in systemic necrosis in tobacco and, when expressed from a potexvirus vector, a similar response is

Table 2. Viral proteins known to function as hypersensitive response elicitors

Virus	Elicitor	Host	R gene
TMV	Coat protein	Tobacco	N'
		Pepper spp.	L^1, L^2, L^3
		Eggplant	$?^1$
	30kDa cell-to-cell movement protein	Tomato	$Tm-2, Tm-2^2$
	Replicase	Tobacco	N
PVX	Coat protein	Potato	Nx
CaMV	Gene VI	Tobacco, *Datura*	$?^1$
TBSV	P19 and p22	Tobacco spp.	$?^1$

[1] ? indicates that corresponding resistance gene has not been identified

observed in eggplant. These studies have confirmed that the ability of the TMV CP to elicit the HR is dependent only upon expression of the CP and no other virus factors. Additionally, the structural characteristics of CP that are required for HR elicitation in these different hosts have been examined. Amino acid substitutions within the CP that alter its overall three-dimensional structure fail to elicit the HR. Additional substitutions show that residues on the right face of the CP four helical bundle may be directly involved in recognition in tobacco (N' gene) and pepper (L^1 gene) (Taraporewala and Culver, 1996; Dardick and Culver, unpublished data). The characteristics of this region are consistent with other known recognition surfaces (Janin and Chothia, 1990) suggesting that this structural region may serve as a receptor binding site for the product of the host-encoded R gene.

Potexvirus X Capsid Protein as Elicitor of Hypersensitive Response

The CP encoded by potexvirus X (PVX) elicits an HR *in Solanum tuberosum* (potato) conferred by the *Nx* gene. Natural PVX isolates that overcome *Nx* exhibited alterations in their CP sequences (Kavanagh *et al.*, 1992). Moreover, a single amino acid substitution was all that was required for PVX to break *Nx* resistance. Additional experiments demonstrated that host recognition could be restored by the introduction of a second site substitution. This is analogous to the ability of specific TMV CP mutations to confer elicitor activity to non-elicitor CPs. These results further illustrate how viruses can evolve structural changes to overcome host defense mechanisms.

Virulence Generated by Different Virus Functions

Tobacco Mosaic Tobamovirus Movement Protein

Two genes, *Tm-2* and *Tm-2²*, in *Lycopersicon esculentum* (tomato) have been identified that control resistance to tobamoviruses. Tomato cultivars carrying either of these genes respond with the HR when inoculated with TMV. Pathogenic determinants of TMV mutants that overcome *Tm-2* and *Tm-2²* resistance were found to reside within the 30kDa cell-to-cell movement protein (Meshi *et al.,* 1989; Weber *et al.,* 1993). Additionally, transgenic expression of the 30kDa movement protein in *Tm-2²* tomato resulted in a necrotic HR-like phenotype. Thus, movement protein alone is sufficient for HR elicitation in tomato.

Tobacco Mosaic Tobamovirus Replicase

The tobacco *N* gene is one of the most effective and durable *R* genes against tobamoviruses. Only one TMV strain, TMV-Ob, is known to overcome resistance provided by the *N* gene. Through domain swapping experiments, Padgett and Beachy (1993) identified that the HR evading determinant in TMV-Ob resided in the helicase domain of the viral replicase. In later experiments, Padgett *et al.* (1997) expressed portions of the TMV replicase gene using a TMV-Ob-based vector and characterized a region spanning 424 amino acids that was responsible for HR elicitation. However, it is not yet clear whether the replicase protein is all that is required or if other viral proteins or processes are involved in elicitation of the HR. As discussed previously, the *N* gene has been cloned and characterized and the predicted protein encoded by *N* has been proposed to be a cytoplasmic protein (Whitham *et al.,* 1994). This correlates well with the known location of the TMV replicase since viral replication occurs in the cytoplasm. Therefore, it is likely that recognition of viral replicase is a result of interaction with *N*-encoded protein.

Cauliflower Mosaic Caulimovirus (CaMV) Gene VI

CaMV typically infects cruciferous plants, however, a strain has been described that replicates in two Solanaceous species, *Datura stramonium* and *Nicotiana bigelovii*. The D4 strain causes systemic symptoms while ordinary CaMV strains induce the HR. By constructing recombinant viruses between D4 and other CaMV strains, Schoelz *et al.* (1987) identified a 496 base pair region in gene VI that determined the host response. CaMV gene VI has been shown to encode a regulatory protein that functions in posttranscriptional transactivation of the major RNA transcript. The gene VI product is primarily found in inclusion bodies and is a principle determinant of system development (Shaw, 1996). Sequence analysis between CaMV strains revealed 20 amino acid substitutions that could account for the ability of D4 to escape recognition.

Multiple Virus Determinants in Tomato Bushy Stunt Virus

The small icosahedral plant virus, tomato bushy stunt virus (TBSV), has a broad host range and induces a variety of different symptoms in various hosts. The ability of two

TBSV-encoded proteins, p19 and p22, to invoke host responses when expressed from a PVX vector has been characterized by Jackson and colleagues and although specific functions have not been ascribed to these proteins, one or both are thought to be involved in the cell-to-cell movement (Shaw, 1996). Their results demonstrated that PVX-expressed p19 caused systemic necrosis in *Nicotiana benthamiana* and *N. clevelandii* but induced necrotic local lesions in *N. tabacum*. In contrast, PVX derivatives expressing p22 induced necrotic local lesions in *N. glutinosa* and *N. edwardsonii* (Scholthof *et al.*, 1995). The ability of these viral proteins to elicit an HR in closely related plant species suggests that these plants contain different *R* genes that respond to distinct TBSV proteins.

PLANT VIRUSES EVOLVE TO EVADE HOST RECOGNITION

The lack of *R* gene durability over long term exposure to pathogens implies that pathogens evolve to avoid host recognition. The finding that single amino acid substitutions within elicitors can drastically effect host responses suggests that small changes in a pathogen's genome are all that are required for the organism to regain its virulence in a particular host. The ability of the PVX CP to elicit *Nx*-mediated resistance provides one example. Cruz *et al.* (1993) demonstrated that a single amino acid exchange in the PVX CP was the only change necessary to confer a resistance breaking phenotype. Furthermore, two second site substitutions were found that could restore elicitor activity. From the predicted structure of the PVX CP, Baulcombe and colleagues speculated that the resistance breaking substitution alters a host recognition site while elicitor substitutions restore this site (Bendahmane *et al.*, 1995).

The molecular mechanisms by which TMV CP substitutions alter host recognition have been investigated. Substitutions in the CP of the TMV U1 strain (normally a non-elicitor of the *N'* gene response) that lead to HR elicitation have been mapped to interface regions between CP subunits (Culver *et al.*, 1994). Moreover, CP mutations have been designed that enhance CP intersubunit interactions by removing repulsive carboxylate interactions that normally help drive virion disassembly (Culver *et al.*, 1995; Lu *et al.*, 1996). When these substitutions are placed in combination with substitutions that lead to HR elicitation, CP elicitor activity is lost (Culver *et al.*, 1995; Dardick *et al.*, unpublished data). Therefore, the quaternary configurations of CP have a drastic effect on host recognition. These findings have prompted the proposal that the TMV U1 strain has evolved a CP quaternary structure that masks the receptor binding site by concealing the recognition surface in an altered oligomeric state (i.e. monomer, dimer, trimer etc). Such a mechanism provides one model by which plant viruses may evolve to evade host recognition.

OTHER VIRUS-HOST INTERACTIONS LEADING TO RESISTANCE

A number of immune responses and HR-like phenotypes have been identified that prevent virus infection. Many of these are the result of unsuitable virus-plant interactions that hinder the replication, cell-to-cell movement, or systemic spread of the infecting virus. Still others may involve unique host defense responses that do not coincide with typical HR characteristics.

In some instances, resistance is provided by the reduced accumulation of viral RNA without obvious host responses. Plants carrying the *Tm-1* gene from *Lycopersicon esculentum* have a reduced capacity to support tomato mosaic tobamovirus (ToMV) replication. Hammamato *et al.* (1997) studied ToMV mutants that overcome *Tm-1* resistance and concluded that conformational changes in the helicase domain of the viral replicase protein were responsible for the *Tm-1* breaking phenotype. This suggests that the *Tm-1* locus encodes a host protein that the virus normally commandeers for its replication machinery; however, in *Tm-1* plants the protein may be altered such that it does not properly interact with the viral replicase, resulting in attenuated viral replication. A similar case has been observed in the interaction of potyvirus Y in *Capsicum annum* (pepper) carrying the y^a gene. Arroyo *et al.* (1996) determined that y^a resistance was a result of impaired cell-to-cell movement. Although the viral elements involved in y^a resistance have not been ascertained, the presence of this gene results in the reduced accumulation and spread of viral RNA.

PVX resistance provided by the *Rx* gene from potato has been described as a rapid resistance response that blocks viral replication at the level of single cells. *Rx* containing plants support undetectable levels of virus replication. From studies in protoplasts, Bendahmane *et al.* (1995) have identified the PVX CP as the viral component involved in *Rx* mediated resistance. Their work suggests the participation of a plant defense response that is elicited by the CP. A similar defense response has also been reported against potato virus Y in potato plants carrying the *Ry* gene, however, the potyviral elicitor has not yet been isolated (Hinrichs *et al.*, 1998). Thus, in some cases plant defense responses may be capable of rapidly limiting virus spread and so seem to provide complete immunity.

APPLICATIONS

Molecular engineering of crop plants will play a vital role in ensuring the availability of food and natural resources to future generations. From transferring resistance to new plant species to the development of broad-spectrum resistance, the farms of the future will likely utilize plants that are genetically altered to resist pathogen infection. In fact, many biotechnology companies are already generating transgenically modified products that increase agricultural yields, alter fruit ripening, incorporate herbicide resistance, and provide novel forms of pathogen resistance. Continuing advances in understanding plant resistance will play a key role in advent and improvement of these technologies.

R Genes can be Transferred across Species Barriers

With the cloning and characterization of *R* genes, several investigators have already shown that resistance can be transferred to new plant species. Baker and colleagues used *Agrobacterium*-mediated transformation to show that the tobamovirus resistance gene, *N*, from *N. glutinosa* is functional in tomato (Whitham *et al.*, 1996). Transgenic tomato plants altered to incorporate the *N* gene showed the same spectrum and level of resistance as tobacco. Additional experiments by Martin and colleagues and Staskawicz and colleagues demonstrated that the bacterial *R* gene *Pto* from tomato is functional in various tobacco species. When expressed as a transgene in *N. benthamiana* and *N. tabacum*, *Pto* conferred a resistance response similar to that observed in tomato (Thilmony *et al.*, 1995; Rommens *et al.*, 1995). Taken together, these findings demonstrate that *R* gene signaling pathways are conserved in different plant species and illustrate the potential utility of transferring *R* genes across species boundaries.

Rational Design of *R* Genes

The discovery that *R* genes share common features has prompted the idea of engineering *R* genes to incorporate new specificities. Using the yeast two-hybrid protein-protein interaction trap, Scofield *et al.* (1996) were able to furnish *avrPto* specificity to the closely related *Fen* gene (*Pto* and *Fen* share over 80% amino acid identity) by constructing *Pto/Fen* chimeras. The ability to transfer recognition specificities illustrates that it is possible to engineer *R* genes with new specificities. Furthermore, structural information on *R* gene-elicitor interactions could conceivably permit the rational design of *R* genes to accommodate specific pathogen elicitors.

Engineering Broad-Spectrum Resistance

Currently, there is limited knowledge on the signaling mechanisms involved in HR-mediated resistance. Although a number of signaling molecules have been identified that are involved in the induction and propagation of defense responses, a comprehensive understanding of the mechanisms that lead to resistance is not yet available. However, because the components involved in resistance to diverse pathogens are similar, it may be feasible to devise strategies that provide broad-spectrum resistance. Presently, potato plants engineered to exhibit high levels of H_2O_2, a molecule associated with HR induction, as well as application of the synthetic compound BTH, an inducer of systemic acquired resistance, have already demonstrated this capability (Wu *et al.*, 1995; Gorlach *et al.*, 1996). Further insights into understanding plant-pathogen interactions will undoubtedly lead to novel pathogen control approaches.

REFERENCES

Arroyo, R., Soto, M. J., Martinez-Sapater, J. M., and Ponz, R. (1996). Impaired cell-to-cell movement of potato virus Y in pepper plants carrying the y^a resistance gene. *Mol. Plant-Microbe Interact.* 9, 314-318.

Baker, B., Zambryski, P., Staskawicz, B., and Dinesh-Kumar, S. P. (1997). Signaling in plant-microbe interactions. *Science* 276, 726-733.

Bendahmane , A., Kohn, B. A., Dedi, C., and Baulcombe, D. C. (1997). The coat protein of potato virus X is a strain-specific elicitor of *Rx-1* mediated virus resistance in potato. *Plant J.* 8, 933-941.

Bennetzen, J., Qin, M., Ingels, S., and Ellingboe, A. (1988). Allele-specific and mutator-associated instability at the *Rp1* locus of maize. *Nature* 332, 369-371.

Berzal-Herranz A., de la Cruz, A., Tenllado, F., Diaz-Ruiz, J. R., Lopez, L., Sanz, A. I., Vaquero, C., Serra, M. T., and Garcia-Luque, I. (1995). In the *Capsicum* L^3 gene-mediated resistance against the tobamovirus is elicited by the coat protein. *Virology* 209, 498-505.

Bevan, M.,Bancraft, I., Bent, E., *et al.* (1998). Analysis of 1.9Mb of contiguous sequence from chromosome 4 of *Arabidopsis thaliana. Nature* 391, 485-488.

Cruz, S. S., and Baulcombe, D. C. (1993). Molecular analysis of potato virus X isolates in relation to the potato hypersensitivity gene *Nx. Mol. Plant-Microbe Interact.* 6, 707-714.

Culver, J. N., and Dawson, W. O. (1989). Point mutations in the coat protein gene of tobacco mosaic virus induce hypersensitivity in *Nicotiana sylvestris. Mol. Plant-Microbe Interact.* 4, 209-213.

Culver, J. N., Stubbs, G., and Dawson, W. O. (1994). Structure function relationship between tobacco mosaic virus coat protein and hypersensitivity in *Nicotiana sylvestris. J. Mol. Biol.* 242, 130-138.

Culver, J. N., Dawson, W. O., Plonk, K., and Stubbs, G. (1995). Site-directed mutagensis confirms the involvement of carboxylate groups in the disassembly of tobacco mosaic virus. *Virology* 206, 724-730.

Dangl, J. (1995). Pièce de Résistance: Novel classes of plant disease resistance genes. *Cell* 80, 363-366.

Dangl, J., and Holub, E. (1997). La Dolce Vita: A molecular feast in plant-pathogen interactions. *Cell* 91, 17-24.

Dardick, C. D., and Culver, J. N. (1997) Tobamovirus coat proteins, Elicitors of the hypersensitive response in *Solanum melongena* (eggplant). Mol. *Plant-Microbe Interact.* 10, 776-778.

de la Cruz, A., Lopez, L., Tenllado, F., Diaz-Ruiz, J. R., Sanz, A. I., Vaquero, C., Serra, M. T., Garcia-Luque, I. (1997). The coat protein is required for the elicitation of the *Capsicum* L^2 gene-mediated resistance against the tobamoviruses. *Mol. Plant-Microbe Interact.* 10, 107-113.

Flor, H. (1971). Current status of the gene-for-gene concept. *Annu, Rev. Phytopathol.* 9, 275-296.

Fraser, R. S. S. (1990). The genetics of resistance to plant viruses. *Annu. Rev. Phytopathol.* 28, 179-200.

Fraser, R. S., and Gerwitz, A. (1987). The genetics of resistance and virulence in plant virus disease genetics and plant pathogenesis. *In* "Genetics and Plant Pathogenesis" (P. R. Day and G. J. Jallis, Eds.). pp. 33-44. Blackwell Scientific Publ. Oxford.

Gorlach, J., Volrath, S., Knauf-Beiter, G., Hengry, G., Beckhove, U., Kogel, K., Oostendorp, M., Staub, T., Ward, E., Kessmann, H., and Ryals, J. (1996). Benzothiadiazole, a novel class of inducers of systemic acquired resistance, activates gene expression and disease resistance in wheat. *Plant Cell* 8, 629-643.

Grant, M. R., Godiard, L., Strauge, E., Ashriled, T., Lewald, J., Sattler, A., Innes, R. W., and Dangl, J. L. (1995). Structure of the Arabidopsis RPM1 gene enabling dual specificity disease resistance. *Science* 269, 843-846.

Hamamoto, H., Watanabe, Y., Kamada, H., and Okada, Y. (1997). Amino acid changes in the putative replicase of tomato mosaic tobamovirus that overcome resistance in *Tm-1* tomato. *J. Gen. Virol,* 78, 461-464.

Heguy, A., Baldari, C., Macchia, G., Telford, J., and Melli, M. (1992). Amino acids conserved in interleukin-1 receptors (IL-1Rs) and the *Drosophila* Toll protein are essential for IL-1R signal transduction, *J. Bio. Chem.* 267, 2605-2609.

Hinrichs, J., Berger, S., and Shaw, J. G. (1998). A hypersensitive response-like mechanism is involved in resistance of potato plants bearing the *Ry(sto)* gene to the potyviruses potato virus Y and tobacco etch virus. *J. Gen. Virol* 79, 167-176.

Janin, J., and Chothia, C. (1990). The structure of protein-protein recognition sites. *J. Biol, Chem,* 265, 16027-16030.

Jones, D., Thomas, C., Hammond-Kosack, K., Balint-Kurti, P., and Jones, J. (1994). Isolation of the tomato *Cf-9* gene for resistance to *Cladosporium fulvum* by transposon tagging. *Science* 266, 789-793.

Kavanagh, T., Goulden, M., Santa Cruz, S., Chapman, S., Barker, I., and Baulcombe, D. C. (1992) Molecular analysis of a resistance-breaking strain of potato virus X. *Virology* 189, 609-617.

Keen, N. T. (1990). Gene-for-gene complementarity in plant-pathogen interactions. *Annu. Rev. Genet.* 24, 447-463.

Kobe, B., and Deisenhofer, J. (1995). A structural basis of the interactions between leucine-rich repeats and protein ligands. *Nature* 374,183-186.

Lawrence, G. J., Finnegan, E. J., Anderson, P. A., and Ellis, J. G. (1995). The L^6 gene for flax resistance is related to the *Arabidopsis* bacterial resistance gene *RPS2* and tobacco viral resistance gene *N*. *Plant Cell* 7,1195-1206.

Lu, B., Stubbs, G., and Culver, J. N. (1996). Carboxylate interactions involved in the disassembly of tobacco mosaic tobamovirus. *Virology* 225, 11-20.

Martin, G., Brommonschenkel, S., Chunwongse, J., Frary, A., Ganal, M., Spivey, R., Wu, T., Earle, E. D., and Tanksley, S. (1993). Map-based cloning of a protein kinase gene conferring disease resistance in tomato. *Science* 262, 1432-1436.

Martin, G. B., Frary, A., Wu, T., Brommonschenkel, S., Chunwongse, J., Earle, E. D., and Tranksley, S. D. (1994). A member of the tomato *Pto* gene faily confers sensitivity to Fenthion resulting in rapid cell death. *Plant Cell* 6, 1543-1552.

Meshi, T., Motoyoshi, T., Maeda, T., Yoshiwoka, S., Watanabe, H., and Okada, Y. (1989). Mutations in the tobacco mosaic virus 30 kD protein gene overcome *Tm-2* resistance in tomato. *Plant Cell* 1, 515-522.

Mindrinos, M., Katagirl, F., Yu, G. L., and Ausubel, M. (1994). The *A. thaliana* disease resistance gene *RPS2* encodes a protein containing a nucleotide-binding site and leucine-rich repeats. *Cell* 78,1089-1099.

Mittler, R., and Lam, E. (1996). Sacrifice in the face of foes: Pathogen -induced programmed cell death in plants. *Trends Microbiol.* 4, 10-15.

Padgett, H. S., and Beachy, R. N. (1993). Analysis of a tobacco mosaic virus strain capable of overcoming *N* gene mediated resistance. *Plant Cell* 5, 577-586.

Padgett, H. S., Watanabe, Y., and Beachy, R. N. (1997). Identification of the TMV replicase sequence that activates the *N* gene-mediated hypersensitive response. *Mol. Plant-Microbe Interact.* 10, 709-715.

Parker, J. E., Coleman, M. J., Szabo, V., Frost, L. N., Schmidt, R., van der Biezen, E., Moores, T., Dean, C., Daniels, M. J., and Jones, J. D. G. (1997). The *Arabidopsis* downy mildew resistance gene *Rpp5* shares similarity to the Toll and Interleukin-1 with receptors with *N* and L^6. *Plant Cell* 9, 879-894.

Rommens, C. M., Salmeron, J. M. Oldroyd, G. E., and Staskawicz, B. J. (1995). Intergenic transfer and functional expression of the tomato disease resistance gene *Pto*. *Plant Cell* 7, 1537-1544.

Salmeron, J. M., Oldroyd, E. D., Rommens, C. M., Scofield, S. R., Kim, H., Lavelle, D. T., Dahlbeck, D., and Staskawicz, B. J. (1996). Tomato *Prf* is a member of the leucine-rich repeat class of plant disease resistance genes and lies embedded within the *Pto* kinase gene cluster. *Cell* 86, 123-133.

Schoelz, J. E., Shephard, R. J., and Dauber, S. D. (1987). Host response to cauliflower mosaic virus (CaMV) in solanaceous plants is determined by a 496bp DNA sequence within gene VI. *In* "Molecular Strategies for Crop Protection" (C. J. Amatzen and C. Ryan, Eds.). pp. 253-265. Alan R, Liss, Inc.

Scholthof, H. B., Scholthof, K. G., and Jackson, A. O. (1995). Identification of tomato bushy stunt virus host-specific determinants by expression of individual genes from a potato virus X vector. *Plant Cell* 7, 1157-1172.

Scofield, S. R., Tobias, C. M., Rathjen, J. P., Chang, J. H., Lavelle, D. T., Michelmore, R. W., and Staskawicz, B. J., (1996). Molecular basis of gene-for-gene specificity in bacterial speck disease of tomato. *Science* 274, 2063-2065.

Shaw, J. G. (1996). Plant Viruses. *In* "Fields Virology" (Bernard N. Fields, Ed.). Lippincott-Raven Publishers, Philadelphia PA.

Sims, J., Acres, B., Grubin, C., McMahan, C., Wignall, J., March, C., and Dower, S. (1989). Cloning the interleukin-1 receptor from human t cell. *Proc. Natl. Acad. Sci. USA* 86, 8946-8950.

Song, W., Wang, G., Chen, L., Kim., L., Holsten, T., Gardner, J., Wang, B., Zhai, W., Zhu, L., Fauquet, C., and Ronald, P. (1995). A receptor kinase-like protein encoded by the rice disease resistance gene, *Xa21*. *Science* 270, 1804-1806.

Staskawicz, B. J., Ausubel, F. M., Baker, B. J., Ellis, J. G., and Jones, J. D. (1995). Molecular genetics of plant disease resistance. *Science* 268, 661-667.

Tang, X. Y., Frederick, R. D., Zhou, J. M., Halterman, D. A., Jia, Y. L., and Martin, G. B. (1996). Initiation of plant disease resistance by physical interaction of *avrPto* and *Pto* kinase. *Science* 274, 2060-2063.

Taraporewala, Z. F., and Culver, J. N., (1996). Identification of an elicitor active site within the three-dimensional structure of the tobacco mosaic virus coat protein. *Plant Cell* 8, 169-178.

Thilmony, R. L., Chen, Z., Bressan, R. A. and Martin, G. B. (1995). Expression of the tomato *Pto* gene in tobacco enhances resistance to *Pseudomonas syringae* pv *tabaci* expressing *avrPto*. *Plant Cell* 7, 1529-1536.

Weber, H., Schultze, S., and Pfitzner, A. J. P. (1993). Two amino acid substitutions in the tomato mosaic virus 30 -Kilodalton movement protein counter the ability to overcome the *Tm-2²* resistance gene in tomato. *J. Virol.* 67, 6432-6438.

Whitham, S., Dinesh-Kumar, S. P., Choi, D., Hehl, R., Corr, C., and Baker, B. (1994). The product of the tobacco mosaic virus resistance gene *N* : Similarity to Toll and the interleukin-1 receptor. *Cell* 78, 1101-1115.

Whitham, S., McCormick, S., and Baker, B. (1996). The *N* gene of tobacco confers resistance to tobacco mosaic virus in transgenic tomato. *Proc. Natl. Acad, Sci.. USA* 93, 8776-8781.

Wu, G., Shortt, B. J., Lawrence, E. B., Levine, E. B., Fitzsimmons, K. C., and Shah, D. M. (1995). Disease resistance conferred by expression of a gene encoding H_2O_2 generating glucose oxidase in transgenic potato plants. *Plant Cell* 7, 1357-1368.

Zhou, J., Loh, Y., Bressan, R. A., and Martin, G. B. (1995). The tomato gene *Pti* 1encodes a serine/ theronine kinase that is phosphorylated by *Pto* and is involved in the hypersensitive response. *Cell* 83, 925-935.

11 MOLECULAR BIOLOGY OF VIROIDS

Ricardo Flores, Marcos de la Peña,
José-Antonio Navarro, Silvia Ambrós, and
Beatriz Navarro

INTRODUCTION

Viroids are the only class of autonomously replicating subviral pathogens with a well defined molecular structure. Their incidental discovery some twenty five years ago, during studies aimed at characterizing the agents of some plant maladies presumably induced by viruses, changed the existing belief that viruses were the smallest inciting agents of infectious diseases. Viroids, single-stranded circular RNAs of 246-399 nucleotides (Flores et al., 1998), a genome size approximately one-tenth that of the smallest known RNA virus, are currently the lowest step of the biological scale. Historically, it is remarkable that tobacco mosaic virus, the first virus discovered almost a century ago, has a plant origin just like potato spindle tuber viroid (PSTVd), the first known subviral pathogen endowed with autonomous replication (Diener, 1971). Although numerous examples of viruses affecting bacterial and animal cells have since been found, viroids remain confined to the plant kingdom.

VIROIDS: THE SIMPLEST GENETIC SYSTEMS

Nucleotide Sequence

The number of the viroid species sequenced is 27 (Table 1). Comparative analyses have revealed the existence of some conserved sequence and structural motifs presumably involved in critical functions, but the most important consequence from them is that viroids, as opposed to viruses, do not appear to code for any protein. This contention is supported by the absence of AUG initiation triplets in PSTVd (Gross et al., 1978), of common open reading frames in closely related viroids, of messenger activity of viroid RNAs in different systems (Davies et al., 1974), and of viroid-specific proteins in infected tissues (Conejero and Semancik, 1977). Moreover, the circular structure of the viroid molecule makes it an unlikely

Table 1. Sequenced viroids with their abbreviations and sizes*

Viroid species	Abbreviation	Size (nt)**
Potato spindle tuber	PSTVd	356,359-360
Mexican papita	MPVd	359-360
Tomato planta macho	TPMVd	360
Citrus exocortis	CEVd	370-375,463
Chrysanthemum stunt	CSVd	354,356
Tomato apical stunt	TASVd	360,363
Iresine 1	IrVd	370
Columnea latent	CLVd	370,372
Hop stunt	HSVd	295-303
Coconut cadang-cadang	CCCVd	246-247,287-301
Coconut tinangaja	CTiVd	254
Hop latent	HLVd	256
Citrus IV	CVd-IV	284
Apple scar skin	ASSVd	329-330
Citrus III	CVd-III	294,297
Apple dimple fruit	ADFVd	306-307
Grapevine yellow speckle 1	GYSVd-1	366-368
Grapevine yellow speckle 2	GYSVd-2	363
Citrus bent leaf	CBLVd	318
Pear blister canker	PBCVd	315-316
Australian grapevine	AGVd	369
Coleus blumei 1	CbVd-1	248,250-251
Coleus blumei 2	CbVd-2	301-302
Coleus blumei 3	CbVd-3	361-362,364
Avocado sunblotch	ASBVd	246-250
Peach latent mosaic	PLMVd	335-338
Chrysanthemum chlorotic mottle	CChMVd	398-399

*Viroids are listed according to Figure 2   ** *nt*, nucleotide

translation template. All the above evidence qualify viroids as non-coding genomes and indicate that differences with viruses go beyond genome size and include fundamental distinctions in function and evolutive origin.

Domains in Proposed Secondary Structure and Elements of Tertiary Structure

Due to intramolecular self-complemenatrity, viroid RNAs display a high degree of secondary structure (Fig. 1). They adopt in most, but not in all cases, *in vitro* rod-like or quasi-rod-like conformations (Gross *et al.,* 1978), that can be divided into five structural domains termed central (C), variable (V), pathogenic (P), and terminal left and right (T_L and T_R, respectively) (Keese and Symons, 1985). During replication viroids can also form alternative metastable conformations with defined hairpins (Qu *et al.,* 1995). Viroid structure *in vivo* is unknown, but support exists for at least a partial rod-like structure in some viroids. Thus, sequence duplications in T_R domain of CCCVd and CEVd occur in a way to preserve the rod-like structure (Haseloff *et al.,* 1982). Within the rod-like structure several conserved motifs can be distinguished: i) the central conserved region (CCR) in the C domain formed by two series of nucleotides flanked by an inverted repeat in upper strand,

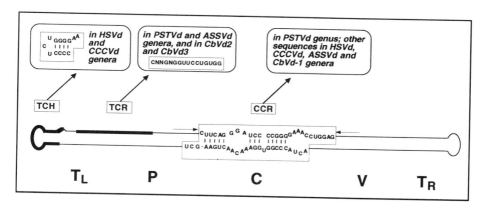

Figure 1. Schematic rod-like structure model for viroids of the PSTVd family. The approximate location of the five domains C, P, V, T$_L$, and T$_R$ are indicated on the bottom of the figure. The core nucleotides of the CCR, TCR and TCH are shown, as well as the presence of these motifs in different members of the family: Arrows indicate flanking sequences which form with the core nucleotides of the CCR upper strand, imperfect inverted repeats.

ii) the terminal conserved region (TCR) in the upper strand of the T$_L$ domain, and iii) the terminal conserved hairpin (TCH), at the end of the T$_L$ domain, conserved in sequence and secondary structure (Fig. 1). The type of CCR and the presence or absence of the TCR and TCH motifs have been used as criteria for allocating most viroids to the first family, *Pospiviroidae,* and within it to different genera (see below). However, the situation is very different in the three viroids that make up the second family, *Avsunviroidae.* ASBVd, PLMVd and CChMVd do not have these conserved motifs and share limited sequence similarity between them, but are endowed with a very peculiar property: their strands of both polarities are able to form hammerhead structures and to self-cleave as predicted by these ribozymes (Prody *et al.,* 1986; Hutchins *et al.,* 1986). Moreover, *in vitro* and probably *in vivo,* PLMVd and CChMVd adopt branched conformations instead of rod-like ones (Hernández and Flores, 1992; Navarro and Flores, 1997).

Elements of tertiary structure have been also detected in viroids, prominent among them is an interaction in the CCR of PSTVd which is also present in the so-called loop E of 5S rRNA (Branch *et al.,* 1985). Fine structural analysis has also revelaed non-canonical base pairs that play a role in stabilizing the whole viroid conformation or specific regions such as the hammerhead structures. Other postulated elements of tertiary structure include a tetraloop in PSTVd (Baumstark *et al.,* 1996) and a pseudoknot-like interaction in PLMVd (Ambrós *et al.,* 1998).

Role of Specific Domains

Of the five structural domains of the rod-like or quasi-rod-like structure of most viroids (Keese and Symons, 1985), some appear to determine functional characteristics. For example, analysis of chimerical viroids, constructed *in vitro* from mild and severe variants of CEVd (Visvader and Symons, 1986) or even from different viroids like CEVd and TASVd (Sano *et al.*, 1992), has shown that symptom severity is regulated by two determinants in the T_L and P domains, and replication/accumulation by a third determinant located in the V and/or T_R domains. Recent results with chimerical viroids derived from CEVd and HSVd support the view that the T_R domain may modulate the efficiency of viroid replication/accumulation (Sano and Ishiguro, 1998).

Strictly conserved sequence motifs such as those forming the CCR, TCR and TCH, may be involved in currently unknown but crucial functions. The best established structural/functional relationship in viroids occurs in ASBVd, PLMVd and CChMVd between the hammerhead structures and RNA self-cleavage.

GENOMIC DIVERSITY

Variability and Viroid Quasispecies

Viroids, like other RNA replicons and particularly RNA viruses, propagate in their hosts as populations of closely-related variants composing what it is known as a quasispecies (Domingo and Holland, 1994). There are numerous examples of minor genetic heterogeneity between sequences coexisting in natural isolates of different viroids including CEVd (Visvader and Symons, 1985), ASBVd (Rakowski and Symons, 1989), PSTVd (Gòra *et al.*, 1994), HSVd (Kofalvi *et al.*, 1997), GYSVd-1 (Koltunow and Rezaian, 1988), CChMVd (Navarro and Flores, 1997) and PLMVd (Hernández and Flores, 1992; Ambrós *et al.*, 1998). The sequence heterogeneity is most probably the result of the inaccurate fidelities of RNA polymerases that lead to mutation rates of 10^{-3} to 10^{-5} substitutions per nucleotide per round of viral RNA replication (Domingo and Holland, 1994). Estimations of such a class are not available for viroids but similar, if not higher, mutation rates can be presumed for them considering that polymerization of viroid RNA is catalyzed by cellular RNA polymerases whose normal templates are double-stranded DNAs (Diener, 1996). Repeated infection of the same individual plant, a feasible situation in hosts such as citrus and fruit trees with a long productive life, may also contribute to sequence diversity. However, analysis of the progenies evolved from inoculations with PLMVd cDNA clones has shown that most of the variability results from the emergence of new variants during replication (Ambrós, Hernández and Flores, unpublished data).

Selection Mechanisms

Although the high mutation rate of RNA polymerases is an essential factor for the genetic divergence of viroids, most of the resulting mutants are filtered through the action of different selection pressures. Some restrictions are imposed by the pathogen itself, this is the case in PSTVd and related viroids regarding the conservation of some sequence (CCR, TCR and TCH) and structural elements (rod-like secondary structure and potential to form certain hairpins). Examples here include PSTVd RNAs with mutations affecting one of these hairpins which when inoculated rapidly revert to the wild-type sequence (Qu *et al.,* 1995), and the evolution of an infectious PSTVd replicon from an *in vitro*-generated non-infectious deletion mutant that occurs via a complementary deletion *in vivo* that restores the rod-like structure (Wassenegger *et al.,* 1994). In other viroids some variability is tolerated in the hammerhead structures in as much as their stabilities are preserved. This is observed in PLMVd, where the additional conservation of a pseudoknot and a branched secondary structure seems also critical (Ambrós *et al.,* 1998). Similarly, the CChMVd polymorphism preserves the stability of some hairpins of the most stable branched conformation, because the mutations affect the loops or because they are compensatory when found in the stems (Navarro and Flores, 1997).

The host and even the tissue in which the viroid propagates also limits the permitted variability. For example, the sequences of CEVd cDNA clones from citron have shown previously unreported mutations in other CEVd variants, and a common pattern of nucleotide exchanges, the 'tomato signature', has been detected in CEVd isolates from tomato tissues (Semancik *et al.,* 1993).

Recombination

Some viroids appear to have emerged from recombination events between two or more parental viroids that presumably coinfected the same plant. The most representative case, CLVd, is made up of a mosaic of sequences from HSVd, PSTVd, TPMVd, and TASVd (Hammond *et al.,* 1989). The chimerical nature of CLVd is also reflected in its host range: CLVd can infect characteristic hosts of HSVd and PSTVd. A second typical example of viroid recombination is AGVd (Rezaian, 1990). Direct evidence supporting a continuous recombination process has come from studies on the long term distribution of the three coleus viroids coinfecting individual plants. Fusion of the right-hand part and the CCR of CbVd 1 with the left-hand part of CbVd 3 produces the recombinant CbVd 2 (Sänger, Henkel and Spieker, unpublished data). Intramolecular rearrangements also occur as illustrated by the already mentioned cases of CCCVd and CEVd. Recombination presumably results from discontinuous transcription by a jumping polymerase with low processivity (Keese and Symons, 1985).

ORIGIN AND EVOLUTION

Viroids as Relics of RNA World

From an evolutionary perspective, viroids are also intriguing systems. Sequence similarities between PSTVd and related viroids, and the termini of transposable elements, led to the proposal that viroids may have evolved from them or from retroviral proviruses by deletion of the internal sequences, although the similarities could be fortuitous. Alternatively, the conservation in viroids of sequences believed to play a critical role in the self-splicing of group I introns, suggested that viroids may represent "escaped" introns. However, a more detailed analysis revealed neither resemblance to group I intron structures of the hammerhead ribozymes mediating self-cleavage in some viroids, nor a common mechanism in both autolytic reactions, making a phylogenetic relationship between viroids and introns unlikely (see for a review Diener, 1996). On the other hand, a number of indications suggest that viroids could be molecular fossils of the RNA world postulated to have existed on the earth before advent of cellular life. These include their small size, circular structure (making genomic tags unnecessary for initiating and terminating transcription), high G+C content increasing their tolerance to error-prone primitive polymerases and, in particular, the presence in some of them of ribozymes (Diener, 1989). This alternative hypothesis means that viroid and viroid-like satellite RNAs may have originated from "free-living" molecules which became intracellular parasites dependent on their host in the case of viroids, or on their helper virus, and therefore probably also on their host, in the case of the viroid-like satellite RNAs.

Phylogenetic Reconstructions

An updated phylogenetic tree of all sequenced viroids is presented (Fig. 2). Even though the recombinations, duplications and rearrangements suffered by viroids could make it difficult to infer phylogenetic relationships, the existence of these processes does not blur the phylogenetic reconstructions. In fact, evidence suggesting a common origin for viroids and viroid-like satellite RNAs has already been presented (Elena *et al.*, 1991). The phylogenetic tree obtained separated typical viroids, autonomously replicating but not self-cleaving through hammerhead ribozymes, from the self-cleaving but not autonomously replicating satellite RNAs. Interestingly ASBVd, the only hammerhead viroid known at that moment, sharing functional similarities with both groups (autonomous replication and self-cleavage), was located between the two clusters suggesting that it might be an evolutionary link between them. Several groups of related viroids can be distinguished, which indicate that phylogenetic reconstructions can be used for classification purposes.

Classification

Following the ICTV guidelines, the 27 viroid species have been classified into two families (Flores *et al.*, 1998). Members of the *Pospiviroidae* family have a CCR but lack self-cleavage mediated by hammerhead ribozymes. Members of the *Avsunviroidae* family do not have a CCR but both polarity strands are able to self-cleave through hammerhead ribozymes. *Pospiviroidae* members are clustered into genera according to the type of CCR and the presence or absence of TCR and TCH, with *Pospi-, Hostu-* and *Cocaviroid* genera being grouped into a subfamily because their CCRs are more closely interrelated than with those of *Apsca-* and *Coleviroid*. Within *Avsunviroidae*, PLMVd and CChMVd form the genus *Pelamoviroid* on the basis of their more similar G+C content, predicted secondary structures of lowest

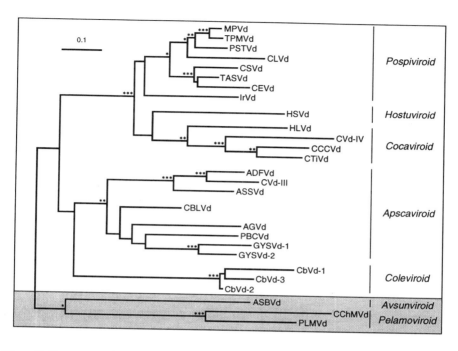

Figure 2. Consensus phylogenetic tree for viroids. ***, **, and *, group monophyletic viroids in more than 95%, 85% and 75% of the 1000 replicates, respectively. Sequence numbering was based on the similarities between the CCR upper strands of *Pospiviroidae* and the upper-left-hand parts of the hammerhead structures of *Avsunviroidae* (Diener, 1989). The alignment was performed with Clustal X 1.64 (gap opening and gap extension penalties 15 and 1, respectively) and minor manual adjustments. Evolutionary distances were estimated according to the Jukes and Cantor model and the phylogenetic tree was constructed by the Neighbor-Joining method with MEGA 1.01. Viroid genera are indicated on the right, and members of *Pospiviroidae* and *Avsunviroidae* families on white and dark backgrounds, respectively.

free energy and morphology of their hammerhead structures, as well as on a physico-chemical criterion such as their insolubility in 2 M LiCl (Navarro and Flores, 1997). The classification of viroids (Fig. 2, right side), is consistent with the phylogenetic tree.

REPLICATION

Rolling Circle Model

The circular structure of the viroid molecule and the detection of multimeric negative RNAs (complementary to the predominant infectious strand to which the plus polarity is conventionally assigned) in PSTVd-infected tomato plants, led to the proposal of a rolling-circle mechanism for viroid replication (Branch and Robertson, 1984). The circular monomeric plus RNA is transcribed into linear multimeric minus strands, which in a second RNA-RNA transcription step generate linear multimeric plus strands that are then processed to unit-length and ligated. This is the asymmetric variant of the model, with a single rolling-circle. However, the finding that dimeric ASBVd RNAs of both polarities self-cleave *in vitro* (Hutchins *et al.*, 1986), like the RNAs of PLMVd and CChMVd (Hernández and Flores, 1992; Navarro and Flores, 1997), provides indirect evidence against either polarity of multimeric RNA serving as a template for these three viroids. They are assumed to follow the alternative symmetric version in which the linear multimeric minus strands are cleaved and ligated to the circular monomeric minus RNA, the template for a second rolling-circle (Branch and Robertson, 1984). The identification in ASBVd-infected tissue of the circular monomeric minus strand (Daròs *et al.*, 1994) directly supports the symmetric pathway for this viroid and, presumably, for PLMVd and CChMVd. By contrast, the impossibility of detecting the circular monomeric minus form in PSTVd-infected tomato (Branch *et al.*, 1988; Feldstein *et al.*, 1998), is consistent with the asymmetric variant operating in PSTVd and other *Pospiviroidae* members.

Transcription, Processing and Ligation, Hammerhead Ribozymes

The three steps of the rolling circle mechanism require the participation of three catalytic activities, RNA-dependent RNA polymerase, RNase and RNA ligase, presumed initially to be host enzymes since viroids do not code for proteins. The identification of the nucleus as the major subcellular accumulation site for PSTVd and related viroids (Harders *et al.*, 1989; Bonfiglioli *et al.*, 1996), and the finding that RNA polymerization is inhibited by nanomolar concentrations of α-amanitin (Mühlbach and Sänger, 1979; Schindler and Mühlbach, 1992), supports the involvement of RNA polymerase II. The situation is very different in ASBVd which accumulates in the chloroplast (Bonfiglioli *et al.*, 1994; Lima *et al.*, 1994) and is insensitive to very high α-amanitin concentrations (Marcos and Flores, 1992; Navarro, Vera and Flores, unpublished data). In this case, and quite likely in PLMVd and CChMVd, polymerization of RNA strands is probably catalyzed by a chloroplast RNA polymerase. An open question here is whether viroid strands have specific sequences for the initiation of transcription. Promoters presumably exist in viroids and some of the conserved motifs in the *Pospiviroidae* family might play such a role. A second question is how cellular RNA polymerases, which under normal conditions transcribe a double-stranded DNA template, are able to act on an RNA template. It can be speculated that the partial double-stranded nature of the viroid molecules serves to recruit one or more transcription factors.

The detailed study of processing of the multimeric replicative intermediates generated in viroid replication, has uncovered a unique mechanism in the three members of the *Avsunviroidae* family. The reaction is autolytic and mediated by hammerhead ribozymes in both polarity strands (Fig. 3). Viroids, therefore, although unable to code for enzymes, can "code" for ribozymes. There is considerable evidence that the hammerhead structures are active not only *in vitro* but also *in vivo*. For example, the 5' termini of some ASBVd and CChMVd linear RNAs

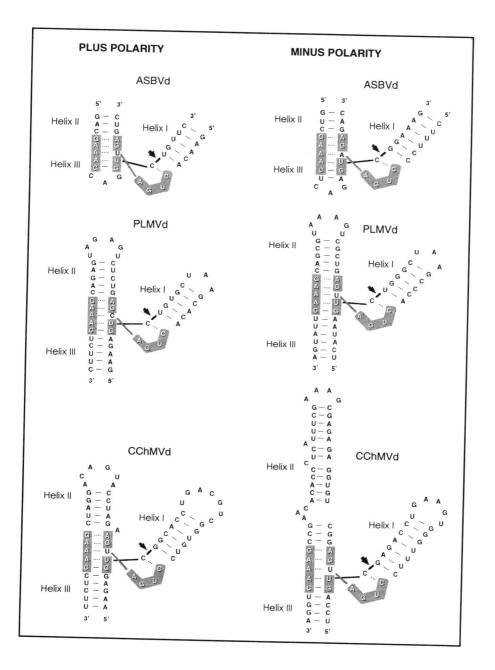

Figure 3. Schematic representation of the hammerhead structures of ASBVd, PLMVd and CChMVd. Letters on a dark background refer to absolutely or highly conserved residues in all natural hammerhead structures. Arrows indicate self-cleavage sites. Watson-Crick base pairs and non-canonical interactions are denoted with continous and broken lines, respectively.

isolated from infected tissues are the same as those resulting from *in vitro* self-cleavage (Daròs *et al.,* 1994; Navarro and Flores, 1997), and the mutations found in PLMVd variants in the sequences forming the hammerhead structures do not alter their thermodynamic stabilities (Hernández and Flores, 1992; Ambrós *et al.,* 1998). Members of the *Pospiviroidae* family can not adopt hammerhead structures, and their RNA processing is generally assumed to be catalyzed by a host RNase, although its nature and the precise processing site remain unknown. Both queries have been tackled by *in vitro* experiments giving inconclusive results because: i) different RNases, including one of fungal origin, are able to catalyze the reaction (Tabler *et al.,* 1992), and ii) although some data point out that the PSTVd processing site is in the CCR upper strand (Baumstark *et al.,* 1997), others obtained with CEVd leave this question open (Rakowski and Symons, 1994). To make the situation even more complex, a recent report suggests that processing of CCCVd RNA is autolytic and mediated by a novel but unknown class of ribozymes, and that the processing site is in CCR lower strand (Liu and Symons, 1998).

RNA ligation most probably occurs in all viroids by a host-encoded RNA ligase. Some *in vitro* data show that monomeric linear PLMVd strands from autolytic processing, can self-ligate to their circular forms. However, the resulting phosphodiester bond is 2'-5' and not 3'-5' (Côte and Perreault, 1997), making the *in vivo* functioning of such a mechanism unlikely.

GENETIC DETERMINANTS OF PATHOGENESIS AND SYMPTOMATOLOGY

Pathogenic Domains

The type and intensity of symptoms are modulated in some viroids of *Pospiviroidae* family by determinants in P and T_L domains (Sano *et al.,* 1992; Sano and Ishiguro, 1998). Data on viroids of *Asunviroidae* are limited. In ASBVd, distinct viroid populations segregate within bleached, variegated and symptomless avocado tissues, with the predominant sites of sequence heterogeneity being the terminal loops of the rod-like structure and, particularly, the right one that is enlarged in the variant associated with bleached tissue (Semancik and Szychowski, 1994). However, the difficulties in achieving reproducible mechanical transmission of ASBVd hampers efforts to obtain direct supportive data. By contrast, reliable bioassays exist for PLMVd and CChMVd based in their natural hosts, but the extreme variability of PLMVd has not permitted pathogenic effect of specific variants to be assigned to a defined structural motif (Ambrós *et al.,* 1998). The situation is less complicated in CChMVd and a molecular determinant of pathogenicity is located in a defined region of RNA molecule(De la Peña, Navarro and Flores, unpublished data).

Symptomatology

Symptom development is not necessarily the outcome of viroid-host plant interaction: some viroids are pathogenic in their natural hosts but others replicate and accumulate to high levels without eliciting visible symptoms. Indeed, viroid reservoirs exist in wild symptomless plants from which, conceivably, they have been transferred to cultivated plants. An example is the recent identification in *Solanum cardiophyllum* plants growing wild in Mexico of MPVd, the putative ancestor of crop viroids such as PSTVd and TPMVd (Martínez-Soriano *et al.,* 1996).

The genetics underlying symptomatology in viroid-host plant interactions is poorly understood, and probably differs in both viroid families if one considers the distinct replication and accumulation sites of their type members. Concerning the viroid, some data have been obtained regarding pathogenic domains like those previously mentioned. Information concerning the plant is scarce and of a circumstantial nature. Although small nuclear or cytoplasmic cellular RNAs were proposed as the primary target, on the basis of their potential to basepair with some viroids, a protein/RNA interaction would seem a more likely candidate for the initial event of the signal transduction pathway which ultimately leads to symptoms. Viroids must certainly interact with proteins during replication and probably in transport and accumulation too, but there is however no direct evidence to support such an interaction. By exploiting the powerful genetic approaches recently developed, this area of research is likely to provide some illuminating insights in the future. In this respect, *in vitro* RNA-ligand screening of a tomato cDNA expression library has led to the isolation of several apparently specific PSTVd-binding proteins, that are being examined by *in vivo* techniques such as the three-hybrid system (Sägesser *et al.,* 1997). On the other hand, PSTVd can activate *in vitro* the mammalian 68-kDa protein kinase with the interesting difference that the effect produced by a severe variant is more pronounced than that corresponding to a mild one. Since activation of P68 elicits a pathway leading to a decrease in protein initiation, it can be speculated that PSTVd might trigger a similar effect on a homologous plant enzyme (Diener *et al.,* 1993).

In this context it should be mentioned that cross-protection phenomena, widely documented in plant viruses, have also been reported in both viroid families and used to detect mild strains of PSTVd and PLMVd even before they were recognized as viroids. These interactions occur at the level of viroid accumulation and symptom expression, and have been observed between strains of the same viroid and between closely-related viroids. The data are consistent with the view that the two coinfecting viroid RNAs compete with different affinities for a limiting host factor.

Host Range

The first distinction to be made here is that with the exception of CCCVd and CTiVd that infect monocotyledons, the other viroids infect dicotyledons. Within the *Pospiviroidae* family very different situations are found represented by CCCVd with a host range restricted to members of the *Palmae* family, and HSVd that is able to infect cucumber, different *Prunus* species (including plum, peach, apricot and almond), pear, citrus and grapevine. The three members of the *Avsunviroidae* family are, however, characterized by very restricted host ranges: they only replicate in their natural hosts or closely-related species. Considering the other partner in the interaction, some plants can host more than one viroid, as illustrated by grapevine and *Citrus* able to harbor naturally at least five different viroids. Essentially nothing is known about the molecular mechanisms governing the host range of viroids. Inoculation of a *Solanum acaule* accession with *Agrobacterium* carrying PSTVd cDNAs has revealed that its resistance to PSTVd is a resistance to mechanical inoculation rather than an immunity to infection (Salazar *et al.,* 1988).

GENETIC DETERMINANTS OF OTHER VIROID FUNCTIONS

Transmission

Viroids are mainly transmitted by agronomic practices that include vegetative propagation and pruning tools (Garnsey and Randles, 1987). Most viroids can be experimentally transmitted by mechanical inoculation. PSTVd and ASBVd are transmitted through the seed of their natural hosts but others, such as PLMVd and PBCVd, are not. Efficient transmission by aphids has only been well documented in TPMVd, and probably occurs under specific ecological conditions in which wild reservoir plants are grown in the vicinity of cultivated tomato plants. These transmission characteristics of different viroids must be based on genetic factors which, however, have not been detected yet.

Viroid Transport in Host Plant

Long-distance transport of viroids has been investigated in detail in PSTVd infecting tomato. The viroid was first detected in the tip and leaves close to it before the onset of symptoms. PSTVd was also translocated to the roots, but not to leaves below the inoculated leaf unless the flux of photosynthetic materials was reversed. These results are consistent with the viroid following the movement of photosynthetic products through the phloem, the same route is also used by most plant viruses (Palukaitis 1987).

Regarding short-distance movement, data obtained from micro-injection studies in tobacco and tomato mesophyll have shown that PSTVd can move from cell-to-cell via plamodesmata and that fusion of PSTVd sequences to a non-mobile RNA promoted the movement of the fusion product. Moreover, the viroid accumulated in the nuclei of the injected and neighboring cells, confirming previous results on its subcellular location site. Interestingly, PSTVd cDNA also appears to mediate plasmodesmatal transport of a plasmid, suggesting that viroid-specific sequence or structural motifs are responsible for this movement (Ding *et al.,* 1997).

The molecular nature of the motifs involved in cell-to-cell or long-distance spread of viroids are unknown. *Agrobacterium*-mediated inoculation of tomato plants with constructs containing PSTVd sequences with mutations in the right terminal loop have shown the confinement of the progeny to specific cell types, namely the vascular cambium and the phloem. This raises the possibility that the lack of systemic spread may be due to a disruption of an essential RNA/protein interaction, although the requirement of specific cell types for replication of these particular PSTVd variants is also possible (Hammond, 1994).

REFERENCES

Ambrós, S., Hernández, C., Desvignes, J. C., and Flores, R. (1998). Genomic structure of three phenotypically different isolates of peach latent mosaic viroid: implications of the existence of constraints limiting the heterogeneity of viroid quasi-species. *J. Virol.* 72, 7397-7406.

Baumstark, T., Schröder, A. R. W., and Riesner, D. (1997). Viroid processing: Switch from cleavage to ligation is driven by a change from a tetraloop to a loop E conformation. *EMBO J.* 16, 599-610.

Bonfiglioli, R., McFadden, G. I., and Symons, R. H. (1994). *In situ* hybridization localizes avocado sunblotch viroid on chloroplast thylakoid membranes and coconut cadang cadang viroid in the nucleus. *Plant J.* 6, 99-103.

Bonfiglioli, R. G., Webb, D. R., and Symons, R. H. (1996). Tissue and intra-cellular distribution of coconut cadang cadang viroid and citrus exocortis viroid determined by *in situ* hybridization and confocal laser scanning and transmission electron microscopy. *Plant J.* 9, 457-465.

Branch, A. D., and Robertson, H. D. (1984). A replication cycle for viroids and other small infectious RNAs. *Science* 223, 450-454.

Branch, A. D., Benenfeld, B. J., and Robertson, H. D. (1985). Ultraviolet light-induced crosslinking reveals a unique region of local tertiary structure in potato spindle tuber viroid and HeLa 5S RNA. *Proc. Natl. Acad. Sci. U. S. A.* 82, 6590-6594.

Branch, A. D., Benenfeld, B. J., and Robertson, H. D. (1988). Evidence for a single rolling circle in the replication of potato spindle tuber viroid. *Proc. Natl. Acad. Sci. U. S .A.* 85, 9128-9132.

Conejero, V., and Semancik, J. S. (1977). Exocortis viroid: alteration in the proteins of *Gynura aurantiaca* accompanying viroid infection. *Virology* 77, 221-232.

Côte, F., and Perrault, J. P. (1997). Peach latent mosaic viroid is locked by a 2',5'-phosphodiester bond produced by *in vitro* self-ligation. *J. Mol. Biol.* 273, 533-543.

Daròs, J. A., Marcos, J. F., Hernández, C., and Flores, R. (1994). Replication of avocado sunblotch viroid: Evidence for a symmetric pathway with two rolling circles and hammerhead ribozyme processing. *Proc. Natl. Acad. Sci. U. S. A.* 91, 12813-12817.

Davies, J. W., Kaesberg, P., and Diener, T. O. (1974). Potato spindle tuber viroid. XII. An investigation of viroid RNA as a messenger for protein synthesis. *Virology* 61, 281-286.

Diener, T. O. (1971). Potato spindle tuber viroid. IV. A replicating, low molecular weight RNA. *Virology* 45, 411-428.

Diener, T. O. (1989). Circular RNAs: Relics of precellular evolution?. *Proc. Natl. Acad. Sci. U. S. A.* 86, 9370 - 9374.

Diener, T. O. (1996). Origin and evolution of viroids and viroid-like satellite RNAs. *Virus Genes* 11, 119-131.

Diener, T. O., Hammond, R. W., Black, T., and Katze, M. G. (1993). Mechanism of viroid pathogenesis: Differential activation of the interferon-induced, double-stranded RNA-activated, M_r 68000 protein kinase by viroid strains of varying pathogenicity. *Biochimie* 75, 533-538.

Ding, B., Kwon, M. O., Hammond, R., and Owens, R. (1997). Cell-to cell movement of potato spindle tuber viroid. *Plant J.* 12, 931-936.

Domingo, E, and Holland, J. J. (1994). Mutation rates and rapid evolution of RNA viruses. *In* "Evolutionary Biology of Viruses" (S. S. Morse, Ed.), pp. 161-184. Raven Press, New York, NY.

Elena, S. F., Dopazo, J., Flores, R., Diener, T. O., and Moya, A. (1991). Phylogeny of viroids, viroid-like satellite RNAs, and the viroidlike domain of hepatitis δ virus RNA. *Proc. Natl. Acad. Sci. U. S. A.* 88, 5631-5634.

Feldstein, P. A., Hu, Y., and Owens, R. A. (1998). Precisely full length, circularizable, complementary RNA: An infectious form of potato spindle tuber viroid. *Proc. Natl. Acad. Sci. U. S. A.* 95, 6560-6565.

Flores, R., Randles, J. W., Bar-Joseph, M., and Diener, T. O. (1998). A proposed scheme for viroid classification and nomenclature. *Arch. Virol.* 143, 623 - 629.

Garnsey, S. M., and Randles, J. W. (1987). Biological interactions and agricultural implications of viroids. *In* "Viroids and Viroidlike Pathogens" (J. S. Semancik, Ed.), pp. 127-160. CRC Press, Boca Raton, FL.

Góra, A., Candresse, T., and Zagórski, W. (1994). Analysis of the population structure of three phenotypically different PSTVd isolates. *Arch. Virol.* 138, 223-245.

Gross, H. J., Domdey, H., Lossow, C., Jank, P., Raba, M., Alberty, H., and Sänger, H. L. (1978). Nucleotide sequence and secondary structure of potato spindle tuber viroid. *Nature* 273, 203-208.

Hammond, R. (1994). *Agrobacterium*-mediated inoculation of PSTV cDNAs onto tomato reveals the biological effect of apparently lethal mutations. *Virology* 201, 36-45.

Hammond, R., Smith, D. R., and Diener, T. O. (1989). Nucleotide sequence and proposed secondary structure of *Columnea* latent viroid: a natural mosaic of viroid sequences. *Nucleic Acids Res.* 17, 10083-10094.

Harders, J., Lukacs, N., Robert-Nicoud, M., Jovin, J. M., and Riesner, D. (1989). Imaging of viroids in nuclei from tomato leaf tissue by *in situ* hybridization and confocal laser scanning microscopy. *EMBO J.* 8, 3941-3949.

Haseloff, J., Mohamed, N. A., and Symons, R. H. (1982). Viroid RNAs of the cadang-cadang disease of coconuts. *Nature* 229, 316-321.

Hernández, C., and Flores, R. (1992). Plus and minus RNAs of peach latent mosaic viroid self cleave *in vitro* via hammerhead structures. *Proc. Natl. Acad. Sci. U. S. A.* 89, 3711-3715.

Hutchins, C. J., Rathjen, P. D., Forster, A. C., and Symons, R. H. (1986). Self-cleavage of plus and minus RNA transcripts of avocado sunblotch viroid. *Nucleic Acids Res.* 14, 3627-3640.

Keese, P., and Symons, R. H. (1985). Domains in viroids: Evidence of intermolecular RNA rearrangements and their contribution to viroid evolution. *Proc. Natl. Acad. Sci. U. S. A.* 82, 4582-4586.

Kofalvi, S. A., Marcos, J. F., Cañizares, M. C., Pallás, V., and Candresse, T. (1997). Hop stunt viroid (HSVd) sequence variants from *Prunus* species: Evidence for recombination between HSVd isolates. *J. Gen. Virol.* 78, 3177-3186.

Koltunow, A. M., and Rezaian, M. A. (1988). Grapevine yellow speckle viroid. Structural features of a new viroid group. *Nucleic Acids Res.* 16, 849-864.

Lima, M. I., Fonseca, M. E. N., Flores, R., and Kitajima, E. W. (1994). Detection of avocado sunblotch viroid in chloroplasts of avocado leaves by *in situ* hybridization. *Arch. Virol.* 138, 385-390.

Liu, Y. H., and Symons, R. H. (1998). Specific RNA self-cleavage in coconut cadang-cadang viroid: Potential for a role in rolling circle mechanism. *RNA* 4, 418-429.

Marcos, J. F., and Flores, R. (1992). Characterization of RNAs specific to avocado sunblotch viroid synthesized *in vitro* by a cell-free system from infected avocado leaves. *Virology* 186, 481-488.

Martínez-Soriano, J. P., Galindo-Alonso, J., Maroon, C. J. M., Yucel, I., Smith D. R., and Diener, T. O. (1996). Mexican papita viroid: Putative ancestor of crop viroids. *Proc. Natl. Acad. Sci. U. S. A.* 93, 9397-9401.

Mühlbach, H. P., and Sänger, H. L. (1979). Viroid replication is inhibited by α-amanitin. *Nature* 278, 185-188.

Navarro, B, and Flores, R. (1997). Chrysanthemum chlorotic mottle viroid: Unusual structural properties of a subgroup of self-cleaving viroids with hammerhead ribozymes. *Proc. Natl. Acad. Sci. U. S. A.* 94, 11262-11267.

Palukaitis P. (1987). Potato spindle tuber viroid: Investigation of the long-distance, intra-plant transport route. *Virology* 158, 239-241.

Prody, G. A., Bakos, J. T., Buzayan, J. M., Schneider, I. R., and Bruening G. (1986). Autolytic processing of dimeric plant virus satellite RNA. *Science* 231, 1577-1580.

Qu, F., Heinrich, C., Loss, P., Steger, G., Tien, P., and Riesner, D. (1995). Multiple pathways of reversion in viroids for conservation of structural domains. *EMBO J.* 12, 2129-2139.

Rakowski, A. G., and Symons, R. H. (1989). Comparative sequence studies of variants of avocado sunblotch viroid. *Virology* 173, 352-356.

Rakowski, A. G., and Symons, R. H. (1994). Infectivity of linear monomeric transcripts of citrus exocortis viroid: terminal sequence requirements for processing. *Virology* 203, 328-335.

Rezaian, M. A. (1990). Australian grapevine viroid — evidence for extensive recombination between viroids. *Nucleic Acids Res.* 18, 1813-1818.

Sägesser, R., Martínez, E., Tsagris, M., and Tabler, M. (1997). Detection and isolation of RNA-binding proreins by RNA-ligand screening of a cDNA expression library. *Nucleic Acids Res.* 25, 3816-3822.

Salazar, L. F., Hammond, R. W., Diener, T. O, and Owens, R. A. (1988). Analysis of viroid replication following *Agrobacterium*-mediated inoculation of non-host species of potato spindle tuber viroid cDNA. *J. Gen. Virol.* 69, 879-889.

Sano, T., Candresse, T., Hammond, R. W., Diener, T. O., and Owens, R. A. (1992). Identification of multiple structural domains regulating viroid pathogenicity. *Proc. Natl. Acad. Sci. U. S. A.* 89, 10104-10108.

Sano, T., and Ishiguro, A. (1998). Viability and pathogenicity of intersubgroup viroid chimeras suggest possible involvement of the terminal right region in replication. *Virology* 240, 238-244.

Schindler, I. M., and Mühlbach, H. P. (1992). Involvement of nuclear DNA-dependent RNA polymerases in potato spindle tuber viroid replication: A reevaluation. *Plant Sci.* 84, 221-229.

Semancik, J. S., and Szychowski, J. A. (1994). Avocado sunblotch disease: A persistent viroid infection in which variants are associated with differential symptoms. *J. Gen. Virol.* 75, 1543-1549.

Semancik, J. S., Szychowski, J. A., Rakowski, A. G., and Symons, R. H. (1993). Isolates of citrus exocortis viroid recovered by host and tissue selection. *J. Gen. Virol.* 74, 2427-2436.

Tabler, M., Tzortzakaki, S., and Tsagris, M. (1992). Processing of linear longer-than-unit-length potato spindle tuber viroid RNAs into infectious monomeric circular molecules by a G-specific endoribonuclease. *Virology* 190, 746 -753.

Visvader, J. E., and Symons, R. H. (1985). Eleven new sequence variants of citrus exocortis viroid and the correlation of sequence with pathogenicity. *Nucleic Acids Res.* 5, 2907-2920.

Visvader, J. E., and Symons, R. H. (1986). Replication of *in vitro* constructed viroid mutants: Location of the pathogenicity-modulating domain in citrus exocortis viroid. *EMBO, J.* 13, 2051-2055.

Wassenegger, M., Heimes, S., and Sänger, H. L. (1994). An infectious viroid RNA replicon evolved from an *in vitro*-generated non-infectious viroid deletion mutant via a complementary deletion *in vivo*. *EMBO J.* 13, 6172 - 6177.

12 MOLECULAR BIOLOGY OF TRANSGENIC PLANTS

Chuni L. Mandahar

INTRODUCTION

Conventional plant breeding methods have been commonly used for introducing natural resistance genes into plants for controlling plant virus diseases. Introduction of one or more genes (transgene) from one to the other plant, which in many cases is not possible by conventional breeding methods because of sexual incompatibility, can also be done by plant transformation methods, based on DNA recombinant technology, to produce the transformed/transgenic plant. Such recombination techniques have proved very useful in several ways: they offer the possibility of genetic diversity, save time by avoiding cross-breeding, allow specific improvements in crops while preserving the desirable characteristics of an inbred line, and create novel types of resistance genes that do not occur in nature

Hamilton (1980) first raised the possibility of imparting resistance, called pathogen-derived resistance (PDR), to plants by transforming them with the genes of a pathogen. Sanford and Johnston (1985) postulated a generalized concept about it. Virus-resistant cultivars are now being developed by introducing pathogen (virus)-derived resistance genes (in the form of their cDNA) into the existing elite commercial inbred hybrids by recombinant technology to produce the transgenic plants which retain all the other characteristics of the parent. Thus, except for the virus-resistance trait, all transgenic lines of ZW-20 squash inbred hybrid had the same growth rate, fruiting time, and susceptibility to powdery mildew, whiteflies, aphids and papaya ringspot virus as their non-transgenic counterparts (Tricoli et al., 1995; Fuchs and Gonsalves, 1995).

The first transgenic plant possessing virus gene-derived, genetically-engineered resistance was the tobacco plant expressing capsid protein (CP) gene of tobacco mosaic virus (TMV) (Powell et al., 1986). Soon after and since then many other reports came in where CP and other viral genes from other plant RNA viruses successfully protected (or cross-protected) the transgenic plant against infection by that (homologous) virus (homologous resistance) as well as against infection by heterologous virus(es) (heterologous resistance). This resistance has been exploited experimentally over a wide scale (reviewed by: Hemenway et al., 1990; Beachy, 1993; Fitchen and Beachy, 1993; Baulcombe, 1994; Lomonossoff, 1995; Baulcombe, 1996; Palukaitis and Zaitlin, 1997; Beachy, 1997; Dempsey et al.,

1998) but practically in only one case (Tricoli *et al.,* 1995; Fuchs and Gonsalves, 1995). Viral genes of DNA plant viruses, like African cassava mosaic virus, beet curly top virus, cauliflower mosaic virus (CaMV), tomato yellow leaf curl virus and some others, have also been incorporated into plants for producing virus-resistant transgenic plants (Frischmuth and Stanely, 1994, 1998; Noris *et al.,* 1996).

Capsid protein gene, generally singly but sometimes in multiple copies, has been mostly employed for developing transgenic resistance. This gene, from at least 35 viruses of 15 different viral groups (Palukaitis and Zaitlin, 1997), gives fairly good control of diseases of annual plants at experimental level. Transgenic squash line ZW-20 is a multiple virus-resistant plant. It contains CP genes of zucchini yellow mosaic virus (ZYMV) and watermelon mosaic virus 2, is completely resistant to both these viruses and has been approved for release by USDA and Environmental Protection Agency of the USA for commercial use (Tricoli *et al.,* 1995; Fuchs and Gonsalves, 1995). Nearly the same is true for papaya while such viral CP-mediated resistant transgenic plants of several other crops are under development. Capsid protein-mediated resistance is narrow to broad-based from being effective against the specific virus from which CP was derived to against several unrelated viruses.

Viral replicase gene, also commonly employed for developing transgenic resistance in plants (Palukaitis and Zaitlin, 1997), imparts nearly full immunity to transgenic plants against infection. But this resistance is generally effective only against the particular virus providing the replicase gene and, consequently, is of a very limited nature. Viral movement protein (MP) gene(s), generally in defective/mutant form but also in native form, has also been used for production of transgenic resistant plants (Beck *et al.,* 1994; Cooper *et al.,* 1995; Tacke *et al.,* 1996; Ares *et al.,* 1998). These plants show broad-spectrum antiviral resistance. Tobacco mosaic virus MP [protein (p) 30] when expressed in mutated form in tobacco plants imparted resistance to it against several viruses belonging to alfamo-, caulimo-, cucumo-, nepo-, tobamo-, and tobraviruses (Cooper *et al.,* 1995). In contrast, non-modified potato virus X and TMV MPs imparted specific protection against that very virus but not against homologous virus (Ares *et al.,* 1998). Other viral sequences used for engineering virus-resistant transgenic plants are entire viral genomes; read-through, modified, mutated, truncated and/or ligated transgenes; antisense RNA; and satellite RNA.

Transgenic plants show no or decreased and/or delayed symptoms upon infection by homologous/heterologous virus(es). This is accompanied by reduced multiplication and accumulation of challenge virus and viral gene product (CP) transcripts in transgenic plants. Movement of challenge virus in transgenic plants is also impaired. Transgene is vertically transmissible because of its being meiotically stable over several generations. Transgenic plants can thus be used as parents for breeding purposes and transgenic resistance is successful under field conditions for several future generations. These two characteristics, decreased symptoms and parent-to-progeny transmissible resistance, make the development and field use of transgenic plants so important.

Pathogen-derived resistance is broadly divisible into two types (Baulcombe, 1996; Beachy, 1997; Ratcliff *et al.,* 1997), viral protein-mediated and RNA-mediated resistance. The viral protein may or may not accumulate during viral protein-mediated resistance, which is generally broad-based, protects the transgenic plants against several viruses, and is more commonly operative in transgenic plants. The second type, the RNA-mediated resistance, seemingly is analogous to the gene silencing process in which expression of an endogenous gene or a transgene is inhibited by the insertion of a homologous sequence or a non-translatable sequence or an intact gene in a sense orientation into the genome of a plant or transgenic plant. It has been dealt in detail in the next chapter (Chapter 13).

Viral CP is a multifunctional protein. It is needed for viral assembly, RNA binding, and for protecting the viral genome. It is also the primary site of and is involved in several other viral functions in different plant viruses. It is involved in movement of some viruses within the host plants, is essential for genome activation, infection and replication of some plant viruses, is directly or indirectly involved in accumulation of viral plus-strand RNA of some viruses, plays a crucial role in virus transmission by insect vectors since it is the recognition site for the vectors, is involved in recognition by host resistance functions (as TMV in *Nicotiana* species with *N* genotype), and has often been used as a transgene for imparting resistance to transgenic plants against virus diseases. It will not be surprising if its last use interferes with its normal role in other functions. Consequently, CP-mediated transgenic plants can pose certain grave risks because they, after infection with a second virus, are like the plants naturally infected with two viruses and can be expected to show heteroencapsidation and recombination (Hull, 1990, 1994; de Zoeten, 1991; Tepfer, 1993; Gibbs, 1994; and Palukaitis and Zaitlin, 1997). Harris (1998) discusses the environmental, regulatory, and risk assessment of transgenic plants.

GENETIC VARIABILITY OF CHALLENGE VIRUS

Recombination

Genetic recombination, formation of chimeric RNA molecules from two segments present on different parental RNAs or present at separate places on the same parental RNA, in plant viruses was first recognized in brome mosaic virus (BMV) on the basis of *in vivo* repair of a deleted genomic RNA (Bujarski and Kaesberg, 1986) (Chapter 5). Later, recombination was detected both in several naturally occurring virus isolates and in experimental virus systems belonging to alfamo- (alfalfa mosaic virus-AMV), bromo- (BMV), caulimo- (CaMV), carmo- (turnip crinkle virus), hordei- (barley stripe mosaic virus), luteo- (potato leaf roll virus-PLRV), and tobravirus (tobacco rattle virus) groups (Chapter 5). Recombination, although not regarded to be common event at present, is considered to be one of the driving forces of plant virus evolution (Dolja and Carrington, 1992; Koonin and Dolja, 1993).

Recombination is of three types: homologous, aberrant homologous, and non-homologous (Simon and Bujarski, 1994) or, in the latest terminology, similarity-essential, similarity non-essential, and similarity-assisted recombination (Nagy and Simon, 1997). Formation of chimeric RNAs can occur by three different mechanisms: RNA cleavage (breakage) and religation (ligation), template switching (copy choice), and breakage-induced template switching (Nagy and Simon, 1997).

Recombination also occurs between an infecting virus RNA and the viral sequences expressed as a transgene in transgenic plants. However, only a few cases have been reported so far. Gal *et al.* (1992) first demonstrated the above type of recombination leading to production of infectious virus following inoculation of a defective virus on a transgenic plant containing the complementary gene as transgene in its genome. Transgenic *Brassica napus* plants containing CaMV translational transactivation gene (gene VI) were inoculated with *Agrobacterium tumefaciens* containing the rest of the CaMV genome. Infectious and replicating CaMV was produced which incited CaMV symptoms in turnip leaves. This happened because of recombination between the transgene and the complementary virus. Both DNA and RNA recombination may have resulted in the formation of infectious CaMV.

The gene VI of CaMV strain D4 determines systemic infection of solanaceous species, including *N. bigelovii,* by this virus. Schoelz and coworkers employed transgenic *N. bigelovii* plants expressing D4 gene VI and inoculated them with various CaMV strains (H13, CM1841, or W260) that are unable to infect systemically the non-transformed *N. bigelovii* plants. All these CaMV strains could systemically infect the transgenic *N. bigelovii* plants because of the recombination between the transgene and the superinfecting CaMV strain H13 (Schoelz *et al.,* 1991), strain CM1841 (Schoelz and Wintermantel, 1993), and strain W260 (Wintermantel and Schoelz, 1996). Thus, CaMV could acquire a copy of its gene VI both from transgenic *Brassica napus* (Gal *et al.,* 1992) as well as transgenic *N. bigelovii* plants (Schoelz and Wintermantel, 1993) containing it as the transgene.

Nicotiana benthamiana plants transformed with 3` two-third part of cowpea chlorotic mottle virus (CCMV) CP were infected with a CCMV deletion mutant lacking the 3` one-third part of CP gene. Recombination between the two mRNAs generated the functional CP so that the deletion mutant, which was unable to infect the transgenic *N. benthamiana* plants, did so successfully (Greene and Allison. 1994). The deletion was restored and four viable recombinant viruses were regenerated through aberrant homologous recombination and they could systemically infect both *N. benthamiana* and cowpea plants. Each recombinant was a distinct variant of wild-type CCMV (Greene and Allison, 1994).

Frischmuth and Stanley (1998) transformed *N. benthamiana* plants with CP constructs of African cassava mosaic geminivirus of genus *Begomovirus*, known earlier as subgroup III. These transgenic plants, when inoculated with a CP deletion mutant, which induces mild systemic symptoms in non-transformed plants, developed typical severe systemic symptoms of this virus because the recombinant progeny possessed wild-type characters.

In studies by Gal *et al.* (1992), and Greene and Allison (1994) deletion mutants defective for systemic infection in their respective hosts were inoculated to transgenic plants and the recombinants between the mutant virus and the transgene could then cause systemic infection. Similarly, CaMV strains H13, CM1841, and W260 fail to systemically infect nontransgenic *N. bigelovii* plants but can infect the transformed *N. bigelovii* plants expressing gene VI of CaMV strain D4 (Schoelz and

Wintermantel, 1993; Wintermantel and Schoelz, 1996). Thus, in all these cases, recombinants acquired enhanced competitiveness and selective advantage (of being able to infect the transgenic plants) (Schoelz and Wintermantel, 1993; Schoelz and Falk, 1993; Falk and Bruening, 1994). A recombinant virus in fact was the predominant virus in transgenic plants inoculated with the CaMV CM1841 strain. Wintermantel and Schloez (1996), in contrast, reported that, even if the wild-type virus could systemically infect the host, the recombinants could be isolated from transgenic plants even then. They referred to such a condition as moderate to weak selection pressure in contrast to the strong selection pressure in which the virus was unable to infect a host/transgenic host unless the recombination event took place as happened in all other cases mentioned above.

The mechanism of above recombination between CaMV and the transgene gene of transgenic plants is based on two template switches occurring during the reverse transcription of viral 35S RNA. The recombination takes place between the two RNA molecules and is mediated by viral reverse transcriptase. During this recombination, the coding region of gene VI of the infecting virus is completely replaced by the gene VI transgene (Wintermantel and Schoelz, 1996).

Rate of recombination between viral gene and transgene in transgenic plants varies even though the RNA and DNA viruses utilize similar mechanisms of recombination. A recombinant form of CCMV was obtained from 3% of the transgenic *N. benthamiana* plants (Greene and Allison, 1994) while a recombinant form of CaMV CM1841 was present in all of the 36% of the infected *N. bigelovii* plants (Wintermantel and Schoelz, 1996). Falk and Bruening (1994) postulated that the potential of recombination to occur between transgene RNA and viral genomic RNA is likely to be equal to the potential for natural RNA- RNA recombination during natural conventional as well as subliminal infections.

It is clearly established that recombination between the transgene and a related virus infecting a transgenic plant does occur. But, does such a recombination lead to the formation of a new recombinant virus, which is more damaging than the infecting virus (parent)? Information on this point is not available yet but it may not be surprising if in some rare instances a new and more virulent virus gets generated. Since generally no resistance is operative in the transgenic plants for viruses that are distantly related to the virus donating the transgene, homologous or nonhomologous recombinants may arise in principle. But there is no such evidence for the time being at least (Palukaitis and Zaitlin, 1997).

Mutation

Wassenegger *et al.* (1994) produced a transgenic tobacco plant by stably incorporating a non-infectious 350 nucleotide long potato spindle tuber viroid (PSTVd) cDNA into its genome. However, the subsequently replicating viroid RNA was found to be a 341 nucleotide long truncated infectious PSTVd. It was produced from the primary transcript of the 350 base pair long cDNA transgene by deletion of 9 nucleotides *in planta*. In addition, five more sequence changes occurred: deletion of one A nucleotide at position 55-60; addition of one U nucleotide in pathogenic domain; transition of G115 to A in central domain, of G254 to A, and of C259 to U in central conserved domain. These five mutations were ascribed to copying errors of RNA polymerase.

HETEROENCAPSIDATION

Assembly of virus particles is a specific process based on recognition and interaction between specific sites on CP molecules and viral genome so that capsid normally encapsulates only its own RNA. However, errors can occur particularly between closely related viruses or different strains of a virus but also even between unrelated viruses so that encapsidation of genome of one virus can take place into the capsid of another virus when these viruses coinfect a host cell. This has been variously called as heteroencapsidation, transcapsidation, heterologous encapsidation or genome masking. It is a well-documented phenomenon (Dodds and Hamilton, 1976; Rochow, 1977; Falk and Duffus, 1981). Such an association between an umbravirus and a helper luteovirus may cause serious plant diseases as the damaging groundnut rosette disease in Africa caused by the dependent groundnut rosette virus and the helper ground rosette assister virus (Reddy *et al.,* 1985). Under unnatural and abnormal *in vitro* conditions, this process can also occur between structurally different and unrelated viruses like encapsulation of elongated barley stripe mosaic virus RNA in capsid of icosahedral BMV (Dodds and Hamilton, 1976) and of elongated TMV RNA in purified cucumber mosaic virus (CMV) CP (Chen and Francki, 1990). However, such abnormal heteroencapsidation has never been reported in nature even though simultaneous infection by both viruses is common.

Does heteroencapsidation of the superinfecting viral genome occur in the constitutive CP synthesized in the CP-mediated transgenic plants? A few such cases have been reported. Encapsidation of genome of CP-defective TMV DT-1 strain and CP-less TMV DT-1G strain in normal TMV U1 CP was possibly the first such report (Osbourne *et al.,* 1990). Potato plant lines Bt6 and Bt10 are resistant to potato virus Y strain N (PVY^N) so that these PVY^N CP-mediated transgenic plants did not synthesize and accumulate CP of this strain. However, CP of strain N accumulated when these transgenic plants were infected by the related PVY strain O (PVY^O) (Farinelli *et al.,* 1992). Immune electron microscopy showed heterologous encapsidatin between the infecting strain O RNA and transgenic PVY^N CP to produce two types of infectious particles: CP of PVY^N encapsidating RNA of PVY^O and CP of PVY^O containing its own RNA. Similarly, transgenic *N. benthamiana* plants, expressing CP of an aphid-transmissible strain of plum pox potyvirus (PPV-D), upon inoculation with an aphid non-transmissible strain of zucchini yellow mosaic potyvirus (ZYMV-NAT) contained virus particles with PPV-CP and ZYMV-NAT RNA as well as particles with ZYMV-NAT CP and ZYMV-NAT RNA (Lecoq *et al.,* 1993).

Heterologous encapsidation of CMV RNA occurred in AMV CP expressed in transgenic tobacco plants (Candelier-Harvey and Hull, 1993). Eleven of the 33 transgenic plants that were examined two weeks after CMV inoculation showed transcapsidation. Beet necrotic yellow vein virus genome, upon infecting the transgenic plants expressing CP of plum pox potyvirus (PPV), gets transcapsidated (Maiss *et al.,* 1994).

Because of the crucial role played by CP in various functions, heteroencapsidation of viral genomes can have several damaging and far-reaching consequences.

Transmission of Non-transmissible Virus Isolates

The CP controls specificity of virus transmission in some cases in part or wholly. Thus, heteroencapsidation of a vector non-transmissible viral genome in the capsid of a vector-transmissible strain (synthesized in a CP-transformed plant) can lead to its transmission by the vector from a transgenic plant expressing CP of the same (or a related) but a vector transmissible virus strain. *Myzus persicae* could acquire and transmit an aphid non-transmissible strain of ZYMV (ZYMV-NAT) from transgenic *N. benthamiana* plants expressing CP of aphid-transmissible strain of PPV but not from *N. benthamiana* control (non-transformed) plants (Lecoq *et al.,* 1993). Thus, transmission of the transcapsidated viral genome could be altered by passage through a transgenic plant. This could have unpredictable effects on host range and/or virulence of a virus.

Complementation

Complementation of defective viruses *in trans* in plants transgenic for the appropriate complementing gene has been commonly demonstrated. It is a common observation that wild-type MP gene used for producing transgenic plants complements a transport deficient mutant when inoculated in that transgenic plant. A MP-frameshift mutant of U1 (common) TMV strain could infect a transgenic tobacco plants expressing the wild-type TMV MP gene resulting in both local and systemic infection but failed to do so in non-transgenic plants (Holt and Beachy, 1991). *Solanum brevidens* was transformed with TMV MP gene. Systemic spread of the *ts* mutant Ls-1, which is defective for transport in plants at 33° C, occurred in these transgenic plants at the non-permissive temperature (Valkonen *et al.,* 1995). Moreover, accumulation of PVY in these transformed *S. brevidens* plants was enhanced compared to its extremely low accumulation in non-transformed plants. This shows that TMV assists in the cell-to-cell spread and/or replication of PVY in transformed *S. brevidens* plants (Valkonen *et al.,* 1995). Gera *et al.* (1995) also got similar results with an MP-deficient TMV mutant which could only move from cell-to-cell in transgenic plants expressing TMV MP gene.

Transgenic tobacco plants were transformed with the cucumber mosaic virus 3a protein, which is the MP of this virus (Kaplan *et al.,* 1995). These transgenic plants complemented the long-distance movement of and systemic infection by CMV deletion mutants that can not move in normal tobacco plants. Transgenic tobacco plants also complemented the long-distance movement of a pseudorecombinant cucumovirus defective for this function in tobacco and also the cell-to-cell but not the long-distance movement of two other related viruses, namely peanut stunt cucumovirus and BMV (Kaplan *et al.,* 1995).

Cell-to-cell movement of defective virus transcripts of red clover necrotic mosaic virus (RCNMV) could be rescued through complementation, rather than recombination, with functional transgenic MP sequences expressed as rnRNA in transgenic *N. benthamiana* plants (Lommel and Xiong, 1991). Transgenic tobacco plants expressing p24, one of the three proteins controlling local movement of potato virus X (PVX), facilitated cell-to-cell movement of a PVX mutant containing a frameshift mutation in p24 (Ares *et al.,* 1998).

Transgenic tobacco plants (*N. tabacum* cv. Xanthi) expressing CP of TMV U1 complemented both the assembly and long-distance spread of defective TMV strains DT-1 and DT-1G due to transcapsidation (Osbourne *et al.,* 1990). Holt and Beachy (1991) found that the TMV CP-frameshift mutant was able to move systemically in transgenic tobacco plants expressing the wild-type TMV CP gene but was unable to do so in non-transformed tobacco plants. This happened due to complementation.

Nicotiana tabacum cv. Samsun NN was transformed with RNA1 of AMV. Prototplasts obtained from this transgenic plant could be infected by the AMV nucleoproteins containing RNAs 2 and 3 because of the complementation between these and the transgene (van Dun *et al.,* 1988). The *A* component of the bipartite tomato golden mosaic virus (TGMV) codes for a viral protein, AL1, which is essential for replication of viral DNA. Several of the transgenic *N. benthamiana* plants that were expressing AL1 ORF complemented a TGMV *A* variant, having a mutation in AL1, upon coinoculation with *B* component of virus (Hanley-Bowdoin *et al.,* 1990). Thus even in absence of TGMV *A* component, single- and double-stranded forms of *B* component were synthesized in leaf discs of the transgenic line.

Extension of Host Range

Synthesis of PVY^N CP in the resistant transgenic potato lines (Farinelli *et al.,* 1992) can be interpreted as extension of host range, caused by a transgene, of a virus to an earlier resistant transgenic plant. Similarly, expression of PSTVd in the resistant tobacco plants generated a new strain PSTVd-NT (by a single nucleotide substitution from C to U at position 259) which is able to infect and multiply upon mechanical inoculation into tobacco plants (Wassenegger *et al.,* 1996). This new strain could only have evolved *de novo* in transgenic tobacco plants and can also be regarded a case of host range extension. Schoelz and Wintermantel (1993) also found that expansion of host range of a virus occurs through complementation and recombination in transgenic plants.

INHERITANCE

Capsid protein gene of white leaf strain of CMV conferred a high level of resistance in transgenic tomato plants against American, Asian, European, and Oceanic strains of both the serogroups of CMV. Study of genetic populations generated by crossing and back-crossing of susceptible cultivars with a homologous CMV-resistant tomato line (TT 5-007-11) showed that a single dominant gene, designated *cmv*, conferred this high level of resistance to the transgenic line. This *cmv* gene contains the CMV CP transgene and the CMV resistance incorporated by it could be used for production of F1 commercial hybrids possessing this resistance (Provvidenti and Gonsalves, 1995).

SYNERGISM

Some unrelated viruses, when present together, show synergistic interaction leading to symptom intensification. Potyvirus-potexvirus and tobamovirus-cucumovirus systems are known to show such reactions. Potyvirus stimulates replication of the potexvirus causing intensification of disease symptoms.

Such a situation may also prevail when a potyvirus/tobamovirus transgenic plant is infected with the second corresponding (potexvirus/cucumovirus) virus. Only one case of synergy has been reported so far. The 5' P-1/helper component - protease genes of a potyvirus, when expressed as a transgene, substituted for the entire potyvirus genome resulting in synergism (Vance *et al.*, 1995). But this was not true of the potyvirus NIb polymerase gene-expressed trangenic plants after PVX infection (Vance *et al.*, 1995) or the transgenic plants expressing TMV 54 kDa gene or defective CMV 2a polymerase gene after infection with the corresponding second (CMV/TMV) virus (Golemboski *et al.*, 1990; Zaitlin *et al.*, 1994). Thus, whether a particular potyvirus/tobamovirus transgene-expressing transgenic plant will exhibit synergistic reaction after infection with the second corresponding virus need to be worked out in each case.

OTHER EFFECTS

Palukaitis and Zaitlin (1997) have visualized some other ways in which transgenes can influence various plant functions. Transgenic weeds showing virus resistance are not likely to act as virus reservoirs for virus spread by vectors to important crop plants. This may decrease the spread of the particular virus to the adjacent non-transgenic crops leading to a possible complete change in epidemiology of that virus disease. Then viral transgenes may influence seed transmissibility of the transgene-donating virus. This may result in increased seed transmissibility in transgenic plants of the virus whose CP (or any other viral) gene is being expressed. Since transcapsidation occurs in CP gene-mediated transgenic plants, there is the possibility of an enhanced seed transmission of a transcapsidated virus.

Spread of transgene-donating virus will be curtailed in transgenic plants. It may leave the area clear for a less competitive virus to spread and become the dominant pathogen in transgenic crops. However, it is not likely to occur since introduction of a gene from one to another species also causes transfer of many linked genes some of which may alter susceptibility of a transgenic plant to unrelated pathogens, including viruses, bacteria, and fungi (Palukaitis and Zaitlin, 1997).

A related phenomenon is the selection of resistance-breaking strains which normally arise under natural conditions (of growing resistant varieties) by one or a few nucleotide changes in an avirulent strain (Mandahar, 1990; Santa Cruz and Baulcombe, 1995). Such a situation may operate with regard to CP-mediated resistance. This in any case can not happen in case of replicase-mediated resistance. However, resistance gene may not be significantly functional against closely related viruses or even strains of the same virus differing in 8-10 nucleotide sequences so that such viruses or strains may replace the viruses against which the resistance is operative. In this way related viruses may get selected and infect the transgenic plant although, strictly speaking, no new resistance -breaking strains have been produced (Palukaitis and Zaitlin, 1997).

All the above possibilities about the induced resistance in weed hosts, increased seed transmissibility, changed virus ecology, and selection of resistance breaking strains by the transgenic plants have as yet no experimental support.

CONCLUSIONS

Cooper *et al.* (1994) sought evidence of RNA-RNA recombination and heteroencapsidation in transgenic tobacco plants cv. Xanthi, expressing the 1.3kb 3` non-coding sequence from a birch isolate of cherry leaf roll nepovirus, after mechanical inoculation with a rhubarb isolate of this virus. No evidence was obtained indicating that recombination and heteroencapsidation are common phenomena. But even then conscious efforts must be made and precautions taken to prevent the occurrence of these phenomenon in transgenic plants. Lecoq *et al.* (1993), Tepfer (1993), Lindbo *et al.* (1994), and Hull (1994) have suggested the possible means of avoiding the potential risks.

One such measure is the modification of the CP gene without compromising its potential for imparting transgenic protection. This appears possible since the area of CP controlling vector specificity has no role in generating transgenic protection (Atreya, C. P., *et al.,* 1990; Atreya, P. L., *et al.,* 1991; Lindbo and Dougherty, 1992a). Modification of this region would prevent the transmission of the recombinant virus by the vectors. A triplet of amino acids (Asp-Ala-Gly) in capsid of potyviruses is essential for vector transmission. Mutations in this triplet of the CP gene, to be used as the transgene, would render the heteroencapsidated virus (if heteroencapsidation does occur) non-transmissible from such transgenic plants (Atreya *et al.,* 1991). In this way, any gene could be tailored to eliminate its potentially risky parts, if known. A still safer method is the use of non-translatable CP gene which also provides protection in some cases (Lindbo and Dougherty, 1992b) or of genes encoding non-structural viral proteins (Lomonossoff, 1992). These precautions will not permit heteroencapsidation to occur.

The frequency of heteroencapsidation during large-scale cultivation of transgenic plants may not increase in general because the synthesis of viral capsid protein in the transgenic plants is lower than its level in normal plants infected by the same virus.

Moreover, infectivity of heteroencapsidated viral genomic RNAs on CP-expressing transgenic plants is less than the corresponding wild-type virus. Heteroencapsidation in CP-mediated transgenic plants may be only a little more than the heteroencapsidation occurring naturally while the importance of such heteroencapsidated virus spread by the vector is likely to be still less (Bruening and Falk, 1994)

However, in certain special situations heteroencapsidation may occur even then. Such a condition can be easily visualized when transgenic plants, expressing CP gene of a luteovirus, are infected by the specific umbravirus. But this and all other cases of heteroencapsidation have one saving grace: the changed feature, that is, the different CP of the virus particles, is restricted to only one generation since the viral genome and hence the viral genetic information remains the same. Consequently, the progeny virus particles will have their own CP enclosing their own RNA. Hence, damage caused will have only a limited impact; in any case, further spread of the virus progeny is not likely to take place if original CP does not play any part in virus transmission.

Frequency of heteroencapsidation and/or recombination needs to be explored more thoroughly to get an idea about whether the use of CP gene for development of transgenic plants can prove a source of potential danger. Nevertheless, as Falk and Bruening (1994) express, the possible advantages of engineered resistance genes are far greater than the possible little danger arising out of the likelihood of generating new and harmful viruses at significantly greater rates than those being generated during the natural course of events.

REFERENCES

Ares, X.,Calamante, G., Cabral, S., Lodge, J., Hemenway, P., Beachy, R. N., and Mentaberry, A. (1998). Transgenic potato plants expressing potato virus X ORF2 protein (p24) are resistant to tobacco mosaic virus and Ob tobamoviruses. *J. Virol.* 72, 731-738.

Atreya, C. P., Raccah, B., and Pirone, T. P. (1990). A point mutation in the coat protein abolishes aphid virus transmissibility of a potyvirus. *Virology* 178, 161- 165.

Atreya, P. L., Atreya, C. D., and Pirone, T. P. (1991). Amino acid substitutions in coat protein result in loss of insect transmissibility of a plant virus. *Proc. Natl. Acad. Sci. USA* 88, 7887-7891.

Baulcombe, D. (1994). Novel strategies for engineering virus resistance in plants. *Curr.Opinion Biotech.* 5, 117-124.

Baulcombe, D. (1996). Mechanisms of pathogen-derived resistance to viruses in transgenic plants.*Plant Cell* 8, 1833-1844.

Bruening, G. M., and Falk, B. W. (1994). Risks in using transgenic plants? Response. *Science* 264, 1651-1652.

Beachy, R. N. (1993). Transgenic resistance to plant viruses. *Semin. Virol.* 4, 327-416.

Beachy, R. N. (1997). Mechanisms and applications of pathogen-derived resistance in transgenic plants. *Curr. Opin. Biotech.* 8, 215-220.

Beck, D. L., Van Dolleweerd, C. J., Lough, T. J., Balmori, E., Voot, D., Andersen, M. T., O' Brien, I. E. M., and Forster, R. L. S. (1994). Disruption of virus movement confers broad-spectrum resistance against systemic infection by plant viruses with a triple gene block. *Proc. Natl. Acad. Sci. USA* 91, 10310-10314.

Bujarski,J. J., and Kaesberg, P. (1986). Genetic recombination between RNA components of a multi-partite plant virus. *Nature* 321, 528-531.

Bujarski, J. J., and Nagy, P. D. (1996). Different mechanisms of homologous and nonhomologous recombination in brome mosaic virus: Role of RNA sequences and replicase proteins. *Semin. Virol.* 7, 363-372.

Candelier-Harvey, P., and Hull, R. (1993). Cucumber mosaic virus genome is encapsidated in alfalfa mosaic virus coat protein expressed in transgenic tobacco plants. *Transg. Res.* 2, 277-285.

Chen, B., and Francki, R. I. B. (1990). Cucumovirus transmission by the aphid *Myzus persicae* is determined solely by the viral coat protein. *J. Gen. Virol.* 71, 939-944.

Cooper, B., Lapidot, M., Heick, J. A., Dodds, J. A., and Beachy, R. N. (1995). A defective movement protein of TMV in transgenic plants confers resistance to multiple viruses whereas the functional analog increases susceptibility. *Virology* 206, 307-313.

Cooper, J. I., Edwards, M. L, Rosenwasser, O., and Scott, N. W. (1994). Transgenic resistance genes from nepoviruses: Efficacy and other properties. *New Zealand J. Crops & Hort. Sci.* 22, 129-137.

de Zoeten, G. A. (1991). Risk assessment: Do we let history repeat itself? *Phytopathology* 81, 585-586.

Dempsey, D. M. A., Silva, H., and Klessig, D. F. (1998). Engineering disease and pest resistance in plants. *Trends Microbiol.* 54, 54-61.

Dodds, J. A., and Hamilton, R. I. (1974). Masking of the RNA genome of tobacco mosaic viruses by the protein of barley stripe mosaic virus in doubly infected barley. *Virology* 59, 418-427.

Dodds, J. A. and Hamilton, R. I. (1976). Structural interactions between viruses as a consequence of mixed infections. *Adv. Virus Res.* 20, 33-86.

Dolja, V. V., and Carrington, J. C. (1992). Evolution of positive-strand RNA viruses. *Semin. Virol.* 3, 315-326.

Falk, B. W., and Bruening, G. (1994). Will transgenic crops generate new viruses and new diseases? *Science* 263, 1395-1396.

Falk, B. W., and Duffs, J. E. (1981). Epidemiology of helper-dependent persistent aphid-transmitted virus complexes.*In* " Plant Disease and Vectors: Ecology and Epidemiology" (K. Maramorosch and K.F. Harris, Eds.), pp. 162-181. Academic Press , New York.

Falk, B. W., Passmore., B. K., Watson, M. T., and Chin, L.- S. (1995). T he specificity and significance of heterologous encapsidation of virus and virus- like RNAs. *In* "Biotechnology and Plant protection :Viral Pathogenesis and Disease Resistance" (D. D. Bills *et al.*, Eds.), pp. 391-415. World Scientific, Singapore.

Farinelli, L., Malnoe, L.,and Collet, G. F. (1992). Heterologous encapsidation of potato virus Y strain O (PVY^O), with the transgenic coat protein of PVY strain N (PVY^N) in *Solanum tuberosum* cv .Bintje. *Bio/Technology* 10, 1020-1025.

Fitchen, J. H.,and Beachy, R. N. (1993). Genetically engineered protection against viruses in transgenic plants. *Annu. Rev. Microbiol.* 47, 739-763.

Frischmuth, T. A., and Stanley, J. (1994). Beet curly top virus symptom amelioration in *Nicotiana benthamiana* transformed with a naturally occurring viral subgenomic DNA. *Virology* 200, 826-830.

Frischmuth, T., and Stanley,J. (1998). Recombination between viral DNA and the transgenic coat protein gene of African cassava mosaic geminivirus. *J . Gen. Virol.* 79, 1265-1271.

Fuchs, M., and Gonsalves, D. (1995). Resistance of transgenic hybrid squash ZW-20 expressing the coat protein genes of zucchini yellow mosaic virus and watermelon mosaic virus2 to mixed infections by both potyviruses. B*io/Technology* 13, 1466-1473.

Gal, S., Pisan, B., Hohn, T.,Grimsley, N., and Hohn, B. 1992. Agroinfection of transgenic plants leads to viable cauliflower mosaic virus by intermolecular recombination. *Virology* 187, 525-533.

Gera, A., Deom, M., Donson, J., Shaw, J. J., Lewandowski, D. J., and Dawson, W. O. (1995). Tobacco mosaic tobamovirus does not require concomitant synthesis of movement protein during vascular transport. *Mol. Plant Microbe Ineract.* 8, 784-787.

Gibbs, M. (1994). Risks in using transgenic plants? *Science* 264, 1650-1651.

Golemboski, D. B., Lomonossoff , G. D., and Zaitln, M. (1990). Plants transformed with a tobacco mosaic virus nonstructural gene sequence are resistant to the virus. *Proc. Natl. Acad. Sci. USA.* 87, 6311-6315.

Greene, A. E., and Allison, R. F. (1994). Recombination between viral RNA and transgenic plant transcripts. *Science* 263: 1423-1425.

Hamilton, R. I. (1980). Defences triggered by previous invaders. *In* "Plant Disease: An Advanced Treatise. Vol. V." (J.G. Horsfall and E.B. Cowling, Eds.), pp. 279-303. Academic Press, New York.

Hanley- Bowdoin, L., Elmer, J. S., and Rogers, S. G. (1990). Expression of functional replication protein from tomato golden mosaic virus in transgenic tobacco plants. *Proc. Natl. Acad. Sci. USA* 87, 1446-1450.

Harris, P. S. (1998). Environmental and regularity aspects of using genetically transformed plants and micro-organisms. *In* "Comprehensive Potato Biotechnology" S. M. Paul Khurana, R. Chandra, and M. D. Upadhya, Eds.), pp. 307-354. Malhotra Publ. House, New Delhi.

Hemenway, C., Haley, L., Kaniewski., W. K., Lawson, E. C., Connell, O' Connell, K. M. , Sanders, P. R., Thomas, P.E., and Tumer, N.E. (1990). Genetically engineered resistance: Transgenic plants. *In* "Plant Viruses. Vol. II. Pathology" (Mandahar, C. L., Ed.), pp. 347-363. CRC Press, Boca Raton, Florida.

Holt, C. A., and Beachy, R. N. (1991). *In vivo* complementation of infectious transcripts from mutant tobacco mosaic virus cDNAs in transgenic plants. *Virology* 181, 109-117.

Hull, R. (1990). The use and misuse of viruses in cloning and expression in plants. *In* "Recognition and Response in Plant Virus Interactions" (Fraser, R. S. S., Ed.), pp. 443-457. NATO ASI, Vol. 441. Springer Verlag, Berlin.

Hull, R. (1994). Risks in using transgenic plants. *Science* 264, 1649-1650.

Kaplan, I. B., Shintaku, M. H., Li, Q., Zhang, L., Marsh, L. E., and Palukaitis, P. (1995). Complementation of virus movement in transgenic tobacco expressing the cucumber mosaic virus 3a gene. *Virology* 209, 188-199.

Koonin, E. V., and Dolja, V.V. (1993). Evolution and taxonomy of positive-strand RNA viruses: Implications of comparative analysis of amino acid sequences. *Crit. Rev. Biochem. Mol. Biol.* 28, 375-430.

Lecoq, H., Ravelonandro, M., Wipf-Scheibel, C., Monsion, M., Raccah, B., and Dunez, J. (1993). Aphid transmission of a non-aphid-transmissible strain of zucchini yellow mosaic potyvirus from transgenic plants expressing the capsid protein of plum pox potyvirus. *Mol. Plant-Microbe Interact.* 6, 403-406.

Lindbo, J. A., and Dougherty, W. G. (1992a). Untranslatable transcripts of the tobacco etch virus coat protein gene sequence can interfere with tobacco etch virus replication in transgenic plants and protoplasts. *Virology* 189, 725-733.

Lindbo, J. A. , and Dougherty, W. G. (1993b). Pathogen-derived resistance to a potyvirus. Immune and resistant phenotypes in transgenic tobacco expressing altered forms of a potyvirus coat protein nucleotide sequence. *Mol. Plant-Microbe Interact.* 5, 144-153.

Lindbo, J. A., Silva-Rosales, L., and Dougherty, W. G. (1993a). Pathogen-derived resistance to potyviruses. Working but why? *Semin. Virol.* 4, 369-379.

Lindbo, J. A., Silva-Rosales, L., Proebsting, W. M., and Dougherty, W. G. (1993b). Induction of a highly specific antiviral state in transgenic plants. Implications for the regulation of gene expression and virus resistance. *Plant Cell* 5, 1749-1759.

Lommel, S. A., and Xiong, Z. (1991). Reconstitution of a functional red clover necrotic mosaic virus by recombinational rescue of the cell-to-cell movement gene expressed in a transgenic plant. *J. Cell. Biochem.* 15 A, 151.

Lomonossoff, G. P. (1992). Virus resistance mediated by a non-structural viral gene sequence. *In* "Transgenic Plants: Fundamentals and Applications" (A. Hiatt, Ed.). pp. 79-91. Marcel Dekker, New York.

Lomonossoff, G. P. (1995). Pathogen-derived resistance to plant viruses. *Annu.Rev. Phytopathol.* 33, 323-343.

Maiss,E., Koenig, R., and Lesemann, D. E. (1994). Heterologous encapsidation of viruses in transgenic plants and in mixed infections. *In* "Third International Symposium on the Biosafety Results of Field Tests of Genetically Modified Plants and Micro-organisms", pp. 129-139. Univ. California Press, Oakland.

Mandahar, C. L. (1990). Variability of plant viruses. *In* "Plant Viruses. Vol. II. Pathology" (C. L. Mandahar , Ed.), pp. 109-151. CRC, Press, Boca Raton, Florida.

Nagy, P. D., and Simon, A. E. (1997). New insights into the mechanisms of RNA recombination. *Virology* 235, 1-9.

Noris, E., Accotto, G. P., Tavazza, R., Brunetti, A., Crespi, S., and Tavazza, M. (1996). Resistance to tomato yellow leaf curl geminivirus in *Nicotiana benthamiana* plants transformed with a truncated viral C1 gene. *Virology* 224, 130-138.

Osbourne, J. K., Sarkar, S.,Wilson, T. M. A. (1990). Complementation of coat protein defective TMV mutants in transgenic tobacco plants expressing TMV coat protein. *Virology* 179, 921-925.

Palukaitis, P.,and Zaitlin, M. (1997). Replicase mediated resistance to plant virus disease. *Adv.Virus Res.* 48, 349-377.

Powell, P. A., Nelson, R. S., De, B., Hoffmann, N., Rogers, S. G., Fraley, R. T., and Beachy, R. N. (1986). Delay of disease development in transgenic plants that express the tobacco mosaic virus coat protein gene. *Science* 232, 738-743.

Provvidenti, R., and Gonsalves, D. (1995). Inheritance of resistance to cucumber mosaic virus in a transgenic tomato line expressing the coat protein gene of the white leaf strain. *J. Hered.* 86, 85-88.

Ratcliff, F., Harrison, B. D., and Baulcombe, D. C. (1997). A similarity between viral defence and gene silencing in plants. *Science* 276, 1558-1560.

Reddy, D. V. R., Murant, A. F., and Ducan, G. H. (1985). Viruses associated with chlorotic rosette and green rosette disease of groundnut in Nigeria. *Ann. Appl. Biol.* 107, 57-64.

Rochow, W. F. (1977). Dependent virus transmission from mixed infections. *In.* "Aphids as Virus Vectors" (K. F. Harris and K. Maramarosch, Eds.), pp. 253- 273 . Academic Press, New York.

Sanford, J. C., and Johnston, S. A. (1985). The concept of parasite-derived resistance - deriving resistance genes from the parasite's own genome. *J. Theorm. Biol.* 113, 395-405.

Santa Cruz, S., and Baulcombe, D. (1995). Analysis of potato virus X coat protein genes in relation to resistance conferred by the genes *Nx, Nb,* and *Rx1* of potato. *J. Gen. Virol.* 76, 2057-2061.

Schoelz, J. E., and Wintermantel, W. M. (1993). Expansion of viral host range through complementation and recombination in transgenic plants. *Plant Cell* 5, 1669-1679.

Schoelz, J. E., and Wintermantel, W. M. (1996). Isolation of recombination viruses between cauliflower mosaic virus and a viral gene in transgenic plants under conditions of moderate selection pressure. *Virology* 223, 156-164.

Schoelz, J. E., Goldberg, K.-B., and Kiernan, J. (1991). Expression of cauliflower mosaic virus (CaMV) gene VI in transgenic *Nicotiana bigelovii* complements a strain of CaMV defective in long-distance movement in non-transformed *N. bigelovii. Mol.Plant-Microbe Interact.* 4, 350-355.

Simon, A. E., and Bujarski, J. J. (1994). RNA-RNA recombination and evolution in virus infected plants. *Annu. Rev. Phytopathol.* 32, 337-362.

Tacke, E., Salamini, F., and Rohde, W. (1996). Genetic engineering of potato for broad-spectrum protection against virus infection. *Nature Biotechnol.* 14, 1597-1601.

Tepfer, M. (1993). Viral genes and transgenic plants. What are potential environmental risks? *Bio/Technology* 11, 1125-1132.

Tricoli, D. M.,Carney, K. J., Russell, P. F., McMaster, J. R., Groff l, D. W. , Hadden, K. C., Himmell, P. T., Hubbard, J. P., Boeshore, M. L., and Quemada, H. D. (1995). Field evaluation of transgenic squash containing single or multiple virus coat protein gene constructs for resistance to cucumber mosaic virus, watermelon mosaic virus 2 and zucchini yellow mosaic vi rus. *Bio/Technology* 13, 1458-1465.

Van Dun, C. M. P., van Vloten-Doting, L., and Bol. J. F. (1988). Expression of alfalfa mosaic virus cDNA 1 and 2 in transgenic tobacco plants. *Virology* 163, 572-578.

Valkonen, J. P. T., Koivu, K., Slack, S. A., and Pehu, E. (1995). Modified resistance of *Solanum brevidens* to potato Y potyvirus and tobacco mosaic tobamovirus following genetic transformation and explant regeneration. *Plant Sci. (Limerick)* 106, 71-79.

Vance, V. B., Berger, P. H., Carrington, J. C., Hunt, A. G., and Shi, X. M. (1995). 5` proximal potyviral sequences mediate potato virus X/potyviral synergistic disease in transgenic tobacco. *Virology* 206, 583-590.

Wassenegger, M., Heimes, S., and Sanger, H. L. (1994). An infectious viroid RNA replicon evolved from an *in vitro* generated non-infectious viroid deletion mutant via a complementary deletion *in vivo. EMBO J.* 13, 6172-6177.

Wintermantel, W. M., and Schoelz, J. E. (1996). Isolation of recombination viruses between cauliflower mosaic virus and a viral gene in transgenic plants under conditions of moderate selection pressure. *Virology* 223, 156-164.

Zaitlin, M., Anderson, T. M., Perry, K. L., Zhang, L., and Palukaitis, P. (1994). Specificity of replicase-mediated resistance to cucumber mosaic virus. *Virology* 291, 200-205.

13 GENE SILENCING

Chuni L. Mandahar

INTRODUCTION

The basic important characteristics of a transgenic plant are the stability of the transgene from generation to generation and the high degree of fidelity in expression of the transgene in each generation. Transgenic plants have, therefore, been widely employed for a variety of experimental purposes including for understanding gene function by engineering over-expression or under-expression of the relevant genes in wild type or mutant plants which generated important data about the role of the respective proteins. The development of transgenic plants also soon led to the efforts to create such transgenic plants that could overproduce a useful product, like vegetables, fruits, and seeds with enhanced food value and protein content, by over-expressing the particular plant protein by having multiple copies of a gene (as homologous transgenes) within a nucleus. Thus, the number of sequences of a particular gene in a transgenic plant was two or more, depending upon the number of transgene copies incorporated into the genome of a transformed plant.

But, surprisingly, the opposite happened in some cases. Instead of producing large amounts of the desired protein product, the high-expressing trangene present in single copy or multiple copies simultaneously inhibited the expression of its some or all transgene copies as well as of plants`s endogenous genes that were homologous to the transgene. Thus, the overproduced transgene became a liability. Often a correlation existed between multiple linked copies of a transgene and poor gene expression due to gene silencing (de Carvalho *et al.*, 1992; Angenent *et al.*, 1993).

Gene silencing is thus the conversion of an actively transcribing gene into a non-transcribing state. It was originally discovered in petunia plants in 1990 as a consequence of the introduction of additional copies of a flower colour gene encoding chalcone synthase (*CHS* gene) or additional copies of a transgene consisting of the corresponding cDNA under the control of cauliflower mosaic virus (CaMV) promoter. When this transgene, in single copy or multiple copies, which was homologous to a part or whole of a host (endogenous) gene, was introduced into a petunia plant, the gene silencing came about (Napoli *et al.*, 1990; Van der Krol *et al.*, 1990). Gene silencing was also later found to occur in transgenic (tobacco) plants in which a virus transgene [capsid protein (CP) of tobacco etch virus (TEV)] had been incorporated (Lindbo *et al.*, 1993). This silencing in transgenic plants occurred because of the sequence similarity between the virus and either the transgene or an endogenous nuclear gene. Gene silencing involves either single transgene locus or interactions between unlinked genes.

Two experimental systems have been commonly used for studying gene silencing. One is the suppression of chalcone synthase expression in flowers of transgenic petunia. It interferes with the *CHS*-catalyzed step during the pathway of anthocyanin pathway so that silencing of *CHS* gene causes the appearance of distinctive and diagnostic colouration patterns in corollas and/or white anthers. The second system is based on CP-mediated resistance of transgenic tobacco plants and related species to infection by tobacco mosaic virus (TMV) and TEV RNA viruses. This resistance of transgenic plants is of two types: constitutive which is also referred to as immunity, and the recovery phenomenon which manifests itself after a transient infection of a plant by a virus.

Silencing affects all the homologous copies (transgenes) of a gene present in plant genome, irrespective of the fact whether they are closely linked on a single DNA molecule or are unlinked loci being located on separate DNA molecules. The former is called *cis*-inactivation and the latter the *trans*-inactivation. The *trans*-inactivation affects sequence repeats (transgenes) occurring at allelic positions or on non-homologous chromosomes. It is either reciprocal in which all gene copies are silenced leading to dominant *trans*-inactivation or non-reciprocal in which only one allele is changed leading to epistatic *trans*-inactivation.

Because of the suppression of gene expression of all the homologous genes, gene silencing is clearly homology-dependent (Matzeke *et al.,* 1994) or repeat-induced (Assad *et al.,* 1993). Moreover, the simultaneous inactivation of gene expression of both the transgene and of the homologous (or partially homologous) host genes in transgenic plants as a consequence of specific sequence duplication by pairing of homologous sequences or, in other words, the simultaneous silencing of both the homologous host genes and the homologous transgenes, is variously called cosuppression, sense suppression, sense cosuppression, or mutual inactivation. It is a commonly observed phenomenon in plants (Meyer, 1995a).

Gene silencing is attracting considerable attention because of several reasons. It controls genome maintenance and plant development. It may also control and coordinate gene expression during developmental changes after aneuploidization and polyploidization within the whole plants and polysomy within single cells. It may regulate gene expression from multigenic families, or may recognize and degrade the potentially harmful transcripts in plant cells. A transgenic plant lacking expression of a transgene and/or of the endogenous homologous gene can also be a useful experimental tool. Moreover, cosuppression is not an artificial process which is only manifested during studies with transgenes. On the other hand, it is to be regarded as a definite phenomenon, which seems to be possibly associated with several existing plant regulatory mechanisms.

It is also emerging as a defense mechanism of plants against virus infection, as the cause of the recovery phenomenon of virus-infected plants, and is the basic reason for the resistance of the transgenic plants to infection by the homologous viruses. Gene silencing is discussed here only with regard to these issues. It is clear from above that RNA viruses carrying sequences that are homologous to a transgene(s) can trigger as well as be targets of gene silencing and cosuppression.

Gene silencing has been reviewed (Jorgensen, 1992, 1995; Finnegan and McElroy, 1994; Flavell, 1994; Matzeke *et al.,* 1995; Matzeke and Matzeke, 1995 a, b; Meyer, 1995a, b; Dougherty and Parks, 1996; Baulcombe, 1996a, b; Baulcombe and English, 1996; Meyer and Sadler, 1996; Taylor, 1997; Jorgensen *et al.,* 1998).

TYPES OF GENE SILENCING

It is two types depending upon when suppression of the homologous genes occurs: transcriptional if it occurs at the time of transcription of the transgene and post-transcriptional if it occurs afterwards. Transcriptional gene silencing is generally associated with methylation of transgene promoter region and there is a complete lack of transcriptional activity as measured by run-on transcription (Park *et al.,* 1996). It requires promoter homology and sometimes induces paramutations, which are the meiotically heritable changes in gene expression. Transcriptional gene silencing causes DNA modifications that remain even after the transgene has segregated (Lindbo *et al.,* 1993; Jorgensen, 1995; Matzke and Matzke, 1995a,b; Park *et al.,* 1996).

The host genes and transgenes continue to be correctly transcribed in the nucleus at about the normal rates whereas the corresponding mRNAs do not accumulate in cytoplasm (de Carvalho *et al.,* 1992; Lindbo *et al.,* 1993; Van Blokland *et al.,* 1994) so that a reduction of both the transgene and host gene mRNAs occurs. Gene silencing is therefore seemingly post-transcriptional, occurs after transgene transcripts have been produced in abundance because of the relatively high transcriptional activity, and is operative at the level of mRNA processing, localization, and/or degradation. Post-transcriptional gene silencing is usually reciprocal causing cosuppression of expression of a transgene and the homologous endogenous or viral genes in transgenic plants, and is generally based on duplication of transcribed sequences. It is of two types: coordinate cosuppression in which coordinate silencing of two or more transgene loci takes place (Jorgensen, 1992) and single-locus silencing in which just a single transgene locus is silenced (Smith *et al.,* 1994; Mueller *et al.,* 1995; English *et al.,* 1996). High levels of resistance to homologous virus was detected in plants in which transcription of transgene was detected in the nucleus, although the corresponding mRNA did not accumulate in cytoplasm (English *et al.,* 1996; Goodwin *et al.,* 1996; Sijen *et al.,* 1996). Because RNA viruses replicate in cytoplasm, it was suggested that virus-induced post-transcriptional gene silencing occurs in cytoplasm (Lindbo *et al.,* 1993; Teresa Ruiz *et al.,* 1998).

Characteristics of gene silencing are: it is heritable indicating that genetic determinants of gene silencing exist in genome; continued transcription along with RNA degradation is one of its hallmark in transgenic plants; increased RNA turnover; is associated with transgene methylation; seems to be reinitiated after each generation, a process called meiotic resetting (de Carvolho *et al.,* 1992; Hart *et al.,* 1992); and is thought to be involved in the resistance of transgenic and non-transgenic plants to virus infection. Methylation of DNA sequences results in decreased transcription and typically manifests itself as a decrease in steady state levels of mRNAs. Both the transgene and host genes could be methylated resulting in their inactivation irrespective of the fact whether these genes have limited or extensive sequence homology between them.

Both the transcriptional and post-transcriptional models may be operative in different situations in triggering the same effect and therefore are not likely to be mutually exclusive (Palauqui *et al.*, 1997)

MECHANISMS OF GENE SILENCING

Four explanations have been put forth for cosuppression (Dougherty and Parks, 1995). An antisense RNA, formed from nuclear transcription of transgene, could suppress the sense transcripts of the endogenous gene and the transgene. Methylation of the transgene and the endogenous gene may inactivate them. Gene expression may be suppressed during DNA-DNA interactions between the endogenous gene and the transgene at the time of transcription. Genes may be suppressed post-transcriptionally.

Transcriptional Gene Silencing

Direct physical association or pairing of alleles causes DNA-DNA pairing between a transgene locus and the endogenous gene. This is suggested to induce a change in the chromatin configuration/methylation of the later (silenced locus) leading to suppression of transcription. During the changed chromatin configuration, chromatin adopts a local configuration that is difficult to transcribe so that both the closely linked repeat transgenes and the endogenous pair of alleles are silenced *in trans* as well as *in cis* leading to transcriptional gene silencing (Jorgensen, 1992; Flavell, 1994; Matzeke *et al.*, 1995).

Post-transcriptional Gene Silencing

Tanzer *et al.* (1997) have proposed a model of post-transcriptional gene silencing. Post-transcriptional gene silencing does not have any effect on mRNA transcription so that a nascent transcript undergoes capping, processing, and is transported from nucleus to cytoplasm where the newly transcribed mature mRNA gets associated with the protein synthesizing machinery. After this, all the events prior to codon recognition (like cap binding, ribosome scanning etc.) occur normally; then translation starts at an appropriate AUG and the polyribosomes are loaded on to mRNA. After initiation of translation, an endonuclease complex in some way recognizes and then cleaves at specific sites the mRNA to be silenced to produce low molecular weight RNAs. Later, non-specific exonucleases further degrade these RNAs. This whole process results in initiation of gene silencing and once initiated further maintenance of the silenced state is independent of the promoter and/or gene copy number. Work on potato virus X (PVX) indicated that initiation of post-transcriptional gene silencing and its maintenance are separate stages. Initiation is completely dependent upon the presence of virus and, once initiated, its maintenance is independent of virus presence (Teresa Ruiz *et al.*, 1998).

Three models have been proposed concerning the mechanism of post-transcriptional gene silencing: threshold model in which post-transcriptional silencing is triggered by the high level of an mRNA beyond a certain threshold, ectopic pairing or aberrant RNA (aRNA) model in which post-transcriptional silencing is triggered by a truncated or altered (aberrant) mRNAs, and a third model comprising of parts of the earlier two models. All the three mechanisms work *in trans* to silence all highly homologous genes.

Threshold Model

The threshold model was proposed by Dougherty and coworkers (Lindbo *et al.*, 1993; Smith *et al.*, 1994; Dougherty and Parks, 1995) while working on resistance of transgenic tobacco plants to TEV infection. It is also called the biochemical switch model or the gene dosage hypothesis: switch model because the particular transcript acts like a switch, switching the silencing on or off depending upon the transcript concentration; and gene dosage model because number of genes present determine the number of transgenic plants silenced.

Threshold model was based on the observation that the nuclear transgenes as also the viral RNA were suppressed even though the transgenes were being expressed and forming transcripts. Something was therefore happening post-transcriptionally, which was inhibiting accumulation of viral mRNAs (transcripts). They suggested on this basis that cosuppression occurs because of the existence of a cytoplasmic concentration-dependent threshold-sensitive system that causes sequence-specific degradation of transcripts that accumulate to a high concentration and beyond a certain threshold. After this critical threshold, the transgene transcripts and the related endogenous gene transcripts undergo sequence-specific degradation. The RNA-dependent RNA polymerase (RdRp) makes this turnover sequence specific while gene determinants of silencing cause the transgene, when expressed under the control of a strong promoter, to be transcribed at a very high level so that its mRNA accumulates above a threshold value. This activates the gene silencing mechanism resulting in specific destruction of all the homologous RNA molecules in cytoplasm.

Other characteristics of this model are (Palauqui *et al.*, 1997): gene silencing occurs far more commonly in homozygous plants than in hemizygous plants; nevertheless, genes of haploid or hemizygous plants, in which only a single copy of the transgene exist, are also silenced indicating that pairing between allelic or ectopic copies is not necessary; transgenes that are expressed in the presence of a 35S promoter with a double enhancer are far more commonly silenced than the transgenes being expressed under the aegis of wild-type 35S promoter.

Cosuppression is gene dosage-dependent. Thus, transgenic plants homologous for the transgene locus show increased frequency and/or intensity of cosuppression than the hemizygous plants (de Carvalho *et al.*, 1992; Angenent *et al.*, 1993). Transgenic plants carrying a combination of unlinked transgene loci exhibit greater cosuppression than the parental lines (Jorgensen *et al.,* 1996; Palauqui and Vaucheret, 1995). The degree and frequency of cosuppression is primarily influenced by the amount of *Chs* mRNA present and reducing the *Chs* transcripts causes a reduction in the degree and frequency of cosuppression (Que *et al.,* 1997). Use of diploid, triploid, and tetraploid plants (with regard to the transgene copies of nitrate reductase gene) of *Nicotiana plumbaginifolia* and *N. tabacum* showed a correlation between the number of *Nia2* genes and the number of plants that were cosuppressed so that greater the number of active *Nia2* genes, the higher was the number of cosuppressed plants (Vaucheret *et al.,* 1997). The rapidity of emergence of viral resistance in transgenic plants correlated with the overall amount of transgenic and TEV mRNAs (Tanzer *et al.,* 1997).

Ectopic Pairing Model

The threshold model explains much of the data on cosuppression. However, several reports show that cosuppression occurs even in the absence of transgene transcription and/or in the absence of detectable accumulation of transgene transcripts. A second model, called ectopic or aberrant RNA model, was proposed to account for these cases (Van Blokland *et al.,* 1994; English *et al.,* 1996; Sijen *et al.,* 1996; Baulcombe and English, 1996; Stam *et al.,* 1997).

Ectopic (non-allelic) pairing model envisages DNA-DNA interaction which involves pairing between cDNA (or RNA) of the transgene present in some host chromosome and the homologous cDNA of RNA of inoculated/infecting virus and/or between endogenous and introduced DNAs. This leads to methylation of specific transgenes and these methylated genes are regarded to produce RNA(s) that is aberrant in some way (aRNA). The aRNAs could be produced by complex transgene loci, particularly the ones containing inverted repeats. Only a very small quantity of the aRNA is produced in addition to the normal mRNA. Moreover, the aRNA is thought to be only a minor portion of the total RNA produced from the receptor locus.

This aRNA moves to cytoplasm, transmits the signal for gene silencing from the DNA to the RNA level, and anomalous features of the aRNAs trigger the sequence-specific degradation in cytoplasm of transcripts (mRNA) produced by all genes homologous to the transgene locus. The target RNA may be the viral RNA or transcripts of the endogenous homologous genes which do not undergo pairing with the transgene locus. This results in post-transcriptional gene silencing (Baulcombe and English, 1996) or, in nut shell, ectopic pairing causes post-transcriptional gene silencing (Smith *et al.,* 1994; Mueller *et al.,* 1995; English *et al.,* 1996).

Third Model

The third model, proposed by Cameron and Jennings (1991) and Metzlaff *et al.* (1997), combines some features of the earlier two models. Cosuppression is triggered by intermolecular (Cameron and Jennings, 1991) or intramolecular

(Metzlaff *et al.*, 1997) base pairing between sense transcripts, instead of the copy RNA (cRNA)-mRNA hybrids produced by the action of an RdRp. With the increase in transcript concentrations, these base pairs could become more common and the putative double-stranded regions trigger the degradation of homologous transcripts. In short, iter- or intramolecular transcript complementarity may trigger cosuppression.

Transmission Signal

The transmission signal is gene-specific, non-metabolic, systemic and mobile (Palauqui *et al.*, 1997). It is widely thought to be an RNA molecule. This RNA signal could be the abnormal transcripts (aRNA) (Baulcombe and English, 1996), transcript fragments generated during RNA degradation (Goodwin *et al.*, 1996; Metzlaff *et al.*, 1997), or cRNA molecules that arise from sense transcripts by reverse transcription of the excess transgene mRNA presumably by endogenous RdRp (Lindbo *et al.*, 1993; Dougherty and Parks, 1995).

The molecules actually triggering cosuppression could be some decay products of cosuppressed transcripts (Que *et al.*, 1997; Tanzer *et al.*, 1997; Vaucheret *et al.*, 1997). It is supported by several observations: accumulation of specific mRNA degradation intermediates occurs during cosuppression (Goodwin *et al.*, 1996; Metzlaff *et al.*, 1997); some fragments of TEV CP mRNA accumulate only in the cosuppressed tissues (Goodwin *et al.*, 1996); and these fragments start accumulating at about the time of cosuppression induction (that is, when transgenic plants start showing signs of virus resistance) (Tanzer *et al.*, 1997). These fragments are the 5`- and 3`-end parts of the transgene CP transcripts degraded by an endonucleolytic cleavage while the transcripts are still in contact with ribosomes so that the degradation intermediates continue to be associated with ribosomes (Tanzer *et al.*, 1997).

Jorgensen *et al.* (1998) favour the cRNA for amplification and transmission of cosuppression state but in the form of a ribonucleoprotein (RNP) complex constituted by cRNA molecules and plant proteins and which might also include a RdRp, an endogenous movement protein, and a double-stranded RNA. Teresa Ruiz *et al.* (1998) suggest that an RNA-containing factor, produced by the transgene, acts as the transmission signal. This factor is envisaged to perform two interrelated functions: the suppression of the transgene and the maintenance of gene suppression even in cells from which the transgene/virus has been eliminated.

The RNA/RNP complex signal moves from cell-to-cell through plasmodesmata, travels through graft union, conceivably moves over long distances through phloem and from phloem into the adjacent tissues (Voinnet and Baulcombe, 1997; Palauqui *et al.*, 1997) to cause systemic acquired silencing (Palauqui *et al.*, 1997) by analogy with the term systemic acquired resistance. Thus cosuppression is a systemic state which effects turnover of all homologous transcripts in the cells/tissues the signal enters. Systemic spread of this signal could, in the most appropriate words of Jorgensen *et al.* (1998), "identify, track, and destroy viral molecules in a sequence-specific manner" in plants to produce resistance in them. Rapid translocation of the signal over long distances in phloem mediates recovery of infected plants from viral infection (Jorgensen *et al.*, 1998), and simultaneously imparts them the capacity to resist infection by the homologous virus (Ratcliff *et al.*, 1997; Covey *et al.*, 1997). This signal thus induces natural systemic defense of plants against virus infection by moving ahead of the infecting virus and so inhibits virus spread, and also causes recovery of virus infected plants. All this comes about by generating anti-viral post-transcriptional gene silencing in the cells/tissues the signal enters (Voinnet and Baulcombe, 1997; Jorgensen *et al.*, 1998).

EFFECTS OF GENE SILENCING

Resistance of transgenic tobacco plants, expressing CP gene of TEV, manifests itself as two distinct phenotypes: immune and recovery phenotypes (Lindbo and Dougherty, 1992b; Dougherty *et al.*, 1994; Goodwin *et al.*, 1996). In the immune phenotype, there is no virus replication, nor accumulation of transgene mRNA, while resistance characteristic exhibits Mendelian inheritance. On the other hand, transgenic tobacco plants showing recovery phenotype initially are susceptible to virus infection, accumulate full-length transgene mRNA but ultimately develop resistance at which stage they contain little transgene mRNA. Further work with the same system showed no such clear-cut distinction between the two phenotypes (Tanzer *et al.*, 1997). Like the recovery phenotype, immune phenotype is also susceptible to TEV infection, accumulates a high level of full-length transgene CP mRNA at the earlier stages of development, and also exhibits meiotic resetting. Thus both these phenotypes seem to have the same mechanisms of virus resistance and gene silencing with one primary difference that presence of TEV is essential for initiating silencing in plants exhibiting the recovery phenotype.

Moreover, immune lines have more transgene copy number (3, 4, or 6 copies), show more transgene transcription, and give low steady-state levels of full-length transgene RNA while recovery lines contain only one or two transgene copies (Goodwin *et al.*, 1996). Transgenic plants containing more transgene copies induced recovery and silencing more rapidly (Tanzer *et al.*, 1997). Thus, phenotype of transgene lines was dependent upon the transgene copy number.

Degradation of transgene/viral RNAs is widely reported (Flavell, 1994; Dougherty and Park, 1995; English *et al.*, 1996). The silenced genes characteristically show decreased levels of full-length mRNA and the formation of specific low molecular weight RNAs. Two sets of discrete silencing-specific mRNA fragments of low molecular weight are found in silenced tissues containing the TEV 2RC transgene or full-length TEV CP transgene (Tanzer *et al.*, 1997). One set lacks poly(A) tails, bears nearly intact 5` ends, and is in association with ribosomes while other set is made of CP mRNA fragments bearing the 3` end. They proposed on this basis that these fragments are produced by transgene/viral RNA degradation in cytoplasm by specific endonucleolytic cleavage of full-length CP mRNA while the CP mRNAs possibly are still in association with polyribsosmes. Some of the remaining low molecular weight RNAs may be the aRNAs or, even it is possible, that the RNA fragments formed as above from full-length mRNA may by themselves act as aRNAs. Similarly, specific degradation of the chalcone synthase mRNA in petunia is postulated to be the cause of silencing (Metzlaff *et al.*, 1997).

Resistance in Transgenic and Non-transgenic Plants

Transgene expression (synthesis of proteins encoded by the transgene) and the transgene-mediated resistance are not correlated in several cases. Resistance of transgenic plants to TEV (Lindbo *et al.*, 1993), PVX (Mueller *et al.*, 1993), potato virus Y (Smith *et al.*, 1994), brome mosaic virus (Kaido *et al.*, 1995) and several other viruses (Baulcombe, 1994; Lomonossoff, 1995; Dougherty and Parks, 1995; English *et al.*, 1996; Baulcombe and English, 1996) was not mediated by the protein product of the transgene. All these transgenic lines, showing post-transcriptional inhibition of transgene mRNA accumulation, were strongly resistant to the corresponding viruses and RNA mediated the resistance. In fact, it is now regarded that RNA-mediated resistance is most likely to apply if resistance causes low levels of transgene-derived mRNA accumulation in transgenic plants.

The above type of virus resistance of transgenic plants and gene silencing are both based on the same homology-dependent mechanism; that is, dependent upon the homology between the transgene and a part of the genome of the infecting virus. Thus, homology-dependent resistance of transgenic plants and homology-dependent gene silencing are related processes (Matzke and Matzke, 1995; Matzeke *et al.*, 1995; Jorgensen, 1995; Baulcombe and English, 1996) and are based on post-transcriptional gene silencing. Therefore, the transgenes confer resistance to transgenic plants against the homologous virus through transgene trans-inactivation at post-transcriptional gene silencing level of gene expression so that virus replication and accumulation in cytoplasm of transgenic plants is greatly inhibited (Lindbo *et al.*, 1993; Smith *et al.*, 1994; Mueller *et al.*, 1995; English *et al.*, 1996; Goodwin *et al.*, 1996; Sijen *et al.*, 1996; Baulcombe, 1996b; Baulcombe and English, 1996; Meyer and Sadler, 1996; Tanzer *et al.*, 1997). It seems from above that many of the cases of pathogen-derived resistance are in reality due to transgene-induced gene silencing operating against viral RNA (Baulcombe, 1996b).

Gene silencing in above-mentioned examples of transgenic lines correlated with the existence of inverted T-DNA repeats. This indicated ectopic pairing between the transgene copies or between the transgene and the homologous host genes leading to production of aberrant unproductive RNA which specifically destroys all the homologous RNA molecules in cytoplasm (Baulcombe and English, 1996; English *et al.*, 1996).

Post-transcriptional gene silencing is also initiated in non-transgenic plants upon virus infection and is thought to be an efficient defense mechanism of the non-transgenic plants (Teresa Ruiz *et al.*, 1998; Ratcliff *et al.*, 1997; Covey *et al.*, 1997; Baulcombe 1996a). Mechanism similar to the transgenic plants appears to be involved in the resistance of the non-transgenic plants as well (Tanzer *et al.*, 1997). Gene silencing, therefore, appears to be an efficient general defense system developed by plants against viruses (Baulcombe 1996a; Pruss *et al.*, 1997) so that it is likely to be induced naturally in virus-infected plants and artificially in transgenic plants as the transgene or its transcript (mRNA) is taken as a virus (Teresa Ruiz *et al.*, 1998).

It appears from above that an interaction of viral RNA with nuclear transgene or with its mRNA initiates gene silencing. On the other hand, Metzlaff *et al.* (1997) suggest RNA double-strandedness as the initiator of gene silencing in petunia. This is supported by the results of work on fungi: transgenes which encode replicating viral RNAs activate post-transcriptional gene silencing more efficiently (Angell and Baulcombe, 1997); and the suggestion that virus RNA initiates gene silencing, independently of the corresponding nuclear gene or its mRNA, by RNA double-strandedness formed transiently as an intermediate stage during viral replication (Teresa Ruiz *et al.*, 1998).

Recovery of Transgenic and Non-transgenic Plants from Disease

As mentioned above, inoculation of viruses on transgenic plants causes cosuppression and post-transcriptional gene silencing of both the transgene(s) and the homologous inoculated virus. However, the transgene is not effective during early stages of growth of the transgenic plants so that it develops normal disease symptoms. But gene silencing becomes functional after sometime in the upper leaves that develop after the virus systemically spreads in the transgenic plant, so that both the transgene and the homologous virus in these leaves become silent. These leaves therefore contain low levels of the transgene RNA, do not show symptoms, are virus-free, and are also resistant to reinfection by the same virus. These transgenic plants have thus "recovered" (Lindbo *et al.*, 1993; Smith *et al.*, 1994). In this way, viruses activate gene silencing in transgenic plants which exhibit homology-dependent resistance that is specific to the isolates of the virus donating the transgene.

The response of transgenic *Brassica napus* plants to systemic infection by CaMV was also identical - initial enhancement and subsequent suppression of virus gene expression which was parallel to the changes in symptomatology (Al-kaff *et al.*, 1998). Co-suppression of virus and transgene was mediated by degradation of homologous RNA. CaMV infection of transgenic *Brassica napus* plants silenced both the viral gene expression as well as transgenes possessing homology to CaMV by post-transcriptional gene silencing mechanism. The GUS transgene homologous to CaMV was co-suppressed in the presence of infectious CaMV during recovery of transgenic plants from the virus.

This type of recovery from virus diseases is also found in non-transgenic plants. CaMV-infected Kohlrabi (*Brassica oleracea gongylodes)* initially [up to 13 days post inoculation (dpi)] developed systemic symptoms from which it completely recovered by loss of virus within a few weeks (Al-Kaff and Covey, 1995). Amelioration of symptoms started about 16 dpi and correlated with rapid destruction of viral polyadenylated RNAs by specific degradation by de-adenylation (Covey *et al.*, 1997).

Severe viral symptoms appeared on inoculated and first leaves of a tomato black ring nepovirus (TBRV)-infected *Nicotiana clevelandii.* However, the upper leaves produced after systemic infection, were symptomatic to begin with but later recovered to become symptomless and contained less virus than the symptomatic leaves (Ratcliff *et al.*, 1997). The recovered leaves, upon inoculation with the same virus, were resistant to viral infection, remained symptom-less, and contained low viral RNA concentration. Thus recovered leaves were resistant to viral infection, resistance was specific to strains related to recovery-inducing virus, and RNA was the target of this recovery (Ratcliff *et al.*, 1997).

The above recovery and subsequent natural resistance of CaMV-infected non-transgenic Kohlrabi plants to CaMV infection and TBRV-infected non-transgenic *N. clevelandii* plants to TBRV infection was due to post-transcriptional gene silencing (Covey *et al.*, 1997; Ratcliff *et al.*, 1997). Thus, recovery from disease symptoms of both the transgenic and non-transgenic plants after initial development of systemic symptoms occurs by the same mechanism - inhibition of virus replication by host due to virus-elicited post-transcriptional gene silencing (Covey *et al.*, 1997; Ratcliff *et al.*, 1997; Taylor, 1997; Meyer and Sadler, 1996; Teresa Ruiz *et al.*, 1998). This indicates that a transgene with homology to the infecting virus might also be silenced after the virus infection. It was found to be so with CaMV infection and the phenomenon was mediated by both transcriptional and post-transcriptional mechanisms, with the latter taking precedence over the former (Al-kaff *et al.*, 1998).

For ectopic pairing (between the RNA of infecting virus and the homologous transgene present as cDNA in some host chromosome present in plant genome inside the nucleus) to start, nuclear barrier between viral RNA and homologous transgene has to be removed to enable the genome of the infecting virus to move to the area where genome of the host is present. This could happen only in dividing cells, during which the nuclear membrane dissolves, to provide opportunity for interaction of the viral RNA with homologous DNA. This could explain the development of recovery phenotype after about three weeks or more of inoculation of the transgenic plants (Baulcombe and English, 1996). The systemic acquired silencing signal has been postulated to move upward from the dividing cells and to actively assist in the production of 'recovery' phenotype (Palauqui *et al.*, 1997). Nepoviruses are able to infect the meristems so that their ability to infect meristems, recovery of infected plants and gene silencing may be interrelated. This could indicate an increased possibility of gene silencing in the event of transgenes being expressed in the meristems, that is, in dividing cells (Ratcliff *et al.*, 1997).

Green Islands

Some dark-green areas, called "green islands," of healthy tissue are scattered among yellow green and white areas in the virus-infected leaves showing mosaic symptoms. These localized areas are resistant to infection by the same virus, contain no or very little infectious virus, and contain no viral protein and double-stranded viral RNA. Resistance of these green islands is due to sequence-specific targeting of viral RNA (Guo and Garcia, 1997) and is also seemingly based on the post-transcriptional gene silencing due to the cell-to-cell transmission of the RNP signal (Jorgensen *et al.*, 1998).

CONCLUSIONS

It is clear from above that post-transcriptional gene silencing has a wide area of action. It can be caused in an infecting virus carrying the homologous gene by a transgene of a transgenic plant; in a transgene of a transgenic plant by an infecting homologous virus; in an endogenous (host) gene by a transgene; in an endogenous (host) gene by an infecting virus; and in an infecting virus by a non-transgenic host.

Cosuppression and gene silencing are manifested as homology-dependent resistance phenomena, provide very high protection to transgenic and non-transgenic plants as well as to green islands against virus infection, and seem to be common phenomena in transgenic plants. Thus both transgenic and non-transgenic plants appear to have successfully evolved cosuppression and gene silencing as natural defenses for suppression of viruses and other foreign genetic elements. Post-transcriptional gene silencing in this way is a genetic disease control mechanism that imparts systemic resistance to plants against virus infection and is also emerging as a normal mechanism of gene expression regulation.

The existence of three models based on conflicting data indicates two possibilities. One is the actual existence of three different models of cosuppression with each being triggered in a different way. The second is that the conflicting data may have originated due to the use of different transgene constructs, different genomic arrangement of transgenes in different studies, and/or due to an overlap between cosuppression phenotypes.

It has not been possible so far to determine whether the RNA threshold and ectopic pairing models are mutually exclusive and whether both the models are correct for different types of transgene loci. On the other hand, it is also possible that all types of transgene loci need aRNA for suppression, that cosuppression is triggered when the aRNA transcripts reach a critical threshold level in all cases, and that more aRNA is produced by inverted repeats than by the single-copy transgenes.

The cosuppression is elicited both by the single copy abundantly expressing transgenes as well as by the inverse repeats of a transgene (Van Blockland *et al.,* 1994; English et. al., 1996; Sijen *et al.,* 1996; Stam *et al.,* 1997). Conceivably, cosuppression in these two situations may possess certain common features even if it is triggered by different types of nucleic acid interactions (Taylor, 1997).

REFERENCES

Al-kaff, N. S., Covey, S. N., Kreike, M. A., Page, A. M., Pinder, R., and Dale, P. J. (1998). Transcriptional and post-transcriptional plant gene silencing in response to a pathogen. *Science* 279, 2113 -2115.

Angenent, G. C., Franken, J., Busscher, M., Colombo, L., and Van Tunen, A. J. (1993). Petal and stamen formation in petunia is regulated by the homeotic gene *fbp1. Plant J.* 4, 101-112.

Angell, S. M., and Baulcombe, D. C. (1997). Consistent gene silencing in transgenic plants expressing a replicating potato virus X RNA. *EMBO J.* 16, 3675 -3684.

Assad, F. F., Tucker, L. K., and Signer, E. R. (1993). Epigenetic repeat-induced gene silencing (RIGS) in *Arabidopsis. Plant Mol. Biol.* 22, 1067-1085.

Baulcombe, D. C. (1994a). Replicase-mediated resistance: A novel type of virus resistance in transgenic plants ? *Trends Microbiol.* 2, 60 - 63.

Baulcombe, D. C. (1994b). Novel strategies for engineering virus resistance in plants. *Curr. Opinion Biotechnol.* 5, 117-124.

Baulcombe, D. C. (1996a). Mechanisms of pathogen-derived resistance to viruses in transgenic plants. *Plant Cell* 8, 1833-1844.

Baulcombe, D. C. (1996b). RNA as a target and an initiator of post-transcriptional gene silencing in transgenic plants. *Plant Mol. Biol.* 32, 79-88.

Baulcombe, D. C., and English, J. J. (1996). Ectopic pairing of homologous DNA and post-transcriptional gene silencing in transgenic plants. *Curr. Opin. Biotechnol.* 7, 173 -180.

Benfey, P. N., Ren, L., and Chua, N.- H. (1989). The CaMV 35S enhancer contains at least two domains which can confer different developmental and tissue-specific expression patterns. *EMBO J.* 8, 2195-2202.

Cameron, F. H., and Jennings, P.A. (1991). Inhibition of gene expression by a short sense fragment. *Nucleic Acids Res.* 19, 469-474.

Covey, S. N., Al-kaff, N. S., Langara, A., and Tumer, D. S. (1997). Plants combat infection by gene silencing. *Nature (London)* 385, 781-782.

de Carvalho, F., Gheysen, G., Kushnir, S., van Montagu, M., Inze, D., and Castresana, C. (1992). Suppression of ß-1, 3 glucanase transgene expression in homozygous plants. *EMBO J.* 11, 2595-2602.

de Carvalho Niebel, F., Frendo, P., van Montagu, M., and Cornelissen, M. (1995). Post-transcriptional cosuppression of *β*-1, 3 -gluconase genes does not affect accumulation of transgene nuclear mRNA. *Plant Cell* 7, 347-358.

Depicker, A., and van Montagu, M. (1997). Post-transcriptional gene silencing in plants. *Curr. Opin. Cell Biol.* 9, 372 -382.

Dougherty, W. G., and Parks, T. D. (1995). Transgene and gene suppression - telling us something new. *Curr. Opin. Cell Biol.* 7, 399-405.

Elmayan, T., and Vaucheret, H. (1996). Single copies of a strongly expressed 35S-driven transgene undergo post-transcriptional silencing. *Plant J.* 9, 282 - 292.

English, J. J., Mueller, E., and Baulcombe, D. E. (1996). Suppression of virus accumulation in transgenic plants exhibiting silencing of nuclear genes. *Plant Cell* 8, 179 -188.

Finnegan, J. and McElroy, D. (1994). Transgene inactivation: Plants fight back. *Biotechnology* 12, 883-888.

Flavell, R. B. (1994). Inactivation of gene expression in plants as a consequence of special sequence duplication. *Proc. Natl. Acad. Sci. USA* 91, 3490 -3496.

Goodwin, J. B., Chapman, K., Parks, T. D., Wernsman, E. A., and Dougherty, W. G. (1996). Genetic and biochemical dissection of transgenic RNA-mediated resistance. *Plant Cell* 8, 95 -105.

Guo, H. S., and Garcia, J. A. (1997). Delayed resistance to plum pox potyvirus mediated by a mutated RNA replicase gene: Involvement of a gene silencing mechanism. *Mol. Plant-Microbe Interact.* 10, 160-170.

Hart, C. M., Fischer, B., Neuhas, J. M., and Meins, F. (1992). Regulated inactivation of homologous gene expression in transgenic *Nicotiana sylvestris* plants containing a defence-related tobacco chitinase gene. *Mol. Gen. Genet.* 235, 179-188.

Iglesias, V. A., Moscone, E. A., Papp, I., Neuhuber, F., Michalowski, S., Phelan, T., Spiker, S., Matzke, M., and Matzke, A. J. M. (1997). Molecular and cytogenetic analyses of stably and unstably expressed transgenic loci in tobacco. *Plant Cell* 9, 1251, 1264.

Jorgensen, R. A. (1992). Silencing of plant genes by homologous transgenes. *AgBiotech. New Infor.* 4, 265 -273.

Jorgensen, R. A. (1995). Cosuppression, flower colour patterns and metastable gene expression states. *Science* 268, 686 - 691.

Jorgensen, R. A., Que, Q., English, J., and Wang, H. Y. (1996). Sense cosuppression of flower colour genes: Metastable morphology-based phenotypes and the prepattern-threshold hypothesis. *In* "Epigenetic Mechanisms of Gene Regulation" (V. E. A. Russo, R. A. Martienssen, and A. D. Riggs, Eds.), pp. 393 - 402. Cold Spring Harbor Lab. Press, Cold Spring Harbor, New York

Jorgensen, R. A., Atkinson, R. G., Forster, R. L. S., and Lucas, W. J. (1998). An RNA-based information superhighway in plants. *Science* 279, 1486-1487.

Kumagai, M. H., Donson, J., Della-Cioppa, G., Harvey, D., Hanley, K., and Grill, L. (1995). Cytoplasmic inhibition of carotenoid biosynthesis with virus-derived RNA. *Proc. Natl. Acad. Sci. USA* 92, 1679-1683.

Lindbo, J. A., and Dougherty, W. G. (1992a). Pathogen-derived resistance to a potyvirus. Immune and resistant phenotypes in transgenic tobacco expressing altered forms of a potyvirus coat protein nucleotide sequence. *Mol. Plant-Microbe Interact.* 5, 144 -153.

Lindbo, J. A., and Dougherty, W. G. (1992b). Untranslatable transcripts of the tobacco etch virus coat protein gene sequence can interfere with tobacco etch virus replication in transgenic plants and protoplasts. *Virology* 189, 725-733.

Lindbo, J. A., Silva-Rosales, L., Procbsting, W. M., and Dougherty, W. G. (1993). Induction of a highly specific antiviral state in transgenic plants. Implications for the regulation of gene expression and virus resistance. *Plant Cell* 5, 1749 -1759.

Lomonossoff, G. P. (1995). Pathogen-derived resistance to plant viruses. *Annu. Rev. Phytopathol.* 33, 323 -343.

Marano, M. R., and Baulcombe, D. C. (1998). Pathogen-derived resistance targeted against the negative strand RNA of tobacco mosaic virus: RNA strand-specific gene silencing. *Plant J.* 13, 537-546.

Matzke, A. J. M., Neuhuber, F., Park, Y.- D., Ambros, P. F., and Matzke, M. A. (1994). Homology-dependent gene silencing in transgenic plants. Epistatic silencing loci contain multiple copies of methylated transgenes. *Mol. Gen. Genet.* 244, 219-229.

Matzke, M. A., and Matzke, A. J. M. (1995a). How and why do plants inactivate homologous (trans)genes ? *Plant Physiol.* 107, 6679 - 6685.

Matzke, M. A., and Matzke, A. J. M. (1995b). Homology-dependent gene silencing in transgenic plants – what does it really tell us. *Trends Genet.* 11, 1-3.

Matzke, M. A., Matzke, A. J. M., and Sheid, O. M. (1995). Inactivation of repeated genes- DNA-DNA interaction. *In* Homologous Recombination and Gene Silencing in Plants (J. Paszkowski, Ed.), pp. 271-307. Kluwer Academic Publishers, Dordrecht.

Meins, F., and Kunz, C. (1995). Gene silencing in transgenic plants - a heuristic autoregulation model. *Curr. Topics Microbiol.- Immunol.* 197, 105 -120.

Metzlaff, M., O'Dell, M., Cluster, P. D., and Flavell, R. B. (1997). RNA- mediated RNA degradation and chalcone synthase A silencing in *Petunia. Cell* 88, 845 - 854, 1-20.

Meyer, P. (1995a). Gene Silencing in Higher Plants and Related Phenomena in Other Eukaryotes. Springer-Verlag, Berlin.

Meyer, P. (1995b). Understanding and controlling transgene expression.*Trends Biotechnol.* 13, 332-337.

Meyer, P., and Saedler, H. (1996). Homology-dependent gene silencing in plants. *Annu. Rev. Plant.Physiol. Plant Mol. Biol.* 47, 23-48.

Mittelsten Scheid, O., Jakovleva, L., Afsar, K., Maluszynska, J., and Paszkowski, J. (1996). A change of ploidy can modify epigenetic silencing. *Proc. Nat. Acad. Sci. USA.* 93, 7114-7119.

Mueller, E., Gilbert, J., Davenport, G., Brigneti, G., and Baulcombe, D. C. (1995). Homology-dependent resistance: Transgenic virus resistance in plants related to homology-dependent gene silencing. *Plant J.* 7, 1001-1013.

Napoli, C., Lemieux, C., and Jorgenson, R. (1990). Introduction of a chimeric chalcone synthase gene into petunia results in reversible co-suppression of homologous genes *in trans. Plant Cell* 2, 279-289.

Palauqui, J. C., and Vaucheret, H. (1995). Field trial analysis of nitrate reductase co-superssion: A comparative study of 38 combinations of transgene loci. *Plant Mol. Biol.* 29, 149-159.

Palauqui, J. C., Elmayan, T., Dorlhac de Borne, F., Crete, P., Charles. C., and Vaucheret, H. (1996). Frequencies, timing, and spatial patterns of co-suppression of nitrate reductase and nitrite reductase in transgenic tobacco plants. *Plant Physiol.* 112, 1442 -1456.

Palauqui, J. C., Elmayan, T., Pollien, J. M., and Vaucheret, H. (1997). Systemic acquired silencing: Transgene-specific post-transcriptional silencing is transmitted by grafting from silenced stocks to non-silenced scions. *EMBO* J. 16, 4738-4745.

Park, Y. D., Papp, I., Moscone, E. A., Iglesias, V. A., Vaucheret, H., Matzke, A. J. M., and Matzke, M. A. (1996). Gene silencing mediated by promoter homology occurs at the level of transcription and results in meiotically heritable alteration in methylation and gene activity. *Plant J.* 9,183-194.

Pruss, G., Ge, X., Shi, X. M., Carrington, J. C., and Bowman Vance, V. (1997). Plant virus synergism: The potyviral genome encodes a broad range pathogenicity enhancer that transactivates replication of heterologous viruses. *Plant Cell* 9, 859-868.

Que, Q., Wang, H., English, J. J., and Jorgensen, R. A. (1997). The frequency and degree of cosuppression by sens chalcone synthase transgenes are dependent on transgenic promoter strength and are reduced by premature nonsense codons in the transgene coding sequence. *Plant Cell* 9, 1357-1368.

Ratcliff, F., Harrison, B. D., and Baulcombe, D. C. (1997). A similarity between viral defence and gene silencing in plants. *Science* 276, 1558-1560.

Ruiz, M. T., Voinnet, O., and Baulcombe, D. C. (1998). Initiation and maintenance of virus induced gene silencing. *Plant Cell* 10, 937-946.

Sijen, T., Wellink, J., Hiriart, J.- B., and van Kammen, A. (1996). RNA- mediated virus resistance: Role of transgene and delineation of targeted regions. *Plant Cell* 8, 2277-2294.

Smith, H. A., Swaney, S. L., Parks, T. D., Wernsman, E. A., and Dougherty, W. G. (1994). Transgenic plant virus resistance mediated by untranslatable sense RNAs: Expression, regulation and fate of non-essential RNAs. *Plant Cell* 6, 1441-1453.

Smith, H. A., Powers, H., Swaney, S., Brown, C., and Dougherty, W. G. (1995). Transgenic potato virus Y resistance in potato: Evidence for an RNA-mediated cellular response. *Phytopathology* 85, 864-870.

Stam, M., de Bruin, R., Kenter, S., Van der Hoorn, R. A. L., Van Blokland, R., Mol, N. M., and Kooter, J. M. (1997). Post-transcriptional silencing of chalcone synthase in *Petunia* by inverted transgene repeats. *Plant J.* 12, 63-82.

Tanzer, M. M., Thompson, W. F., Law, M. D., Wernsman, E. A., and Uknes, S. (1997). Characterisation of post-transcriptionally suppressed transgene expression that confers resistance to tobacco etch virus infection in tobacco. *Plant Cell* 9, 1411-1423.

Taylor, C. B. (1997). Comprehending cosuppression. *Plant Cell* 9, 1245-1249.

Tenllado, F., Garcia-Luque, I., Serra, M. T., and Diaz Ruiz, J. R. (1995). *Nicotiana benthamiana* plants transformed with the 54- kDa region of pepper mild mottle tobamovirus replicase gene exhibit two types of resistance response against viral infection. *Virology* 211, 170-183.

Teresa Ruiz, M., Voinnet, O., and Baulcombe, D. C. (1998). Initiation and maintenance of virus-induced gene silencing. *Plant Cell* 10, 937-946.

Van Blokland, R., Vandergeest, N., Mol, J. N. M., and Kooter, J. M. (1994). Transgene mediated suppression of chalcone synthase expression in *Petunia hybrida* results from an increase in RNA turnover. *Plant J.* 6, 861- 877.

Van der Krol, A. R., Mur, L. A., Beld, M., Mol. J. N. M., and Stuitje, A. R. (1990). Flavonoid genes in petunia: Addition of a limited number of gene copies ma lead to a suppression of gene expression. *Plant Cell* 2, 291-299.

Vaucheret, H., Nussaume, L., Palauqui, J. C., Quillere, I., and Elmayan, T. (1997). A transcriptionally active state is required for post-transcriptional silencing (cosuppression) of nitrate reductase host genes and transgenes. *Plant Cell* 9, 1495 -1504.

Voinnet, O., and Baulcombe, D. C. (1997). Systemic signalling in gene silencing. *Nature (London)* 389, 553.

Wassenegger, M., Heimes, S., Riedel, L., and Sanger, H. L. (1994). RNA-directed *de novo* methylation of genome sequences in plants. *Cell* 76, 567-576.

SUBJECT INDEX

ATPase activity 34, 49-50,70,73
AUG initiation codon 13, 30-31, 33, 39-40, 76, 100-102,104,107, 204-206, 225, 258
Aberrant homologous recombination 243-244
Aberrant RNA 101, 259-260, 263, 266
Aberrant RNA model of gene silencing 259-260
Actin filaments 163, 168,170-172
African cassava mosaic virus 185-186, 242, 244
Agrobacterium 221, 236
Agrobacterium tumefaciens 244
Alfalfa mosaic alfamovirus 2, 5-6, 10, 13, 5-17, 65, 102, 121, 124, 126, 131-132, 202, 243, 246, 248
Alfamovirus 2, 5-6, 63, 65,
Alfamovirus(es) 64-66, 164, 166, 169, 173, 242-243
Allexvirus 47, 75-77
Allexvirus(es) 76
Alphavirus 47, 75-77
Alpha-like viruses/supergroup 8-9, 16, 47, 61, 75, 82,105-106, 108-109, 111-113, 125
Alphacryptovirus 2, 6, 48, 81
Alphavirus(es) 106, 108, 113, 125
Ambisense coding strategy/RNA/virus 5, 79-80, 82, 134
Antisense RNA 242, 258
Aphid transmission factor 37-38, 185
Aphid vector/transmission by 12, 29, 37, 50, 55-58, 63, 109, 184-185, 187-188, 236
Aphis gossypii 148
Aphthovirus 112-113
Apple chlorotic leafspot trichovirus 3, 5-6, 8, 74
Apple dimple fruit viroid 226
Apple scar skin viroid 226
Apple stem grooving

capillovirus 2, 5, 6, 74
Apple stem pitting foveavirus 76-77
Apscaviroid 230
Arabidopsis thaliana 202, 213-214, 216
Arterivirus(es) 110, 112
Arthropod vector 183
Assembly of viruses 17-18, 37, 50-51, 72, 76-77, 81, 150, 190, 201, 243, 246, 248,
Assembly origin 17, 51, 61, 67, 175
Astrovirus(es) 110
Aureusvirus 59
Australian grapevine viroid 226, 229
Avenavirus 59
Avocado sunblotch viroid 226, 228, 232, 234, 236
Avsunviroidae 227, 230-232, 235

Bacilliform DNA virus 29
Back mutation 124
Badnavirus 29, 33, 37
Badnavirus(es) 29, 33-34, 37, 39, 40
Bamboo mosaic virus 3
Banana bunchy top virus 29, 31-32, 36
Barley mild mosaic virus 191
Barley stripe mosaic hordeivirus 2, 5, 7, 11, 68, 70, 103, 106-107, 109, 193-195, 243, 246
Barley yellow dwarf luteovirus 2, 5, 7, 11-14, 57-58, 107, 109, 187
Barley yellow mosaic bymovirus 2, 5-6, 49, 51
Bean common mosaic bymovirus 193
Bean golden mosaic virus 31
Beet 3 alphacryptovirus 2, 6, 81
Beet cryptic alphacryptovirus-1 6
Beet curly top curtovirus 31, 36, 39, 148-149,152, 186, 242, 247
Beet necrotic yellow vein benyvirus 2, 5, 7, 9, 11-12, 14-16, 70, 72-73, 104, 12-113, 126, 132, 134, 146, 190, 247

Beet soil-borne pomevirus   69-72, 104
Beet virus Q 14, 71
Beet western yellows luteovirus 12, 122, 188
Beet western yellows virus ST9-associated RNA 57-58
Beet yellows closterovirus 2, 6, 11, 14, 62, 110, 113, 122
Begomovirus 29, 31, 35, 185, 188, 244
Begomovirus(es) 30-31, 34-36, 38-39
Bemisia tabaci 31,188
Benyvirus 66, 70, 72, 190
Benyvirus(es) 51, 102, 104, 113
Betacryptovirus 48, 81
Betacryptovirus(es) 1
Biochemical switch model of gene silencing 259
Biological cloning 144
Bipartite genome/viruses 29, 31, 35-36, 51-52, 61-63, 67, 71-72, 104, 112, 164, 172, 190, 193, 202, 248
Brassica napus 244, 264
Brassica oleracea gongylodes 264
Broad bean mottle virus 2, 132, 134-135
Broad bean necrosis virus 71
Brome mosaic bromovirus 2, 6, 9-11, 13-18, 64-66, 102, 106, 124-128, 130-131, 134, 172, 202-207, 243,246, 248, 263
Brome streak mosaic rymovirus 3, 5, 8
Bromoviridae 2, 47, 57, 63-66, 106, 193
Bromovirus 2, 5-6, 63-65
Bromovirus(es) 15, 64-65, 103, 106, 124-126, 131-132, 134, 164, 166, 169, 171, 173, 175, 202-204, 206, 243
Buchnera 187, 188
Bunyaviridae 3, 48, 77-79, 185
Bunyavirus(es) 78, 136, 186
Bymovirus 2, 5-6, 49, 51, 75, 104, 190-191
Bymovirus(es) 51, 104, 112, 190-191

Callose 163, 211
Cap 5, 9, 12-13, 49, 56, 59, 62-63, 66-68, 71-75, 78, 100-102, 114, 258
Cap-dependent translation 102, 113

Cap-independent translation and translational enhancer 13, 51, 58, 102, 107
Cap snatching 48, 78
Capillovirus 2, 5-6, 47, 74-75
Capillovirus(es) 75, 112
Capsicum annum 220
Capsid/capsid protein   6, 9-18,   30, 35-39, 49-77, 79, 81, 103-104, 106-110, 124, 127, 147, 149, 164, 170, 172-173, 175, 202-207, 216-217, 219-220, 241-244, 246-251, 255-256, 261-262
Capsid-binding sites 17-18
Cardamine lilacina 146-147
Carlavirus 2, 5-6, 47, 75-77
Carlaviruses 9-11, 75-77, 102, 104, 110, 112-113, 164
Carmovirus 2, 5-6, 58-60, 193
Carmovirus(es) 10, 14, 16, 55-56, 58, 106, 108-09, 164, 192, 203, 207, 243
Carmovirus-like viruses/supergroup 2, 8-9, 14, 16, 47-48, 56-57
Carnation mottle carmovirus 2, 5-6, 14, 60
Carnation ringspot dianthovirus 2, 5-6, 14, 60
Carrot mottle mimic umbravirus 57
Carrot red leaf virus 58
Cassava vein mosaic virus 33-34, 38
Cauliflower mosaic caulimovirus 29, 33, 36-40, 125, 151-152, 165, 185, 217-218, 242, 244, 255
Caulimoviridae 29, 32, 36, 39
Caulimovirus 29, 32, 36
Caulimovirus(es) 29, 32-33, 39, 164, 166, 169-170, 184-185, 242-243
Cell-to-cell movement 9-10, 36, 50, 53, 55-57, 59, 61, 64, 67-68, 70, 72-73, 76, 99, 106, 108, 124-125, 153, 161-177, 201, 206, 217-220, 222, 236, 247-248, 261, 265
Chalcone synthase gene 255, 260, 263
Chaperone 171
Chaperonin 12
Chenopodium amaranticolor 124, 206
Chenopodium hybridum 124, 202-203,

206
Chenopodium quinoa 202, 205-206
Cherry leaf roll nepovirus 7, 250
Chinese cabbage 202
Chlorotic lesions 201-203, 206
Chymotrypsin-like proteinase 11, 47, 50, 52-56, 110-112
Chrysanthemum chlorotic mottle viroid 226-230, 232, 234
Chrysanthemum stunt viroid 226
Chytrid fungi/chytridiomycetes 190-191
Circulative virus transmission 56, 183-185, 187
Cis-activity/cleavage/elements 11-16, 50, 53-54, 58-59, 61, 63-64, 66-68, 71, 73, 77, 82, 103, 106-107, 111-113
Cis-inactivation 256, 258
Citrus III viroid 226
Citrus IV viroid 226
Citrus bent leaf viroid 226
Citrus exocortis viroid 226, 228-229, 234
Citrus tatter leaf capillovirus 5
Citrus tristeza closterovirus 2, 14, 62, 103
Cladosporium fulvum 214
Closteroviridae 2, 47, 62, 103-104, 125
Closterovirus 2, 5-6, 62, 104
Closterovirus(es) 10-11, 14, 62-63, 106, 110-114
Clover yellow mosaic potexvirus 3, 132
Coat protein - see Capsid protein
Cocaviroid 230
Cocksfoot mottle sobemovirus 8, 53-54, 110
Coconut cadang cadang viroid 226, 229, 234-235
Coconut foliar decay virus 29, 32, 38
Coconut tinangaja viroid 226, 235
Coleviroid 230
Coleus blumei 1 viroid 226
Coleus blumei 2 viroid 226
Coleus blumei 3 viroid 226
Columnea latent viroid 226, 229
Commelina yellow mottle virus 33, 38
Comoviridae 2-3, 47, 52-53, 193
Comovirus 2, 5-6, 52-53

Comovirus(es) 13, 53, 100, 102, 104, 107-108, 112, 164, 166, 169, 172
Compensatory mutations 124
Complementation between functions/strains 125, 127, 152-154, 164, 247-248
Consensus sequences 108-109, 112
Copy choice model 135, 244
Coronaviruses 110, 112, 114, 125
Cosuppression 256-261, 264, 266
Cotton leaf curl begomovirus 146, 148-149, 151
Cowpea chlorotic mottle bromovirus 10, 131-132, 202-203, 244-245
Cowpae mosaic comovirus 2, 5, 6, 9-11, 13, 15, 52-53, 107, 165, 169, 172-173
Crinivirus 62, 104
Cross-protection 235,241
Cucumber mosaic cucumovirus 2-3, 5-6, 10-11, 15-16, 65, 122-124, 131-132, 144, 146, 148-153, 193, 202-204, 207, 246-249
Cucumber necrosis tombusvirus 10, 191-192
Cucumovirus 2, 5-6, 63-66, 193
Cucumovirus(es) 15, 64, 104, 106, 122, 125, 131, 134, 151-152, 164, 166, 203-204, 242, 248-249
Curcurbita pepo 202
Curtovirus 29, 31, 36, 186
Curtoviruses 34-36, 38
Cymbidium ringspot tombusvirus 16, 132
Cysteine proteinases 50, 110-112
Cysteine-rich protein 103, 105-106, 194
Cytorhabdovirus 1, 16, 48, 78

DNA viruses 29-41
DNA binding activity/domain 35
DNA-dependent RNA polymerases 66, 123
DNA recombinant technology 241
Datura stramonium 217-218
Defective interfering RNAs 1-3, 123, 126, 132, 134, 152
Deletion mutation/mutant 121, 123,

134-135, 151-152
Desmodium species 146
Desmotubule 162-163
Dianthovirus 2, 5-6, 59-61
Dianthovirus(es) 61, 71, 104, 106, 109-110, 164, 166, 169
Digitaria streak virus 38
Divided genome – see Multipartite genome/viruses
Double-stranded DNA genome/template/virus(es) 29, 32-33, 36-37, 39, 144, 166, 169, 173, 228, 232, 248, 261
Double-stranded RNA genome/virus(es) 48, 80-81, 110, 128-129, 166, 264-265
Drosophila 214-215

Ectopic 36
Ectopic pairing 259-260, 263, 265-266
Ectopic pairing model of gene silencing 260
Enamovirus 2, 5-6, 48, 53, 55-56
Endoplasmic reticulum 9, 36, 50, 66-67, 162-163
Endosymbiotic bacterium 187-188
Evolution/evolutionary relationship 47, 82, 99, 102-103, 106, 108, 113-114, 121-122, 124-126, 132, 134-136, 143-154, 191, 216, 229-231, 243

Faba bean necrotic yellows virus 29, 31-32, 36
Fabavirus 52
Fabavirus(es) 1
Fijivirus 2, 48, 80-81
Foot and mouth disease *Aphthovirus* 113, 186
Founder effect 144, 146-150
Foveavirus 47, 75-77
Frameshift/frameshift mutation 6-8, 13-14, 49, 53, 55-58, 61, 63, 76, 104, 109-111, 113-114, 123, 134, 216, 247-248
Fungal transmission factor/of viruses

10-12, 51, 59, 71, 109, 190-193
Furovirus 2, 5, 7, 66, 69, 71, 190
Furovirus(es) 10-11, 101, 104, 108-109, 126, 132, 134, 164, 166, 176

Geminiviridae 29, 30, 34, 38, 185
Geminivirus(es) 29-31, 34, 36, 38-39, 125, 148, 150, 152, 164, 167, 169-170, 172-173, 185-186, 244
Gene dosage hypothesis 259
Gene expression 30, 34, 36, 38-39, 68, 99-113, 194, 255-258, 263-264, 266
Gene-for-gene hypothesis/interaction 211-221
Gene silencing 243, 255-266
Genetic distance 154
Genetic diversity/variability 121-135, 143-156, 241, 243-244
Genetic stability 103, 145-148, 154
Genetically-engineered plant/resistance 241-250
Genetic determinants 193-195, 202-206, 234-238
Genome activation 50, 64
Genome masking 246
Genome organisation of DNA viruses 30-34
Genome organisation of RNA viruses 48-81, 104
Genome-linked protein 5, 8-9, 11, 47-56, 58, 175
Gomphrena globosa 202-203
Grapevine fanleaf nepovirus 9, 11
Grapevine trichovirus B 5, 8
Grapevine virus A 74
Grapevine yellow speckle 1 viroid 226, 228
Grapevine yellow speckle 2 viroid 226
Green islands 265-266
GroEL protein 55-56, 187-188
Groundnut rosette assister virus 246
Groundnut rosette disease/umbravirus 4-5, 8, 246
Guanylyltransferase activity/domain 9, 47-48, 61-62, 64, 66, 68, 71-75, 78, 81

Hammerhead ribozymes 227, 230, 232-234

Hammerhead structures 227-234

Haplotypes 146-148

Heat shock protein 63, 102, 166, 172, 187

Helicase 6, 9, 47-48, 50, 52, 61-62, 64, 66, 68, 70-76, 104-106, 108-110, 112-113, 128, 130, 166-167, 218, 220

Helper component proteinase 10, 12, 49-51, 112-113, 173, 175-176, 194-195

Heteroencapsidation/heterologous encapsidation 11, 243, 246-247, 250-251

Heterologous mixture of strains 121, 144-147, 151, 228, 234

Heterologous resistance 241

Heterozygosity index 145, 147

Homologous recombination 126-127, 129-130, 134-135, 243-245

Homologous resistance 241

Homology-dependent gene expression/mechanism/resistance 256, 263-264, 266

Hop latent viroid 226

Hop stunt viroid 226, 228-229, 235

Hordeivirus 2, 5, 7, 66, 68, 70-73, 193

Hordeivirus(es) 68, 71-73, 103-104, 106, 126, 164, 243

Horseradish curly top virus 31

Host factors for pathogen recognition 213-216

Host passage effects 144

Host range 10-11, 37, 122, 125, 131, 136, 192, 203-204, 229, 235-236, 247-248

Host specificity determinants 64, 131

Hostuviroid 230

Hypersensitive response 61, 66-67, 124, 202, 204, 211-212, 214-221

Idaeovirus 2, 5, 7, 47, 63-65, 104

Ilarvirus 2, 5, 7, 63-65

Ilarvirus(es) 7, 10, 64, 66, 104

Immunity 220, 236, 256, 262

Inclusion body 37, 173-174

Insertion mutation 121, 123, 127, 134-135

Intergenic region 2, 16, 30-31, 33-34, 55, 78-79, 150

Interleukin-1 receptor 214-215

Internal initiation of gene expression/translation 12-13, 15-16, 76, 104, 107-108, 110

Internal ribosome binding/entry site 13, 53, 67, 107-108

Intron 30-31, 33-34, 38, 40, 230

Ipomovirus 51

Iresine 1 viroid 226

Kennedya rubicunda 146

Kennedya yellow mosaic tymovirus 146-147, 150

Kinase(s) 11, 166, 171, 174, 213-215, 235

Kinesin 170

Laboratory stocks of viruses 144-145, 147-148, 151, 151-153

Leaky stop codon/leaky scanning gene expression 40, 100-101, 104, 107-109, 113

Lettuce big vein virus 191

Lettuce infectious yellows crinivirus 14, 62, 103

Lettuce mosaic virus 193

Leucine-rich repeat 213-215

Leucine zipper 213

Lily virus X 101

Lipaphis erysimi 184

Local lesion 124, 201-204, 206, 212, 219

Long-distance movement 10-11, 53, 56-57, 59, 61, 64, 66-67, 72-73, 76, 108, 161, 163, 174-176, 201

Luteoviridae 185, 187

Luteovirus 2, 5, 7, 48, 57-58, 187

Luteovirus(es) 9-12, 14, 17, 19, 48, 53, 55-58, 102, 106-107, 109-110, 122, 123, 126, 186-188, 243, 246, 251

Lychnis ringspot hordeivirus 68
Lycopersicon esculentum 217, 220

Machlomoviridae 3
Machlomovirus 3, 5, 7, 59-60
Macluravirus 51
Maize chlorotic mottle machlomovirus 3, 5, 7, 60
Maize rough dwarf fijivirus 2
Maize stripe tenuivirus 3
Maize streak virus 31, 38
Marafivirus 3, 5, 7, 47, 74
Marafivirus(es) 10, 74-75, 112
Mastrevirus 29, 31, 34
Mastrevirus(es) 30-31, 35, 38-39
Meiotic resetting 257, 262
Melon necrotic spot carmovirus 193
Methyl transferase 6, 9, 49, 104-106, 108-110, 113
Methyl/guanylyltransferase 47-48, 61-62, 64, 66, 68, 71-75, 78, 81
Mexican papita viroid 226
Microtubule 10, 63, 170-171
Milk vetch dwarf virus 29
Mononegavirales 48
Monocomponent/monopartite genome/ viruses 29, 31, 36, 103-104, 106, 114, 202-203
Movement protein 6-8, 10-11, 16, 30-31, 35-39, 49, 53, 55-59, 61, 64, 66-68, 71-73, 75-76, 79, 101, 103-106, 125, 131, 164-176, 202, 206, 216-218, 242, 247
Muller's ratchet 153
Mutation 9, 13, 15-17, 36, 40, 50, 75, 99, 121-126, 129, 134, 144, 149, 152-154, 165, 168-169, 175, 189, 194, 206-207, 214, 217, 219, 228-229, 234, 236, 245-246, 248, 250
Mutation rate 122, 228
Multicomponent/multipartite genome/ viruses 103-104, 106, 114, 202
Myndus taffini planthopper 29
Myosin 163, 171
Myzus persicae 148, 187-188, 247

Nanovirus 29, 31-32, 36, 38, 185
Nanovirus(es) 29, 31-32, 36, 38-39
Necrotic lesion 218-219, 201-204, 206
Necrovirus 3, 5, 7, 59-60, 191
Necrovirus(es) 10, 164
Negative selection 151-152, 154
Negative-sense RNA plant viruses 48
Nematode pathogen/vector 11-12, 68, 188-189, 201, 211, 213
Nepovirus 3, 5, 7, 52, 188, 193
Nepovirus(es) 11, 104, 110, 112, 126, 164, 169, 172, 188-190, 250, 264-265
Nicotiana benthamiana 125, 127, 202, 206, 219, 221, 244-248
Nicotiana bigelovii 218, 244-245
Nicotiana clevelandii 219, 265
Nicotiana edwardsonii 202-203, 219
Nicotiana glauca 147, 153
Nicotiana glutinosa 214, 219, 221
Nicotiana plumbaginifolea 260
Nicotiana sylvestris 124, 202, 204, 216
Nicotiana tabacum 202, 219, 221, 248, 260
Non-AUG initiation codons 76, 101
Non-circulative virus transmission 183-184
Non-homologous recombination 126-127, 129-130, 243, 245
Non-propagative virus transmission 183, 187
Nuclear shuttle protein 30, 35-36, 39
Nucleorhabdovirus 3, 5, 7, 48, 78
Nucleotide binding activity/sites 9, 214
Nucleotide diversity 145-146, 148-149, 151
Nucleotidyl transferase 128

Oat blue dwarf marafivirus 3, 5, 7, 74-75
Oat chlorotic stunt virus 59
Oleavirus 63-65
Olive latent virus 6
Olpidium bornovanus 190-193
Olpidium brassicae 190-192
Origin of assembly – see Assembly

origin
Origin of replication 34
Origin of viroids 229-230
Oryzavirus 3, 7, 48, 80-81

Panicovirus 59-60
Panicum mosaic panicovirus 3, 60
Papain-like proteinases 11, 50-51, 62, 72-76, 110-112
Papaya ringspot virus 241
Paramutation 257
Pararetroviruses 29, 39, 40
Parsnip yellow fleck sequivirus 3, 5, 8, 51
Partitiviridae 2, 48, 81
Pathogen-derived resistance 136, 241-242, 263
Pathogenesis-related genes/proteins 211, 215
Pathogenic domain of viroids 226-227, 234-235
Pathogenicity determinants 31, 36, 50, 68, 234
Pea early browning tobravirus 12, 68, 131, 189, 194
Pea enation mosaic enamovirus 2, 5-6, 13-14, 55-57
Pea seed-borne mosaic potyvirus 13, 193-195
Peach latent mosaic viroid 226-230, 232, 234-236
Peanut clump pecluvirus 14, 69-70, 72, 104, 107-109, 132, 191
Peanut stunt cucumovirus 131, 248
Pear blister canker viroid 226, 236
Pecluvirus 66, 69, 72, 190-191
Pelamviroid 230
Pepper mild mottle tobamovirus 147, 149
Peranospora parasitica 216
Petunia 255, 263-264
Phloem-limited viruses 55, 167
Phytoreovirus 3, 7, 48, 80-81, 185
Picorna-like supergroup/picornavirus(es) 6-8, 9, 13, 47-48, 52, 54, 107-108, 110, 111-113, 125

Pinwheel inclusion body 173
Plasmodesmata 10, 35-36, 50, 61, 64, 67, 76, 79, 163-166, 168-174, 176, 195, 236, 261
Plasmodesmata size exclusion limit 10, 36, 50, 61, 64, 67, 76, 163-165, 168-171, 173-174, 176, 183
Plasmodiophoromycetes 190-191
Plum pox potyvirus 3, 9, 15, 127, 246-247
Poa semilatent hordeivirus 68
Point mutation 121-122, 134-135
Polerovirus 48, 53-56, 187-188
Polerovirus(es) 55-56, 122
Poliovirus 9, 122
Pollen transmission of viruses 183, 192-194
Poly(A) end/tail 2, 5, 7, 12, 15-16, 49, 51-53, 68, 78, 100-102, 262
Polymerase(es) – see RNA-dependent RNA polymerase
Polymerase stuttering 123, 129
Polymyxa betae 72, 190
Polymyxa graminis 51, 190
Pomovirus 51, 66, 69, 71, 190
Pomovirus(es) 71-72
Pospiviroidae 227, 230-235
Post-transcriptional gene silencing 257-260, 262-266
Potato carlavirus M 2, 5, 6, 77, 110
Potato leaf roll polerovirus 2, 7, 9, 14, 54-56, 122, 126, 187, 243
Potato mop top pomovirus 71-72, 104, 190-191
Potato spindle tuber viroid 225-229, 231-232, 234-237, 245, 248
Potato potexvirus X 3, 5, 8-9, 13, 15-16, 76-77, 108, 124, 217, 219-220, 242, 248-249, 258, 263
Potato potyvirus Y 3, 5, 8-9, 49, 51, 127, 151-152, 220, 246-248, 263
Potato virus A 151
Potexviridae 47, 75-77
Potexvirus 3, 5, 8, 47, 75, 77, 217
Potexvirus(es) 9-10, 75-77, 101-102, 108, 132, 164, 166, 170-171, 217
Potyviridae 2-3, 47, 49, 51, 104, 190,

193
Potyvirus 3, 5, 8, 49, 51, 104, 193, 195
Potyvirus(es) 9-10, 12-13, 49-51, 55, 102, 108, 110, 112-113, 127, 146, 164, 173, 175, 184-185, 189, 194, 220, 246-247, 249-250
Propagative virus transmission 183
Protease(s) 37-38, 48-56, 62, 72-76, 186, 194, 249
Proteinases 11, 49-50, 62, 110, 112-114, 117, 166, 173
Proteolytic cleavage/processing 6-8, 11, 37, 76, 100, 110, 174, 188
Prune dwarf ilarvirus 7
Pseudoknot 13-15, 66-67, 71, 73, 102, 109, 227, 229
Pseudorecombinant(s) 131-132, 188, 202-203, 206, 248
Pseudorecombination 124, 131
Pseudorevertants 124

Quasi-species 121, 145, 228

RNA binding domain 38, 165, 167
RNA-dependent RNA polymerase 1, 6-8, 9, 12, 14-17, 47-50, 52-53, 55-59, 61, 63-64, 66, 68, 71-75, 78-82, 99, 103-106, 108-110, 113-114, 122-123, 127, 130, 144, 175, 259, 261
Random amplified polymorphic DNA 146
Raspberry bushy dwarf idaeovirus 2, 5, 7, 63-65
Raspberry ringspot virus 202
Readthrough domain/protein 6-8, 10-12, 14, 17, 49, 51, 55-56, 58-59, 66, 68, 71-72, 100, 104-105, 108-109, 114, 187-188
Reassortants 152
Receptors of viruses in vector 184, 186-187, 189
Recombination 9, 55, 99, 103, 123-131, 134-136, 143-154, 216, 229, 241, 243-245, 248, 250-251
Recombinosome 131

Recovery phenomenon 256, 264-265
Red clover necrotic mosaic dianthovirus 11, 14, 17, 106, 170, 174, 202, 248
Red Queen theory 136
Reoviridae 2, 3, 48, 80
Reovirus(es) 10, 80, 151-152
Repeat-induced gene silencing 256
Replicase/replicase gene/recognition site 15, 54, 59, 68, 73, 76, 78-79, 99, 101, 103-104, 106-110, 113-114, 122-123, 126, 129-132, 134-135, 170, 175, 194, 206-207, 216-220, 242
Replicase-mediated resistance 250, 242
Replicase role in long-distance virus movement 175
Replicase stuttering 129
Replication-associated gene/protein 30-31, 34-36, 39
Replication-enhancer protein 30-31, 34-35, 39
Resistance against disease 132, 139, 154, 174, 195, 211-221, 236, 241-242, 244-245, 248-250, 261-263, 265-266
Resistance genes of plants 150, 211-215
Retrovirus(es) 29, 39, 103, 108, 110, 125, 229
Reverse genetics 121
Reverse mutation 124
Reverse transcriptase-polymerase chain reaction 147
Reverse transcription/transcriptase 37-39, 245, 261
Rhabdoviridae 3, 48, 77-78
Rhabdovirus(es) 10, 78, 132, 150
Rhopalosiphum padi 187
Ribonuclease protection analysis 145-149
Ribosome shunting 39
Ribosomal frameshifting – see Frameshift
Rice dwarf phytoreovirus 3, 7, 80-81, 146, 185, 188
Rice grassy stunt tenuivirus 3, 80
Rice ragged stunt oryzavirus 3, 7, 202
Rice stripe tenuivirus 3, 5, 8, 79-80
Rice tungro bacilliform virus 33, 38-40
Rice tungro spherical virus 4-5, 8, 38,

51, 58, 112
Rice yellow mottle sobemovirus 8, 53
Rolling-circle mechanism of replication 231-232
Rubivirus(es) 112
Ryegrass mosaic rymovirus 49
Rymovirus 3, 5, 8, 49, 51

Salivary glands 183, 187
Satellite RNA(s) 1-4, 6-8, 12, 123, 126, 129, 144-145, 148-150, 152-153, 207, 230, 242
Satellite tobacco mosaic virus 123, 144-145, 147, 149
Satellite tobacco necrosis virus 102, 144
Schizaphis graminum 187
Seed transmission 68, 183, 192-195, 236, 249-250
Selection of virus/viroid strains 143-153, 228
Sequiviridae 3-4, 10, 47, 51-52, 185
Sequivirus 3, 8, 51-52
Sequivirus(es) 112
Serine proteinases 11, 50, 54-56, 110-112
Shallot virus X 77, 101
Signal transduction 213-215, 235
Similarity-assisted recombination 243
Similarity essential recombination 243
Similarity non-essential recombination 243
Sindbis virus 122
Single lesion passage 145
Sitobion avenae 187
Size exclusion limit of plasmodesmata – see Plasmodesmata
Sobemovirus 3, 5, 8, 48, 53-54
Sobemo-like supergroup 8-9, 16, 47-48, 53-57, 111
Sobemovirus(es) 11, 53-55, 58, 101, 107, 110-112
Soil-borne wheat mosaic furovirus 2, 5, 7, 14, 69-71, 101, 108-109, 132, 134, 190-191
Solanum acaule 236

Solanum brevidens 247
Solanum cardiophyllum 235
Solanum melongena 216
Solanum tuberosum 217
Sonchus yellow net rhabdovirus 3, 5, 7, 78, 132
Southern bean mosaic sobemovirus 3, 5, 8, 53-54, 107
Soybean dwarf luteovirus 2, 58
Soybean mosaic virus 195
Spongospora subterranea 190
Stem-loop region/structure 16-17, 31-32, 39, 52-54, 58-59, 61, 63, 66, 73
Strawberry mild yellow edge-associated virus 101
Strain mixture – see Heterologous mixture of strains
Stylets of aphids 12, 37, 184
Subgenomic promoter 16-17, 54, 66-67, 74, 103, 106, 130
Subgenomic RNA 52, 54, 56-59, 61, 66, 68, 71, 73-75, 77, 100, 103-104, 106, 204-205, 207
Subterranean clover stunt virus 29, 31, 36
Suppression of leaky stops/stop codon – see Readthrough
Suppression mutation 124
Symbiotic relationship 53, 55
Symptom determinants/mutants 12, 59, 64, 72, 124-125, 132, 144, 194, 201-207, 228, 234-236
Synergism 249
Systemic acquired resistance 211, 221, 261
Systemic acquired silencing 261, 265
Systemic virus movement – see Long-distance movement

tRNA-like structure/functions of 5, 13-15, 64, 66-68, 71-73, 82, 102
Temperature-sensitive mutant 122, 126, 130, 163
Template switching 129, 244-245
Temporal fashion/regulation of gene expression 106, 112

Tenuivirus 3, 8, 48, 77, 79-80,

Tenuivirus(es) 16, 79

Threshold model of gene silencing 259-260

Thrips vector 136, 186

Tobacco etch potyvirus 9-11, 13, 16, 124, 184, 255-256, 259-263

Tobacco leaf curl virus 151

Tobacco mild green mosaic virus 3, 145-147, 149-153

Tobacco mosaic tobamovirus 3, 5, 8-15, 17, 59, 66-71, 75, 102, 106, 108, 124-125, 144-145, 153, 163-168, 170-171, 173-176, 193, 202, 204, 207, 212, 214, 216-219, 241-243, 246-249, 256

Tobacco necrosis necrovirus 3, 5, 7, 60, 191-192

Tobacco rattle tobravirus 3, 5, 8, 12, 67-70, 108-109, 131, 189, 243

Tobacco ringspot virus 193

Tobacco streak ilarvirus 2, 5, 7, 65

Tobacco stunt virus 191

Tobacco vein mottling potyvirus 9, 184

Tobacco yellow dwarf virus 38

Tobamovirus 3, 5, 8, 66, 69, 150, 193, 216, 218

Tobamovirus(es) 13-15, 61, 63-64, 66-69, 72, 101-104, 106, 108, 122, 124, 126, 147, 150, 153, 164-167, 170, 173, 175, 203-204, 214, 216, 218, 220-221, 242, 249

Tobravirus 3, 5, 8, 66-67, 69, 188

Tobravirus(es) 12, 14-15, 67-68, 72, 103-104, 106, 108, 126, 131, 164, 166, 188-189, 242-243

Tomato apical stunt viroid 226, 228-229

Tomato aspermy cucumovirus 125, 153, 202-203

Tomato black ring nepovirus 3, 5, 7, 52, 264-265

Tomato bushy stunt tombusvirus 3, 5, 8, 11, 60, 126, 132, 175, 192, 217-219

Tomato golden mosaic virus 248

Tomato leaf curl virus 31, 36

Tomato mosaic virus 220

Tomato planta macho viroid 226

Tomato pseudo-curly top virus 31, 36

Tomato spotted wilt tospovirus 3, 5, 8, 78-79, 132, 185-186

Tomato yellow leaf curl virus 31, 36, 188, 242

Tombusviridae 2-3, 48, 58-61, 82, 191-192

Tombusvirus 3, 8, 58-60, 191-192

Tombusvirus(es) 14, 59, 106-109, 125-126, 132, 164

Tospovirus 3, 5, 8, 48, 78-79, 185

Tospovirus(es) 16, 78-79, 132, 136, 152, 164, 169, 172

Transcapsidation 246-249

Transcription gene/factor/initiation/start site 16-17, 35, 39, 48, 61-62, 66, 78-79, 101, 104, 106, 114, 134, 215, 229-232, 255, 257-258, 260, 262

Transcriptional activator gene 30, 35, 39

Transcriptional control of gene expression 104, 110

Transcriptional gene silencing 257-258, 265

Transcriptional repressor gene/protein 34

Transgene 217-218, 221, 241-246, 249-251, 255-266

Transgenic/transformed plants 36-37, 127, 136, 164, 167-169, 174, 187, 204, 221, 241-251, 255-257, 259-266

Transgenic resistance 154, 164, 221-222, 241-242, 244-245, 248-250, 256-257, 260-263

Trans-inactivation 256, 263

Transition mutation 122, 125

Translation 1, 11-14, 33, 37, 39-41, 51, 53, 57-58, 63, 66-67, 71, 77, 79, 99-119, 123, 164, 167, 170-171, 176, 194, 205, 226, 258

Translational enhancers 102-103

Translational transactivation gene/protein 37-38, 40, 244

Transmission determinants/factors of viruses and viroids 10-12, 35, 37-38, 50-51, 56-59, 63-64, 68, 71-72, 149, 151, 153-154, 183-200, 234, 236, 243,

247, 249-251
Transport of viruses and viroids – see
  Virus movement
Transversion mutation 122
Trichovirus 3, 8, 47, 74-75
Trichovirus(es) 75, 102, 166
Tricornaviruses 131
Triple gene block 6-10, 70, 72-73, 76-
  77, 101, 104-107, 109, 164, 168, 171,
  194
Tritimovirus 51
Tubiviridae 47, 66, 70, 72
Tubules 36, 165, 168-173
Turnip crinkle carmovirus 2, 10, 17, 61,
  124-127, 129-130, 132, 202, 207, 243
Turnip mosaic potyvirus 13, 184
Turnip yellow mosaic tymovirus 3, 5, 8-
  9, 11, 14-15, 17, 73-74, 146, 202, 206
Tymovirales 48, 73-74, 77
Tymovirus 3, 5, 8, 47, 73-74
Tymovirus(es) 11, 15, 17, 61, 73-75,
  106, 112-113, 146

Umbravirus 4, 5, 8, 48, 56-57, 188
Umbravirus(es) 56-57, 188, 246, 251
Untranslated 5' and/or 3' regions of
  RNA 5, 12-13, 15-18, 51, 53, 58, 64,
  67, 71, 73-74, 77, 100-103, 127, 194,
  206

Variability/variation 121-135, 143-153,
  228-229, 234, 243
Varicosavirus 191
Vector specificity 11, 35, 63, 80, 184,
  186-189
Vector-assisted seed transmission 193
Vesicular stomatitis virus 125, 150
Vicia faba 203
Viroids 1, 3, 12, 144, 150, 152-153,
  225-236, 245
Virus assembly – see Assembly of
  viruses
Virus factors in gene-for-gene
  interaction 216-219
Virus movement/transport in plants

10, 50, 55-56, 61, 63-64, 107, 125,
  161-176, 194, 217, 235-236
Virus population 121-122, 149-154
Virus transmission – see Transmission
Vitivirus 47, 74-75
Vitivirus(es) 75

Waikavirus 4, 8, 51-52
Waikavirus(es) 52, 112
Watermelon mosaic virus 2 242
Wheat dwarf virus 38
Wheat streak mosaic rymovirus 147,
  149
White clover 2 betacryptovirus 81
White clover mosaic potexvirus 10, 15
Whitefly vector 31, 186, 188
Wound tumor phytoreovirus 3, 7, 13,
  146, 151

Xanthomonas oyzae 214

Zoospores 109, 190-192
Zucchini yellow mosaic virus 184-185,
  242, 246-247